# Unifying Themes in Complex Systems IV

# Springer Complexity

Springer Complexity is an interdisciplinary program publishing the best research and academic-level teaching on both fundamental and applied aspects of complex systems – cutting across all traditional disciplines of the natural and life sciences, engineering, economics, medicine, neuroscience, social and computer science.

Complex Systems are systems that comprise many interacting parts with the ability to generate a new quality of macroscopic collective behavior the manifestations of which are the spontaneous formation of distinctive temporal, spatial or functional structures. Models of such systems can be successfully mapped onto quite diverse "real-life" situations like the climate, the coherent emission of light from lasers, chemical reaction-diffusion systems, biological cellular networks, the dynamics of stock markets and of the internet, earthquake statistics and prediction, freeway traffic, the human brain, or the formation of opinions in social systems, to name just some of the popular applications.

Although their scope and methodologies overlap somewhat, one can distinguish the following main concepts and tools: self-organization, nonlinear dynamics, synergetics, turbulence, dynamical systems, catastrophes, instabilities, stochastic processes, chaos, graphs and networks, cellular automata, adaptive systems, genetic algorithms and computational intelligence.

The two major book publication platforms of the Springer Complexity program are the monograph series "Understanding Complex Systems" focusing on the various applications of complexity, and the "Springer Series in Synergetics", which is devoted to the quantitative theoretical and methodological foundations. In addition to the books in these two core series, the program also incorporates individual titles ranging from textbooks to major reference works.

# New England Complex Systems Institute

NECSI

**President**
Yaneer Bar-Yam
New England Complex Systems Institute
24 Mt. Auburn St.
Cambridge, MA 02138, USA

For over 10 years, The New England Complex Systems Institute (NECSI) has been instrumental in the development of complex systems science and its applications. NECSI conducts research, education, know-ledge dissemination, and community development around the world for the promotion of the study of complex systems and its application for the betterment of society.

NECSI was founded by faculty of New England area academic institutions in 1996 to further international research and understanding of complex systems. Complex systems is a growing field of science that aims to understand how parts of a system give rise to the system's collective behaviors, and how it interacts with its environment. These questions can be studied in general, and they are also relevant to all traditional fields of science.

Social systems formed (in part) out of people, the brain formed out of neurons, molecules formed out of atoms, and the weather formed from air flows are all examples of complex systems. The field of complex systems intersects all traditional disciplines of physical, biological and social sciences, as well as engineering, management, and medicine. Advanced education in complex systems attracts professionals, as complex systems science provides practical approaches to health care, social networks, ethnic violence, marketing, military conflict, education, systems engineering, international development and terrorism.

The study of complex systems is about understanding indirect effects. Problems we find difficult to solve have causes and effects that are not obviously related. Pushing on a complex system "here" often has effects "over there" because the parts are interdependent. This has become more and more apparent in our efforts to solve societal problems or avoid ecological disasters caused by our own actions. The field of complex systems provides a number of sophisticated tools, some of them conceptual helping us think about these systems, some of them analytical for studying these systems in greater depth, and some of them computer based for describing, modeling or simulating them.

NECSI research develops basic concepts and formal approaches as well as their applications to real world problems. Contributions of NECSI researchers include studies of networks, agent-based modeling, multiscale analysis and complexity, chaos and predictability, evolution, ecology, biodiversity, altruism, systems biology, cellular response, health care, systems engineering, negotiation, military conflict, ethnic violence, and international development.

NECSI uses many modes of education to further the investigation of complex systems. Throughout the year, classes, seminars, conferences and other programs assist students and professionals alike in their understanding of complex systems. Courses have been taught all over the world: Australia, Canada, China, Colombia, France, Italy, Japan, Korea, Portugal, Russia and many states of the U.S. NECSI also sponsors postdoctoral fellows, provides research resources, and hosts the International Conference on Complex Systems, discussion groups and web resources.

# New England Complex Systems Institute Book Series

**Series Editor**

NECSI

Dan Braha
New England Complex Systems Institute
24 Mt. Auburn St.
Cambridge, MA 02138, USA

**New England Complex Systems Institute Book Series**

The world around is full of the wonderful interplay of relationships and emergent behaviors. The beautiful and mysterious way that atoms form biological and social systems inspires us to new efforts in science. As our society becomes more concerned with how people are connected to each other than how they work independently, so science has become interested in the nature of relationships and relatedness. Through relationships elements act together to become systems, and systems achieve function and purpose. The study of complex systems is remarkable in the closeness of basic ideas and practical implications. Advances in our understanding of complex systems give new opportunities for insight in science and improvement of society. This is manifest in the relevance to engineering, medicine, management and education. We devote this book series to the communication of recent advances and reviews of revolutionary ideas and their application to practical concerns.

# Unifying Themes in Complex Systems IV

*Proceedings of the Fourth International Conference on Complex Systems*

**Edited by Ali Minai and Yaneer Bar-Yam**

Ali A. Minai

Univeristy of Cincinnati
Department of Electrical and
Computer Engineering, and Computer Science
P.O. Box 210030, Rhodes Hall 814
Cincinnati, OH 45221-0030
USA
Email: Ali.Minai@uc.edu

Yaneer Bar-Yam

New England Complex Systems Institute
24 Mt. Auburn St.
Cambridge, MA 02138-3068
USA
Email : yaneer@necsi.org

This volume is part of the
New England Complex Systems Institute Series on Complexity

ISBN 978-3-642-09308-1 e-ISBN 978-3-540-73849-7



Springer is a part of Springer Science+Business Media
springer.com

# CONTENTS

## PART III: Applications

# INTRODUCTION

The mysteries of highly complex systems that have puzzled scientists for years are finally beginning to unravel thanks to new analytical and simulation methods. Better understanding of concepts like complexity, emergence, evolution, adaptation and self-organization have shown that seemingly unrelated disciplines have more in common than we thought. These fundamental insights require interdisciplinary collaboration that usually does not occur between academic departments. This was the vision behind the first International Conference on Complex Systems in 1997; not just to present research, but to introduce new perspectives and foster collaborations that would yield research in the future.

As more and more scientists began to realize the importance of exploring the unifying principles that govern all complex systems, the Fourth ICCS attracted a diverse group of participants representing a wide variety of disciplines. Topics ranged from economics to ecology, particle physics to psychology, and business to biology. Through pedagogical, breakout and poster sessions, conference attendees shared discoveries that were significant both to their particular field of interest, as well as the general study of complex systems. These volumes contain the proceedings from that conference. Even with the fourth ICCS, the science of complex systems is still in its infancy. In order for complex systems science to fulfill its potential to provide a unifying framework for various disciplines, it is essential to establish a standard set of conventions to facilitate communication. This is another valuable function of the conference; it allowed an opportunity to develop a common foundation and language for the study of complex systems.

These efforts have produced a variety of new analytic and simulation techniques that have proven invaluable in the study of physical, biological and social systems. New methods of statistical analysis led to better understanding of polymer formation and complex fluid dynamics; further development of these methods has deepened our understanding of patterns and networks. The application of simulation techniques such as agent-based models, cellular automata, and Monte Carlo simulations to complex systems has increased our ability to understand or even predict behavior of systems that once seemed completely unpredictable.

The concepts and tools of complex systems are of interest not only to scientists, but to corporate managers, doctors, political scientists and policy makers. The same rules that govern neural networks apply to social or corporate networks, and professionals have started to realize how valuable these concepts are to their individual fields. The ICCS conferences have provided the opportunity for professionals to learn the basics of complex systems and share their real-world experience in applying these concepts.

# Fourth International Conference on Complex Systems: Organization and Program

## Organization:

### Host:

New England Complex Systems Institute

### Partial financial support:

National Science Foundation
Nationa Institute for General Medical Sciences, NIH
American Institute of Physics
Perseus Press
University of Chicago Press
World Scientific

### Conference Chair:

Yaneer Bar-Yam - NECSI *

### Executive Committee:

Larry Rudolph - MIT *
Ali Minai - University of Cincinnati
Dan Braha
Helen Harte
Gunter Wagner

Temple Smith

## Program Committee:

Yaneer Bar-Yam - NECSI
Philippe Binder - University of Hawaii
Dan Braha - MIT
Helen Harte - NECSI Organization Science Program
Sui Huang - Harvard University
Michael Jacobson - Research Consultant
Mark Klein - MIT
Seth Lloyd - MIT *
David Meyer - UCSD
Ali Minai - University of Cincinnati
Lael Parrott - University of Montreal
Jeff Stock - Princeton University *
David Sloan Wilson - Binghamton University

## Organizing Committee:

Philip W. Anderson - Princeton University
Kenneth J. Arrow - Stanford University
Michel Baranger - MIT *
Per Bak - Niels Bohr Institute
Charles H. Bennett - IBM
William A. Brock - University of Wisconsin
Charles R. Cantor - Boston University *
Noam A. Chomsky - MIT
Leon Cooper - Brown University
Daniel Dennett - Tufts University
Irving Epstein - Brandeis University *
Michael S. Gazzaniga - Dartmouth College
William Gelbart - Harvard University *
Murray Gell-Mann - CalTech/Santa Fe Institute
Pierre-Gilles de Gennes - ESPCI
Stephen Grossberg - Boston University
Michael Hammer - Hammer Co
John Holland - University of Michigan
John Hopfield - Princeton University
Jerome Kagan - Harvard University *
Stuart A. Kauffman - Santa Fe Institute
Chris Langton - Santa Fe Institute
Roger Lewin - Harvard University
Richard C. Lewontin - Harvard University

Albert J. Libchaber - Rockefeller University
Seth Lloyd - MIT *
Andrew W. Lo - MIT
Daniel W. McShea - Duke University
Marvin Minsky - MIT
Harold J. Morowitz - George Mason University
Alan Perelson - Los Alamos National Lab
Claudio Rebbi - Boston University
Herbert A. Simon - Carnegie-Mellon University
Temple F. Smith - Boston University *
H. Eugene Stanley - Boston University
John Sterman - MIT *
James H. Stock - Harvard University *
Gerald J. Sussman - MIT
Edward O. Wilson - Harvard University
Shuguang Zhang - MIT

## Session Chairs:

Albert-Laszlo Barabasi
Itzhak Benenson
Bruce Boghosian
Jeff Cares
Irving Epstein - Brandeis University
Dan Frey - MIT
Charles Goodnight
Helen Harte
Sui Huang
James Kaput
Les Kaufman
Mark Kon
Jason Redi
Dwight Reed
Larry Rudolph - MIT
Anjali Sastry - MIT
Hiroki Sayama
Temple Smith
David Sloan Wilson

* NECSI Co-faculty
† NECSI Affiliate

# Subject areas: Unifying themes in complex systems

## The themes are:

**EMERGENCE, STRUCTURE AND FUNCTION:** substructure, the relationship of component to collective behavior, the relationship of internal structure to external influence.

**INFORMATICS:** structuring, storing, accessing, and distributing information describing complex systems.

**COMPLEXITY:** characterizing the amount of information necessary to describe complex systems, and the dynamics of this information.

**DYNAMICS:** time series analysis and prediction, chaos, temporal correlations, the time scale of dynamic processes.

**SELF-ORGANIZATION:** pattern formation, evolution, development and adaptation.

## The system categories are:

**FUNDAMENTALS, PHYSICAL & CHEMICAL SYSTEMS:** spatio-temporal patterns and chaos, fractals, dynamic scaling, non-equilibrium processes, hydrodynamics, glasses, non-linear chemical dynamics, complex fluids, molecular self-organization, information and computation in physical systems.

**BIO-MOLECULAR & CELLULAR SYSTEMS:** protein and DNA folding, bio-molecular informatics, membranes, cellular response and communication, genetic regulation, gene-cytoplasm interactions, development, cellular differentiation, primitive multicellular organisms, the immune system.

**PHYSIOLOGICAL SYSTEMS:** nervous system, neuro-muscular control, neural network models of brain, cognition, psychofunction, pattern recognition, man-machine interactions.

**ORGANISMS AND POPULATIONS:** population biology, ecosystems, ecology.

**HUMAN SOCIAL AND ECONOMIC SYSTEMS:** corporate and social structures, markets, the global economy, the Internet.

**ENGINEERED SYSTEMS:** product and product manufacturing, nano-technology, modified and hybrid biological organisms, computer based interactive systems, agents, artificial life, artificial intelligence, and robots.

# Program:

## Sunday, June 9, 2002

**PEDAGOGICAL SESSION: StarLogo and SIMP/STEP - Ali Minai -** Session Chair

**Mark Smith** - Application to Medical Management

**Eric Heller** - Quantum Chaos

**Charles Bennett** - Quantum Information Processing

**Greg Chaitin** - Algorithmic Complexity

**Bud Mishra** - Biomedical Systems

**Usama Fayyad** - Data Mining

**Dan Schrag** - Climate Change

**Special Memorial in Honor of Herbert A. Simon and Claude E. Shannon**

## Monday, June 10, 2002

**Yaneer Bar-Yam** - Welcome

**EMERGENCE - Irving Epstein** - Session Chair

**Philip Anderson** - Emergence of Complex Systems

**John Sterman** - Social Systems

**Dwight Read** - Cultural Anthropology

**Rita Colwell** - Biocomplexity

**DESCRIPTION AND MODELING - Larry Rudolph** - Session Chair

**John Casti** - Agent Based Modeling

**Jeffrey Kephart** - Agent Economies

**Seth Lloyd** - Networks

**Mitchell Feigenbaum** - Universality and the Dynamics of Chaos

**PARALLEL SESSIONS**

**Konstantin Kovalchuk**

**N. Oztas, T. Huerta, & P. J. Robertson** - Mapping the Field: Complexity Sciences in Organization and Management

**Engineering**

**A. Das, M. Marko, A. Probst, M. A. Porter & C. Gershenson** - Neural Net Model for Featured Word Extraction

**C. Gershenson, M. A. Porter, A. Probst & M. Marko**

**Complexity**

**Carlos E. Puente** - More Lessons From Complexity. The Origin: The Root of Peace

**Carlos Gershenson** - Complex Philosophy

**Sergei Victorovich Chebanov**

**Menno Hulswit**

**Denys Lapinard**

**Vidyardhi Nanduri**

**Konstantin L Kouptsov** - Using a Complex Systems approach to undo Brainwashing and Mind Control

**Physical Systems & Formal Methods**

**Mark R. Tinsley & Richard J. Field** - Dynamic Instability in Tropospheric Photochemistry: An Excitability Threshold

**Carlos E. Puente** - Treasures Inside the Bell

**Pierre Sener**

**Md. Shah Alam** - Algebra of Mixded Number

**John Maweu** - Self Organized Criticality in State Transition Systems

**H. N. Mhaskar**

**Tools**

**Edward A Bach** - SIMP/STEP: A Platform for Fine-Grained Lattice Computing

# Tuesday, June 11, 2002

**EDUCATION - Jim Kaput** - Education

**Robert Devaney** - Chaos In The Classroom

**Jerry Sussman** - Formalizing Science

**Robert Savit** - Minority Game

## PARALLEL SESSIONS

### Learning Theory - Mark Kon - Session Chair

**Hrushikesh Mhaskar** - When is approximation by Gaussian networks necessarily a linear process?

**Art Werschulz** - Overview of Information-based Complexity

**Alan Waxman**

**George Cybenko**

**Lee Jones**

**Hillol Kargupta**

**Marc Parham**

### Altruism - David Sloan Wilson - Session Chair

**Sanjay Jain** - Emergence, Growth and Collapse of Cooperative Organizational Structure in Evolving Networks

**Jeff Carpenter**

**Gregory Velicer** - Altruism and Social Conflict in the Bacterium Myxococcus Xanthus

### Modular Systems

**Jochen Triesch** - Towards Understanding Self-organized Information Flow in the Cortex

**Jennifer Hallinan** - Iterative Diffusion of Vectors for the Detection of Modularity in Complex Networks

**Richard A. Watson** - Compositional Evolution: Evolvability, Modularity, and Symbiosis

### Market and Industry Dynamics - Anjali Sastry - Session Chair

**Jeho Lee** - Reconsideration of the Winner-Take-All Hypothesis

**Yuri Yegorov** - The Transitions in Industry, Research and Society: Dynamic Equilibrium Approach

**Pierpaolo Andriani**

**David W. Peterson** - An Adaptive Endogenous Theory of Technological and Economic Growth

### Arts

**Paula Matthusen** - In Memory of an Anthill: Complexity Theory and Compositional Processes

**Philip Galanter** - On four modes of artistic engagement with complexity

**Ellen Levy** - Initial Conditions to Final Results in the Complex Systemsof Art and Science

**Ivar Hagendoorn** - Emergent Patterns in Dance Improvisation and Choreography

**R.P. Taylor, B. Spehar, C.W.G. Clifford & B.R. Newell** - The Visual Complexity of Pollock's Dripped Fractals

**Military Systems - Jeff Cares** - Session Chair

**Alfred Brandstein** - The Role of Analysis in the Brave New World

**Ray Christian**

**Gary Horne**

**John Q. Dickman** - Complex Systems Research and Information Age Warfare

**Management - Helen Harte** - Session Chair

**Eve Mitleton-Kelly** - Organizational Complexity

**Harold E. Klein** - Designing Organizations to be Responsive to the Complex Changing Environment

**Yasmin Merali** - The Concept of Emergent Ecologies and Persistence in Cyberspace

**Brigitte Fleeman** - Sensemaking of a Change Intervention With Insights From Complexity Science

**Barry G. Silverman**

**Tessaleno C. Devezas** - Aggregate Output Model Describing the Limit-Cycle Behavior of a Logistic Growing Process in Socioeconomic Systems

**Bio-Medical - Sui Huang** - Session Chair

**Bud Mishra**

**I. Rouzine & J. M. Coffin** - Realistic Model of Cellular Immune Response Against Human Immunodeficiency Virus

**Solomon Gilbert Diamond** - Measuring Hypnosis: Relating Mental State to Systematic Physiological Changes

**D. E. Creanga, E. Lozneanu and J. C. Sprott** - Computational Analysis in Temporal Series Describing Kidney Excretory Function

**Christopher A. Shaw, Jason M.B. Wilson & Iraj Khabazian** - Reverse Engineering Neurological Diseases

**Elena Bystrova, Evgenia Bogomolova, Ludmila Panina, Anton Bulianitsa & Vladimir Kurochkin** - Fungal Colony Patterning as an Example of Biological Self-Organization

**James Holland Jones**

**Patterns In Physical Systems - Bruce Boghosian** - Session Chair

**Eugenio Degroote** - Flame Spreading Over Liquid Fuels: A General Model

**Jacques Lewalle** - Self-organization in Navier-Stokes turbulence

**Timothy Field**

**Masaharu Kuroda & Francis C. Moon** - Local Complexity and Global Nonlinear Modes in Large Arrays of Elastic Rods in an Air Cross-Flow

**Xiang San Liang & Allan R. Robinson** - A Multiscale Interactive Dynamical Analysis for Oceanic Flows Intermittent in Space and Time

**Benjamin Skellett** - Classical Dynamics of Magnetically Coupled Spins

# Wednesday, June 12, 2002

## NATURAL AND ENGINEERED SYSTEMS

**Ali Minai** - Natural and Engineered Systems

**Rodney Brooks** - Robots and Beyond

**William Cotton** - Air Traffic Control

**Edward Lorenz** - Meteorological Prediction

**Allan R. Robinson** - Multiscale Interactions In The Sea

## SPECIAL SESSIONS

**Special Talk**

**Yu-Chi Ho** - No-Free-Lunch Theorem and Computer Security

**Funding**

**Susan M. Fitzpatrick** - Mcdonnell Foundation

**Education and Psychology**

**Iris Stammberger** - Contemporary Models of Creative Cognition: an Ethnographic Review of the Impact of Complexity, Evolutionary and Computational Theories

**Len Troncale** - Stealth Studies in Complex Systems that Completes Science GE Requirements at Most Universities

**Steve Hassan** - Mind Control

**Organizations**

**Eric Bonabeau** - Co-Evolving Business Models

**Edgar Peters** - Complexity and Efficient Markets

**Ken O'brien** - Organizations

## PARALLEL SESSIONS

### Thread A

#### Education

**Michael J. Jacobson** - Complex Systems in Education: Integrative Conceptual Tools and Techniques for Understanding the Education System Itself

**Michael Connell** - Neuroscience and Education–Bridging the Gap

**Val Bykoski** - Complex Models and Model-Building Automation

**Plamen Petrov** - *The* Game (Introduction to Digital Physics)

**Len Troncale** - An Open Source Computer-Based Tool for Research and Education in the Systems Sciences

**Brock Dubbels** - Building a Network to the People

### Thread B

#### Cellular Automata & Dynamical Systems

**Arnold Smith & Stephanie Pereira** - Continuous and Discrete Properties in Self-Replicating Systems

**Mathieu S. Capcarrere** - Emergent Computation in CA: A Matter of Visual Efficiency?

**Howard A. Blair** - Unifying Discrete and Continuous Dynamical Systems

**Douglas E. Norton** - Epsilon-Pseudo-Orbits and Applications

**H. Sabelli, L. Kauffman, M. Patel & A. Sugerman** - Bios: Mathematical, Cardiac, Economic and Meteorological Creative Processes Beyond Chaos

**Atin Das** - Nonlinear Data Analysis of Experimental [EEG] data and Comparison with Theoretical [ANN] Data

### Thread C

#### Urban and Global Change

**Itzhak Benenson & Erez Hatna**

**Eric Allison** - Self Organization in Cities: Case Studies in Neighborhood Historic Preservation

**Networks**

**Fabrice Saffre** - RAn (Robustness Analyser)
**Laszlo Gulyas**
**Xuenan Li** - The Complexity of the Growing Network

**Thread D**

**Multi-Agent Systems**

**Fred M. Discenzo, Francisco P. Maturana & Dukki Chung** - Managed Complexity in An Agent-based Vent Fan Control System Based on Dynamic Re-configuration
**Iqbal Adjali, David Collings & Paul Marrow** - Modelling Third Generation Mobile Spectrum Auctions
**Sorinel Oprisan**

**Knowledge Management**

**Vinod Subramanian, Rajkumar Arumugam & Ali A. Minai** - Self-Organization of Connectivity and Geographical Routing in Large-Scale Sensor Networks
**John Trimble & Harry Keeling** - Coping with Complexity in Knowledge Management
**Zann Gill** - Webtank

**BANQUET SESSION**

**Marvin Minsky** - Society Of Mind

## Thursday, June 13, 2002

**BIOCOMPLEXITY** - **Temple Smith** - Session Chair

**Chris Johnson** - Visualizing Biological Systems

**Brad Smith** - NMR Imaging Of Biological Dynamics

**Geoffrey West** - Scaling In Biology

**Albert-Laszlo Barabasi** - Scale Free Networks

**PARALLEL SESSIONS**

**Complexity In Engineering** - **Dan Frey** - Session Chair

**Dan Hastings**

**David Wallace**

**Dave Darmofal**

**Mark Klein, Hiroki Sayama, Peyman Faratin & Yaneer Bar-Yam** - A Complex Systems Perspective on Collaborative Design

**Social Interactions - Dwight Reed** - Session Chair

**Robert L. Goldstone & Benjamin C. Ashpole** - The Distribution of Agents to Resources in a Networked Multi-player Environment

**Peyman Faratin** - A Multiagent Simulation Model of the Emergence and Dynamics of Negotiation in Complex Contracting Games

**Keith Warren, Elena Irwin, Brian Roe & William Sean Newsome** - The Sum of the Parts: Two Studies of Interpersonal Influence on Aggressive Behaviors in Small Group Settings

**George J. Gumerman, Alan Swedlund, Jeffery S. Dean, Joshua Epstein & Robert Axtell** - Evolving Social Complexity in the Prehistoric American Southwest

**Corey Lofdahl** - On Trade and the Environment as a Complex System

**Thomas J. Wheeler** - Interdisciplinary Conceptual Model Blending

**Jeffrey C. Schank** - Cycle Variability, Follicle Competition, and Female Mate Choice

**Games On Networks - Albert-Laszlo Barabasi** - Session Chair

**Sui Huang** - Gene network topology and dynamics in mammalian cell fate regulation

**Benjamin Shargel** - Optimization of Robustness and Connectivity in Complex Networks

**K. Josef, J. Saranak & K. W. Foster** - Laboratory Controlled Model System for Study of Complexity Models that Apply to the Signaling Network of a Single Biological Cell

**Adrian Li Mow Ching** - User and Service Interaction Dynamics

**Derek Raine** - The Complexity Of Canonical Power Law Networks

**Urban Systems - Itzhak Benenson** - Session Chair

**Takeshi Arai & Tetsuya Akiyama** - A CA Based Two-Stage Model of Land Use Dynamics in Urban Fringe Areas: An Application to the Tokyo Metropolitan Region

**Fabiano Sobreira & Joana Barros** - City of Slums: Self-Organisation Across Scales

**Laszlo Gulyas & Yuri Mansury**

**Gary G. Nelson & Peter M. Allen** - Self-Organizing Geography: Scaled Objects and Regional Planning in the U.S.

**D. Borri, G. Concilio & E. Conte** - Managing Urban Traffic Dynamics by a Multi-Agent DSS

**Gulie Benguigui & Danny Czamanski**

**Distributed Robots and Computing - Jason Redi** - Session Chair

**Radhika Nagpal** - Programmable Pattern-Formation and Scale-Independence

**Regina Estkowski, Michael Howard & David Payton** - Amorphous Predictive Nets

**Ian W. Marshall, Chris Roadknight & Lionel Sacks** - The Role of Complex Systems in the Management of Pervasive Computing

**Sanjay Sarma**

**Theodoros Michalareas**

**Evolution - Hiroki Sayama** - Session Chair

**Janet Wiles, James Watson, Bradley Tonkes & Terrence Deacon** - Evolving Complex Integrated Behaviour by Masking and Unmasking Selection Pressures

**Erik Rauch** - The Relationship Between Measures of Fitness and Time Scale in Evolution

**Josh Mitteldorf, S. Chandu Ravela, Robert Bell, Dominic L. Boccelli, David H. Croll & Deva Seetharam** - Virtual Stability as a Conceptual Principle Underlying the Directionality of Evolutionary Processes

**Burton Voorhees** - Virtual Stability: A Principle of Complex Systems

**Luis Mateus Rocha** - Indirect Encoding of Phenotypes in Evolutionary Agents

**Daniel W. Miller** - The Complexity of Homeodynamic Psychophysiological Systems

**Complex Systems Curriculum Open Discussion - James Kaput** - Session Chair

**Saint Joseph College** - Ronald Degray, Shyamala Raman

**University Of Alaska** - Cheryl Wright, Kim Peterson, Jerzy Maselko, and Jim Liszka

# Friday, June 14, 2002

**SPECIAL DAY ON EVOLUTION - Les Kaufman and Charles Goodnight** - Session Chairs

**Terrence Deacon** - Evolution and Mind

**David Sloan Wilson** - Darwin's Cathedral

**Joel Peck** - Sex and Altruism

**Raffaele Calabretta** - Modularity

**Mike Wade:** - Gene Interactions

**Jason Wolf** - Maternal Effects

**Lisa Meffert** - Evolutionary Bottlenecks

**Hiroki Sayama** - Beyond The Gene Centered View

# Publications:

## Proceedings:

Conference proceedings (this volume)
On the web at http://necsi.org/events/iccs/2002/proceedings.html
Video proceedings are available to be ordered through the New England Complex Systems Institute.

## Journal articles:

Individual conference articles were published online at http://interjournal.org

## Web pages:

**The New England Complex Systems Institute**
http://necsi.org

**The First International Conference on Complex Systems (ICCS1997)**
http://www.necsi.org/html/ICCS_Program.html

**The Second International Conference on Complex Systems (ICCS1998)**
http://www.necsi.org/events/iccs/iccs2program.html

**The Third International Conference on Complex Systems (ICCS2000)**
http://www.necsi.org/events/iccs/iccs3program.html

**The Fourth International Conference on Complex Systems (ICCS2002)**
http://www.necsi.org/events/iccs/iccs4program.html

**The Fifth International Conference on Complex Systems (ICCS2004)**
http://www.necsi.org/events/iccs/openconf/author/iccsprogram.php

**The Sixth International Conference on Complex Systems (ICCS2006)**
http://www.necsi.org/events/iccs6/index.php

**NECSIWiki**
http://necsi.org/community/wiki/index.php/Main_Page

**InterJournal** - The journal of the New England Complex Systems Institute
http://interjournal.org

# Part I:

# Methods

# Chapter 1

# A Complex Systems Perspective on Collaborative Design


**Mark Klein**
Massachusetts Institute of Technology
m_klein@mit.edu

**Peyman Faratin**
Massachusetts Institute of Technology
peyman@mit.edu

**Hiroki Sayama**
New England Complex Systems Institute
sayama@necsi.org

**Yaneer Bar-Yam**
New England Complex Systems Institute
yaneer@necsi.org


## 1. The Challenge: Collaborative Design Dynamics

Collaborative design is challenging because strong interdependencies between design issues make it difficult to converge on a single design that satisfies these dependencies and is acceptable to all participants. Current collaborative design processes are typically characterized by (1) multiple iterations and/or heavy reliance on multi-functional design reviews, both of which are expensive and time-consuming, (2) poor incorporation of some important design concerns, typically later life-cycle issues such as environmental impact, as well as (3) reduced creativity due to the tendency to incrementally modify known successful designs rather than explore radically different and potentially superior ones.

This article examines what complex systems research can do to help address these issues, by informing the design of better computer-supported collaborative design technology. We will begin by defining a simple model of collaborative design, discuss some of the insights a complex systems perspective has to offer, and suggest ways to better support innovative collaborative design building on these insights.

## 2. Defining Collaborative Design

A design (of physical artifacts such as cars and planes as well as behavioral ones such as plans, schedules, production processes or software) can be represented as a set of *issues* (sometimes also known as *parameters*) each with a unique value. If we imagine that the possible values for every issue are each laid along their own orthogonal axis, then the resulting multi-dimensional space can be called the *design space*, wherein every point represents a distinct (though not necessarily good or even physically possible) design. The choices for each design issue are typically highly *interdependent*. Typical sources of inter-dependency include shared resource (e.g. weight, cost) limits, geometric fit, spatial separation requirements, I/O interface conventions, timing constraints etc.

*Collaborative* design is performed by multiple participants (representing individuals, teams or even entire organizations), each potentially capable of proposing values for design issues and/or evaluating these choices from their own particular perspective (e.g. manufacturability).

Some designs are better than others. We can in principle assign a *utility* value to each design and thereby define a *utility function* that represents the utility for every point in the design space. The *goal* of the design process can thus be viewed as trying to find the design with (close to) the optimal (maximal) utility value,.

The key challenge raised by the collaborative design of complex artifacts is that the design spaces are typically huge, and concurrent search by the many participants through the different design subspaces can be expensive and time-consuming because design issue interdependencies lead to conflicts (when the design solutions for different subspaces are not consistent with each other). Such conflicts severely impact design utility and lead to the need for expensive and time-consuming design rework.

## 3. Insights from Complex Systems Research

A central focus of complex systems research is the dynamics of distributed networks, i.e. networks in which there is no centralized controller, so global behavior emerges solely as a result of concurrent local actions. Such networks are typically modeled as multiple nodes, each node representing a state variable with a given value. Each node in a network tries to select the value that optimizes its own utility while maximizing its consistency with the influences from the other nodes. The global utility of the network state is simply the sum of node utilities plus the degree to which all the influences are satisfied. The dynamics of such networks emerge as follows: since all nodes update their local state concurrently based on their current context (at time T), the choices they make may no longer be the best ones in the new context of node states (at time T+1), leading to the need for further changes.

Is this a useful model for understanding the dynamics of collaborative design? We believe that it is. It is straightforward to map the model of collaborative design presented above onto a network. We can map design participants onto nodes, where each participant is trying to maximize the utility of the choices it is responsible for, while ensuring its decisions will satisfy the relevant dependencies (represented as the links between nodes). As a first approximation, it is reasonable to model the utility of a design as the local utility achieved by each participant plus a measure of how well all the decisions fit together. Even though real-world collaborative design clearly has top-down elements, the sheer complexity of many design artifacts means that no one person is capable of keeping the whole design in his/her head and centralized control of the design decisions becomes impractical, so the design process is dominated by concurrent local activities. The remainder of this paper will be based on this view of the collaborative design process.

How do such distributed networks behave? Let us consider the following simple example, a network consisting of inter-linked binary-valued nodes. At each time step, each node selects the value for itself that is the same as that of the [majority of the] nodes it is linked to. We can imagine using this network to model a real-world situation wherein six subsystems are being designed and we want them to use matching interfaces. The network has converged onto a *local* optimum (no node can increase the number of influences it satisfies by a local change), so it will not reach as a result a *global* optimum (where all the nodes have the same value). (Figure 1):

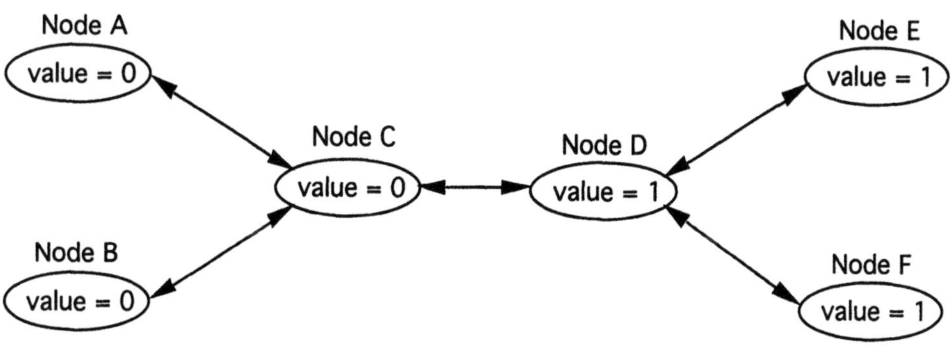


Figure 1: A simple network.

Generally speaking, networks may not always converge upon the global optimum, and in some cases (as we shall see with *dynamic attractors*), a network may not converge at all. *Insights into whether and how global optima can be found in networks represent the heart of what complex systems research offers to the understanding of collaborative design.*

We will discuss these insights in the remainder of this section. The key factor determining network dynamics is the nature of the influences between nodes. We will consider two important distinctions: whether the influences are *linear* or not, and whether they are *symmetric* or not. We will then discuss subdivided network topologies, and the role of learning. Unless indicated otherwise, the material on complex systems presented below is drawn from [1].

## 3.1. Linear vs. Non-Linear Networks

If the value of nodes is a linear function of the influences from the nodes linked to it, then the system is linear, otherwise it is non-linear. Linear networks have a single *attractor*, i.e. a single configuration of node states that the network converges towards no matter what the starting point, corresponding to the global optimum. This means we can use a 'hill-climbing' approach (where each node always moves directly towards increased local utility) because this always move the network towards the global optimum.

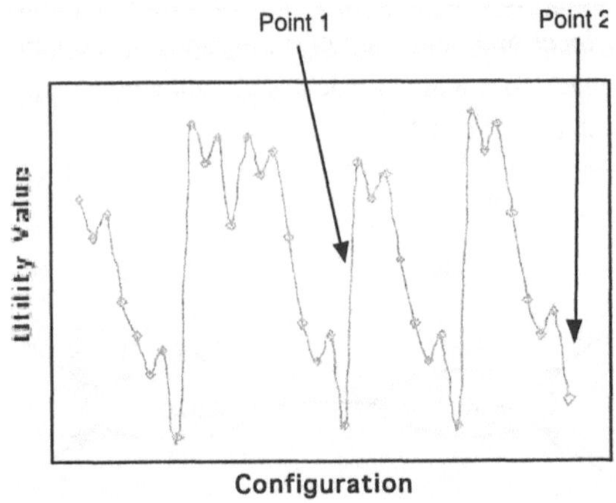


Figure 2. A multi-optimum utility function.

Non-linear networks, by contrast, are characterized by having multiple attractors and multiple-optima utility functions, like that shown in Figure 2 above.

A key property of non-linear networks is that search for the global optima can *not* be performed successfully by pure hill-climbing algorithms, because they can get stuck in local optima that are globally sub-optimal. Consider, for example, what would happen if the system started searching at Point 1 in Figure 2 above. Hill-climbing would take it to the top of the local optimum, which is substantially lower than optima in other regions of the utility function. Hill-climbing would do even more poorly if it started at Point 2.

One consequence of this reality is a tendency to stick near well-known designs. When a utility function has widely separated optima, once a satisfactory optimum is found the temptation is to stick to it. This design conservatism is exacerbated by the fact that it is often difficult to compare the utilities for radically different designs. We can expect this effect to be especially prevalent in industries, such as commercial airlines and power plants, which are capital-intensive and risk-averse, since in such contexts the cost of exploring new designs, and the risk of getting it wrong, can be prohibitive.

A range of techniques have emerged that are appropriate for finding optima in multi-optima utility functions, all relying on the ability to search past valleys in the utility function. Simulated annealing, for example, endows the search procedure with a tolerance for moving in the direction of lower utility that varies as a function of a virtual 'temperature'. At first the temperature is high, so the system is as apt to move towards lower utilities as higher ones. This allows it to range widely over the utility function and possibly find new higher peaks. Since higher peaks are also typically wider ones, the system will tend to spend most of its time in the region of high peaks. Over time the temperature decreases, so the algorithm increasingly tends towards pure hill-climbing. While this technique is not provably optimal, it has been shown to get close to optimal results in most cases.

Annealing, however, runs into a dilemma when applied to systems with multiple actors. Let us assume that at least some actors are self-interested 'hill-climbers', concerned only with directly maximizing their local utilities, while others are 'annealers', willing to accept, at least temporarily, lower local utilities in order to increase the utility in other nodes. Simulation reveals that while the presence of annealers always increases *global* utility, annealers always fare *individually* worse than hill-climbers when both are present [2]. The result is that globally beneficial behavior is not individually incented.

How do these insights apply to collaborative design? Linear networks have been used successfully to model *routine* design [3], involving highly familiar requirements and design options, as for example in automobile brake or transmission design [4]. Today's most challenging and important collaborative design problems (e.g. concerning software, biotechnology, or electronic commerce) are, however, *not* instances of routine design. They typically involve *innovative* design, radically new requirements, and unfamiliar design spaces. It is often unclear as a result where to start to achieve a given set of requirements. There may be multiple very different good solutions, and the best solution may be radically different than any that have been tried before. For such cases non-linear networks seem to represent a more accurate model of the collaborative design process.

This has important consequences. Simply instructing each design participant to optimize its own design subspace as much as possible (i.e. 'hill-climbing') can lead to the design process getting stuck in local optima that may be significantly worse than radically different alternatives. Design participants must be willing to explore alternatives that, at least initially, may appear much worse from their individual perspective than alternatives currently on the table. Designers often show greater loyalty to producing a good design for the subsystem they are responsible for, than to conceding to make someone else's job easier, so we need to find solutions for the dilemma identified above concerning the lack of individual incentives for such globally helpful behavior. We will discuss possible solutions in the section below on "How We Can Help".

## 3.2. Symmetric vs. Asymmetric Networks

Symmetric networks are ones in which influences between nodes are mutual (i.e. if node A influences node B by amount X then the reverse is also true), while asymmetric networks do not have this property. Asymmetric networks (with an exception to be discussed below) add the complication of *dynamic* attractors, which means that the network does not converge on a *single* configuration of node states but rather cycles indefinitely around a relatively small *set* of configurations. Let us consider the simplest possible asymmetric network: the 'odd loop' (Figure 3):

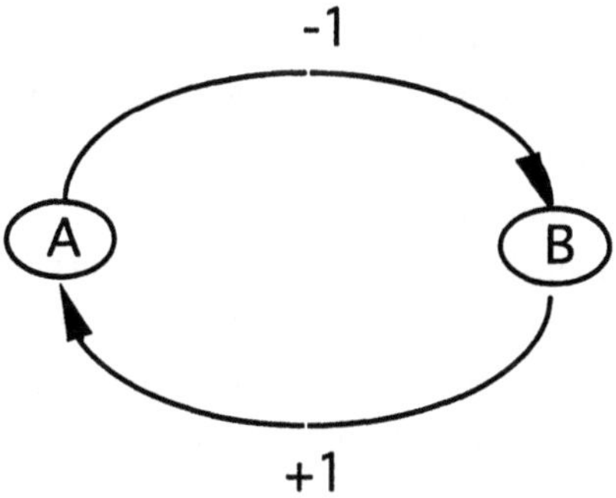


Figure 3. The simplest possible asymmetric network – an 'odd loop'.

This network has two links: one which influences the nodes to have the same value, the other which influences them to have opposite values. Imagine we start with node A having the value 1. This will influence node B to have the value –1, which will in turn influence node A towards the value –1, which will in turn cause node B to flip values again, and so on *ad infinitum.*

Current collaborative design practice is characterized by asymmetric influence loops likely to produce dynamic attractors and therefore non-convergent dynamics. Feedback from later product life cycle perspectives such as manufacturability and transportability, for example. tends to be weaker and slower than influences from design to these perspectives.

## 3.3. Subdivided Networks

Another important property of networks is whether or not they are sub-divided, i.e. whether they consist of sparsely interconnected 'clumps' of highly interconnected nodes. When a network is subdivided, node state changes can occur within a given clump with only minor effects on the other clumps. This has the effect of allowing the network to explore more states more rapidly. This effect is in fact widely exploited in design communities, where it is often known as *modularization.* This involves intentionally creating subdivided networks by dividing the design into subsystems with pre-defined standardized interfaces, so subsystem changes can be made with few or any consequences for the design of the other subsystems. The key to using this approach successfully is defining the design decomposition such that the impact of the subsystem interdependencies on the global utility is relatively low, because the standardized interfaces rarely represent an optimal way of satisfying these dependencies. In most commercial airplanes, for example, the engine and wing subsystems are designed separately, taking advantage of standardized engine mounts to allow the airplanes to use a range of different engines. This is not the optimal way of relating engines and wings, but it is good enough and simplifies the design process considerably. If the engine-wing interdependencies were crucial, for example if standard engine mounts had a drastically negative effect on the airplane's aerodynamics, then the design of these two subsystems would have to be coupled much more closely in order to produce a satisfactory design.

## 3.4. Imprinting

One common technique used to speed network convergence is *imprinting*, wherein the network influences are modified when a successful solution is found in order to facilitate quickly finding (similar) good solutions next time. A common imprinting technique is reinforcement learning, wherein the links representing influences that are satisfied in a successful final configuration of the network are strengthened, and those representing violated influences weakened. The effect of this is to create fewer but higher optima in the utility function, thereby increasing the likelihood of hitting such optima next time.

Imprinting is a crucial part of collaborative design. The configuration of influences between design participants represents a kind of 'social' knowledge that is generally maintained in an implicit and distributed way within design organizations, in the form of individual designer's heuristics about who (i.e. which individual or design group) should talk to whom when about what. When this knowledge is lost, for example due to high personnel turnover in an engineering organization, the ability of that organization to do complex design projects is compromised. It should be noted, however, that imprinting reinforces the tendency we have already noted for organizations in non-linear design regimes to stick to tried-and-true designs, by virtue of making the previously-found optima more prominent in the design utility function.

# 4. How Can We Help?

Once the design of a complex artifact has been distributed to many players, encouraging proper influence relationships and local search strategies is the primary tool available to design managers, and should therefore be supported by computer-supported collaborative design technology. This can occur in several ways. Such technology can help monitor the influence relationships between design participants. One could track the volume of design-related exchanges or (a more direct measure of actual influence) the frequency with which design changes proposed by one participant are accepted as is by other participants. This can be helpful in many ways. Highly asymmetric influences could represent an early warning sign of non-convergent dynamics. Detecting a low degree of influence by an important design concern, especially one such as environmental impact that has traditionally been less valued, can help avoid utility problems down the road. A record of the influence relationships in a successful design project can be used to help design similar future projects. Influence statistics can also be used to help avoid repetitions of a failed project. If a late high-impact problem occurred in a subsystem that had a low influence in the design process, this would suggest that the influence relationships should be modified in the future. Note that this has the effect of making a critical class of normally implicit and distributed knowledge more explicit, and therefore more amenable to being preserved over time (e.g. despite changes in personnel) and transferred between projects and even organizations.

Computer-supported collaborative design technology can also help assess the degree to which the design participants are engaged in routine vs innovative design strategies. We could use such systems to estimate for example the number and variance of design alternatives being considered by a given design participant. This is important because, as we have seen, a premature commitment to a routine design strategy that optimizes a given design alternative can cause the design process to miss other alternatives with higher global optima. Tracking the degree of innovative exploration can be used to fine-tune the use of innovation-enhancing interventions such as incentives, competing design teams, introducing new design participants, and so on.

## 5. References


[1] Bar-Yam, Y., *Dynamics of complex systems*. 1997, Reading, Mass.: Addison-Wesley. xvi, 848.

[2] Klein, M., et al., *Negotiating Complex Contracts*. 2001, Massachusetts Institute of Technology: Cambridge MA USA.

[3] Brown, D.C. *Making design routine*. in *Proceedings of IFIP TC/WG on Intelligent CAD*. 1989.

[4] Smith, R.P. and S.D. Eppinger, *Identifying controlling features of engineering design iteration*. Management Science, 1997. **43**(3): p. 276-93.

# Chapter 2

# RAn (Robustness Analyser)

**Fabrice Saffre**
BTexact Technologies
Adastral Park, Antares 2 pp 5
Martlesham IP5 3RE
United Kingdom
fabrice.saffre@bt.com

## 1. Introduction

In this paper, we present RAn, a simple and user friendly software tool for analysing topological robusntess of random graphs. The underlying model and the methodology are briefly described, then RAn's potential as a research and design tool is illustrated through a basic practical example.

Robustness of complex networks has been extensively discussed in the scientific literature for the last few years. Several authors have pointed out that different topologies would react differently to node failure and/or broken links (see e.g. Albert et al., 2000; Cohen et al., 2000) and that mathematical techniques used in statistical physics could effectively be used to describe their behaviour (see e.g. Callaway et al., 2000). It has also been demonstrated that most artificial networks, including the Internet and the World Wide Web, can be described as complex systems, often featuring "scale-free" properties (see e.g. Albert et al., 1999; Faloutsos et al., 1999; Tadic, 2001).

In this context, it is becoming increasingly obvious that the robustness of a wide variety of real distributed architectures (telecommunication and transportation networks, power grids etc.) is essentially a function of their topology, and could therefore be evaluated on the basis of their blueprint. Similarly, several alternative designs could be compared before their actual implementation, in order, for example, to balance redundancy costs against increased resilience.

## 2. The model

Efficient quantification and comparison requires selecting a consistent set of global variables that are considered a suitable summary of network behaviour under stress. In a previous work (Saffre and Ghanea-Hercock, 2001), we found that the decay of the average relative size of the largest component <*S*> could effectively be modelled using a basic non-linear equation of the form:

$$\langle S \rangle = \frac{X}{X + e^{\beta x}} \qquad [1a]$$

where $X$ and $\beta$ are constants, while $x$ is the fraction of nodes which have been removed from the original network. Depending on the topology, a better fitting can sometimes be obtained for a slightly different expression:

$$\langle S \rangle = \frac{X}{X + x^{\beta}} \qquad [1b]$$

Equations [1a] and [1b] obey a very similar logic though, and their relative efficiency in describing the system's behaviour can actually be used as a first indication to discriminate between 2 "qualitatively" different categories of architecture.

In any case, if expression [1a] or [1b] give a satisfactory approximation of the decay of a specific network's largest component, then the corresponding $X$ and $\beta$ global variables are all that is required to quantify its resilience to cumulative node failure. For increased clarity, it might be preferable to use an adjusted value of $X$:

$$X_c = \frac{\ln(X)}{\beta} \qquad [2a]$$

or

$$X_c = \sqrt[\beta]{X} \qquad [2b]$$

for [1a] and [1b] respectively. $X_c$ is then the value of $x$ for which the average relative size of the largest component is equal to 0.5, that is: the critical fraction of "missing" nodes above which, on average, less than 50% of the surviving elements are still interconnected. Finally, the value of $\beta$ itself roughly indicates the slope of the curve around this critical value.

## 3. Operation

The 3 global variables mentioned in the previous section ($\beta$, $X$ and $X_c$) are automatically computed by RAn, after performing a statistical analysis on data produced using Monte Carlo simulation techniques. Network structure and simulation parameters have to be specified by the user. A Graphical User Interface (GUI) allows these be entered/modified very easily (see Fig. 1). After a properly formatted topology file has been generated for the network to analyse, the user may launch RAn to perform robustness tests.

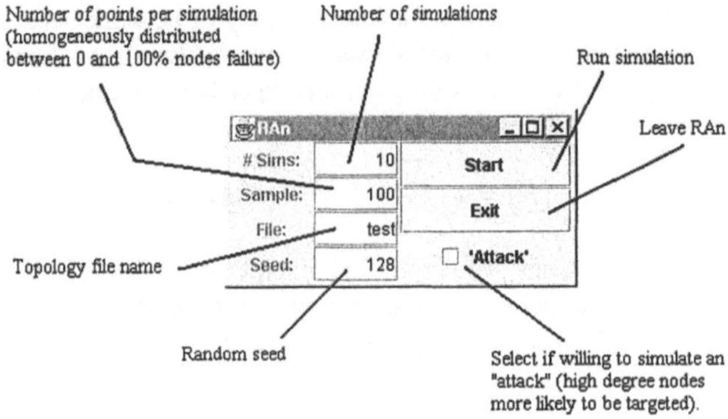


**Figure 1:** RAn Graphical User Interface (GUI).

The total duration of the simulation process is obviously highly dependent on parameter values and network size. As an order of magnitude, analysing the resilience to random failure of a network up to ten thousand nodes large (with a similar number of connections) is typically done in less than 2 minutes on a standard PII Desktop PC.

After the simulation phase is over, RAn analyses the data and the results are summarised as a series of automatically generated files. In addition to these, RAn also provides a graphical summary, as illustrated in Fig. 2.

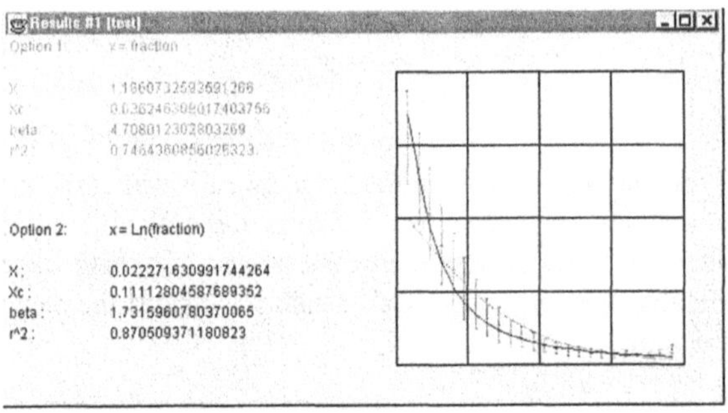


**Figure 2:** Results window displayed by RAn after the robustness analysis is completed. It includes values for all global variables and a graph showing simulation data (average +/- standard deviation) as well as both fitting curves. In this example (1000 nodes, 999 links, scale-free) expression [1b] (option 2) provides better fitting (brittle network).

If the "Attack" option is selected in the GUI (see Fig. 1), RAn makes the assumption that the attacker possesses partial information about network topology, and follows a "best guess" strategy in order to chose which node to target next. This is modelled by attributing to each surviving node a probability of being selected that is linearly proportional to its degree *k*:

$$P_i = \frac{(k_i + 1)}{\sum_{j=1}^{n} (k_j + 1)} \qquad [3]$$

Equation [3] is obviously a variant of the preferential attachment rule presented by Barabasi et al. (1999). Unfortunately, $P_i$ has to be recalculated after each attack in order to take into account the changing probability distribution caused by the elimination of one of the nodes. This increased complexity is the reason why testing a network's resilience for directed attack is considerably more intensive (and time consuming) than for random failure. However, this could probably be improved by optimising the algorithm.

Finally, it might be worth mentioning that the "Attack" scenario, because of its stochastic nature, could also be used to model special forms of accidental damage where connectivity level is involved. For example, it is conceivable that in a network where congestion is a cause for node failure, key relays (high degree nodes) would also be more likely to suffer breakdown, which could easily be modelled using expression [3].

## 4. Example scenario

This section consists in a scenario emphasising how RAn could be used as a design tool when planning network architecture. The planned network is a relatively large 3000 nodes system. The cheapest way to have them all interconnected (from a strictly topological point of view!) would involve 2999 links. They could all be arranged in a single "star" or in a closed "loop", but more realistic architectures would probably involve inter-connected sub-domains of different size and/or topology. Because it has been shown that most networks belong to this category, we did "grow" a scale-free network of the appropriate size (3000 nodes, one link per node except the 1st one) to use as the basic blueprint. Obviously, the process of generating such blueprint would be different if a real system was being designed, because it would have to take into account many other parameters (node type and capability, geographical location, connection type...). However, this makes no difference for RAn, as long as the corresponding topology is translated into the appropriate file format.

Assuming that this topology is actually the blueprint for a real architecture, the network designer could use RAn to compute statistics about its resilience to node failure, in terms of the cohesion of its largest component (initially including all nodes). After the complete process is over (simulation + analysis took just under 1 min for this example), RAn displays the following results window:

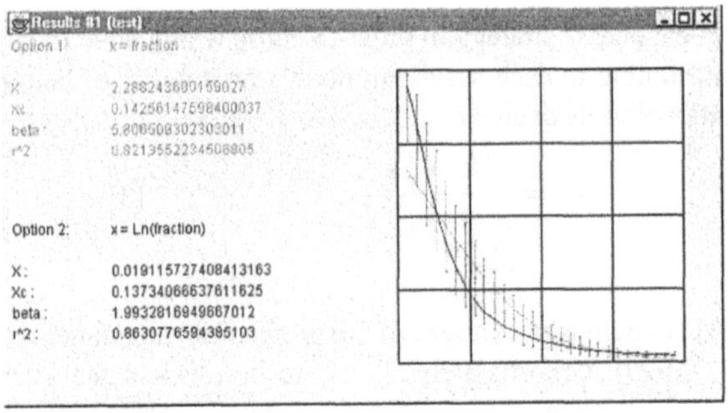


**Figure 3:** Analysis results for a model scale-free network (3000 vertices, 2999 edges).

Because this sort of network is basically a tree-like hierarchical structure with no built-in redundancy (1 link per node), it is not very robust to node failure. Indeed, RAn finds that, on average, removing only about 14% of all vertices (equivalent to severing all their links) is enough to reduce the size of the largest component to 50% of the surviving population ($X_c \sim 0.14$). So in effect, RAn tells the designer that if 500 nodes out of 3000 are malfunctioning, chances are the largest sub-set of relays that are still interconnected contains less than a half of the 2500 surviving nodes. In other words, it is likely that in this situation, around 1250 otherwise perfectly operational nodes are in fact cut from (and, obviously, unable to exchange any information with) the core of the network.

As can be expected, testing the same architecture for "attack" gives even more concerning results. In this scenario, killing only about 2% of the population (but this time selecting preferentially highly connected nodes) is enough to reach the same situation. So when applied to a typical scale-free architecture, RAn correctly and automatically predicts the type of network behaviour described by Albert et al. (2000), with the added advantage of summarising it by a set of global variables.

In the hypothesis that the designer wants to increase the robustness of the planned network, alternative blueprints could be produced, then analysed using RAn in order to compare their performance against that of the original, "cheapest", structure. For example, a straightforward way of increasing robustness is to add at least some backup links, so that alternative routes are available between nodes in case the primary (presumably most efficient) path becomes unavailable due to node failure(s). In our example, the designer could want to test the influence of doubling the total number of connections (raising it to 5999 links). As shown on Fig. 4a, this has a rather spectacular effect on robustness:

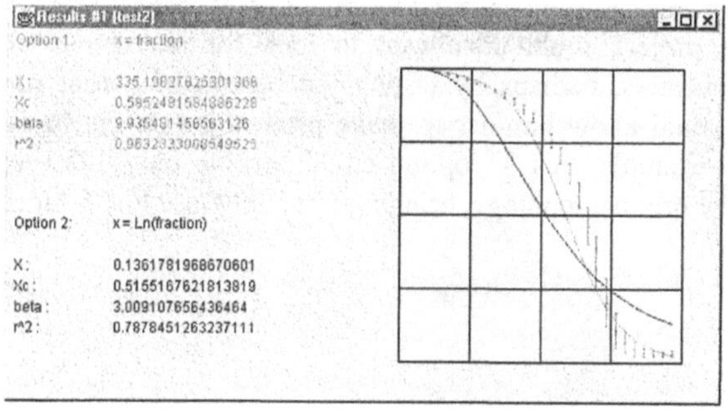


**Figure 4a:** Analysis results for a higher connection density (3000 vertices, 5999 edges).

With 3000 new connections added to the original blueprint, the network becomes much more resilient to node failure: it now takes about 60% nodes missing before more than a half of the surviving population is cut from the largest component. It is also clear that option 1 (expression. [1a]) now gives a much better fitting than option 2 (expression [1b]), suggesting a "qualitative" change in network behaviour. Moreover, RAn provides additional information in the form of the evolution of the standard deviation around the average value. Indeed, until up to 50 percent nodes have failed, the relative size of the largest component appears extremely stable from one simulation to the other, unlike in the original architecture (meaning the reaction of the network to cumulative stress has become more predictable). Finally, the ability of the network to withstand directed attack is even more dramatically increased, as shown on Fig. 4b. Indeed, instead of requiring the removal of only 2% of the nodes, it is now necessary to kill up to 40% to break the largest component, even though the most highly connected vertices are still specifically targeted.

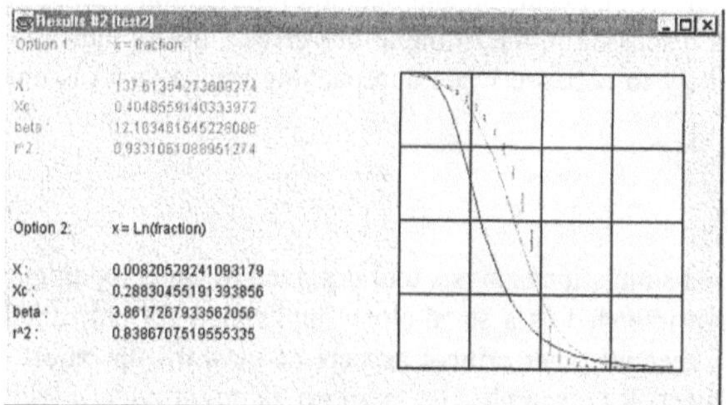


**Figure 4b:** same network as in Fig. 4a, this time submitted to directed attack.

However, simply doubling the number of links could be regarded as an unacceptable solution because of financial considerations. The network designer in charge of the project could then want to look for alternative, cheaper ways of improving robustness, perhaps by testing the benefit of partial route redundancy. Again, RAn would allow him/her to make projections on the basis of yet another blueprint. For example, this 3[rd] option could involve only 1000 extra-connections compared to the original topology, bringing it to 3999 (see Fig. 5 for results).

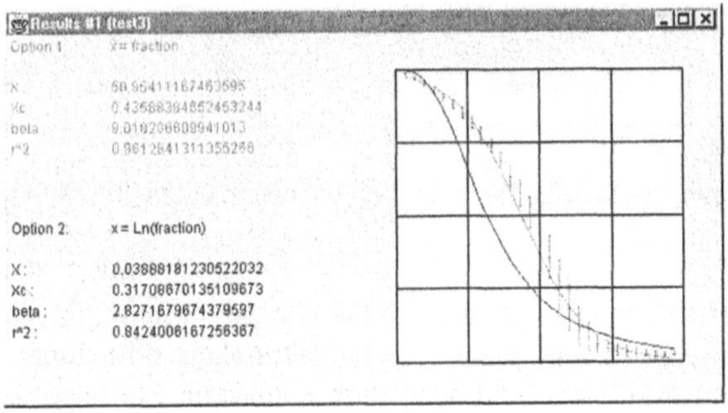


**Figure 5:** Analysis results for intermediate connection density (3000 vertices, 3999 edges).

Obviously, there is no free lunch: robustness is not increased in the same proportion as before. However, considering that only 33% extra links were created instead of 100%, this solution might in fact be a better one. Indeed, the critical size $X_c$ is shifted to ~0.44, meaning a factor 3 was gained compared to the original blueprint, against a factor 4 when *doubling* the number of connections, so the last choice is certainly more cost-effective!

Of course, those results by themselves are not very meaningful, since they relate to a constructed model, but they demonstrate that RAn is potentially a powerful tool for the network designer. Indeed, provided that the appropriate topology files are available, obtaining all this valuable and detailed information (including the value of $\beta$, which was not discussed in the example but gives a useful indication of how fast the network is likely to collapse when approaching critical size) is only a matter of minutes.

## 5. Conclusion

RAn is a combined simulation/analysis tool designed to study topological robustness. It is obviously not intended as a stand-alone application for network analysis, as it doesn't take into account other critical aspects of network operation like traffic or routing management for example. Its purpose is to provide a suitable way of estimating the speed and profile of the largest component's decay under cumulative node failure, a necessary first step in assessing a system's ability to withstand damage.

However, developing strategies to increase cohesion of the largest component (a task that can effectively be conducted using RAn to test alternative designs) is not sufficient to guarantee quality of service. This would also require being able to adapt network operations to a changing topology (which is an entirely different problem). It cannot be denied though that maintaining all nodes within the largest component is a first and necessary condition in order to increase network robustness. In that respect, RAn could be a most valuable tool for the network designer, even though it would require being combined with others to generate an accurate and practical simulation of a realistic architecture.

## References

Albert R., H. Jeong, and A.-L. Barabasi (1999), "*Diameter of the World-Wide Web*", Nature **401**, pages 130-131.

Albert R., H. Jeong, and A.-L. Barabasi (2000), "*Error and attack tolerance of complex networks*", Nature **406**, pages 376-382.

Barabasi A.-L., R. Albert and H. Jeong (1999), "*Mean-field theory for scale-free random networks*". Physica A **272**, pages 173-187.

Callaway D.S., M.E.J. Newman, S. H. Strogatz, and D.J. Watts, "*Network Robustness and Fragility: Percolation on Random Graphs*" (2000), Phys. Rev. Letters **85**, pages 5468-5471.

Cohen R., K. Erez, D. ben-Avraham and S. Havlin (2000), "*Resilience of the Internet to random breakdowns*", Phys. Rev. Letters **85**, pages 4626-4628.

Faloutsos M., P. Faloutsos, and C. Faloutsos (1999), "*On Power-Law Relationships of the Internet Topology*", ACM SIGCOMM '99, Comput. Commun. Rev. **29**, pages 251-263.

Tadic B. (2001), "*Dynamics of directed graphs: the world-wide Web*", Physica A **293**, pages 273-284.

Saffre F. and R. Ghanea-Hercock (2001), "*Robustness in Complex Network: a simplified model*", International Conference on Dynamical Networks in Complex Systems.

# Chapter 3

# Invariant Subsets of the Search Space, and the Universality of a Generalized Genetic Algorithm.

**Boris Mitavskiy**
School of Medicine, University of Sheffield
Sheffield, S10 2JF, United Kngdom
B.Mitavskiy@sheffield.ac.uk

In this paper we shall give a mathematical description of a general evolutionary heuristic search algorithm which allows to see a very special property which slightly generalized binary genetic algorithms have comparing to other evolutionary computation techniques. It turns out that such a generalized genetic algorithm, which we call a binary semi-genetic algorithm, is capable of encoding virtually any other reasonable evolutionary heuristic search technique.

## 1 Introduction

In this paper we shall describe a mathematical framework which allows to see some special properties which binary genetic algorithms have comparing to other evolutionary computation techniques. It turns out that a slightly generalized version of a binary genetic algorithm can encode virtually any "reasonable" heuristic search algorithm. (see corollary 5.6 and corollary 5.7.) This seems interesting at least from a philosophical point of view, for it says something about the special properties of the reproduction mechanisms occurring in nature. Moreover, it has been pointed out in [7] that such generalizations may actually be useful for practical purposes. In fact, the binary embedding theorem and its corollaries (see theorem 5.5 and corollary 5.6) provide both, sufficient and necessary conditions when a given evolutionary algorithm can be embedded into (encoded by) a binary semi-genetic algorithm. These conditions depend only on the nature of the family of the reproduction transformations, and are completely independent of any particular structure on the search space. Theorem 5.5 classifies all such encodings in terms of the invariant subsets of the search space. This may be useful for practical purposes, to simulate a given evolutionary heuristic search algorithm on a computer.

By an evolutionary heuristic search algorithm we mean a heuristic search technique used to solve optimization problems which mimics the basic natural evolution cycle: the natural selection, (or the survival of the fittest) reproduction, and mutation. The precise mechanism is outlined in the following sections.

# 2 Notation

$\Omega$ is a finite set, called a *search space*.

$f : \Omega \to (0, \infty)$ is a function, called a *fitness* function. The goal is to find a maximum of the function $f$.

$\mathcal{F}_q$ is a collection of $q$-ary operations on $\Omega$. Intuitively $\mathcal{F}_q$ can be thought of as the collection of reproduction operators: some $q$ parents produce one offspring. In nature $q = 2$, for every child has two parents, but in the artificial setting there seems to be no special reason to assume that every child has no more than two parents.

$\mathcal{M}$ is a collection of unary operations on $\Omega$. Intuitively these are asexual reproduction, or mutation operators.

# 3 how does a heuristic search algorithm work?

A population $P = \begin{pmatrix} x_1 \\ x_2 \\ \vdots \\ x_m \end{pmatrix}$ with $x_i \in \Omega$ is selected randomly.

**Evaluation:**

Individuals of $P$ are evaluated:

$$\begin{pmatrix} x_1 \\ x_2 \\ \vdots \\ x_m \end{pmatrix} \begin{matrix} \to \\ \to \\ \vdots \\ \to \end{matrix} \begin{matrix} f(x_1) \\ f(x_2) \\ \vdots \\ f(x_m) \end{matrix}$$

**Selection:**

A new population

$$P' = \begin{pmatrix} y_1 \\ y_2 \\ \vdots \\ y_m \end{pmatrix}$$

is obtained where $y_i = x_j$ with probability $\frac{f(x_j)}{\Sigma_{l=1}^{m} f(x_l)}$.

In other words, all of the individuals of $P'$ are these of $P$, and the expectation of the number of occurrences of any individual of $P$ in $P'$ is proportional to the number of occurrences of that individual in $P$ times the individual's fitness value. In particular, the fitter the individual is, the more copies of that individual are likely to be present in $P'$. On the other hand, the individuals having relatively small fitness value are not likely to enter into $P'$ at all. This is designed to imitate the natural survival of the fittest principle.

**Partition:**

The individuals of $P'$ are partitioned into pairwise disjoint tuples for mating according to some probabilistic rule: For instance the tuples could be

$$Q_1 = \begin{pmatrix} y_{i_1^1} \\ y_{i_2^1} \\ \vdots \\ y_{i_{q_1}^1} \end{pmatrix} \quad Q_2 = \begin{pmatrix} y_{i_1^2} \\ y_{i_2^2} \\ \vdots \\ y_{i_{q_2}^2} \end{pmatrix} \dots \; Q_j = \begin{pmatrix} y_{i_1^j} \\ y_{i_2^j} \\ \vdots \\ y_{i_{q_j}^j} \end{pmatrix} \dots$$

**Reproduction:**

Replace every one of the selected $q_j$-tuples $Q_j = \begin{pmatrix} y_{i_1^j} \\ y_{i_2^j} \\ \vdots \\ y_{i_{q_j}^j} \end{pmatrix}$ with the $q_j$-tuples

$$Q' = \begin{pmatrix} T_1(y_{i_1^j}, y_{i_2^j}, \dots, y_{i_{q_j}^j}) \\ T_2(y_{i_1^j}, y_{i_2^j}, \dots, y_{i_{q_j}^j}) \\ \vdots \\ T_{q_j}(y_{i_1^j}, y_{i_2^j}, \dots, y_{i_{q_j}^j}) \end{pmatrix}$$

for some randomly selected $q_j$-tuples of transformations $(T_1, T_2, \dots, T_{q_j}) \in (\mathcal{F}_{q_j})^{q_j}$. This gives us a new population

$$P'' = \begin{pmatrix} z_1 \\ z_2 \\ \vdots \\ z_m \end{pmatrix}$$

**Mutation:**

Finally, with small probability we replace $z_i$ with $F(z_i)$ for some randomly chosen $F \in \mathcal{M}$. This, once again, gives us a new population $P''' = \begin{pmatrix} w_1 \\ w_2 \\ \vdots \\ w_m \end{pmatrix}$

Upon completion of mutation start all over with the initial population $P'''$. The cycle is repeated a certain number of times depending on the problem.

# 4 a couple special heuristic search algorithms:

The search space of every one of the following heuristic search algorithms is $S = \{0, 1\}^n$.

**Binary Genetic Algorithm:**

For every subset $M \subseteq \{1, 2, \dots, n\}$, let

$$L_M(\mathbf{a}, \mathbf{b}) = (x_1, x_2, \dots, x_i, \dots, x_n)$$

where $\mathbf{a} = (a_1, a_2, \dots, a_n)$ and $\mathbf{b} = (b_1, \dots, b_n) \in S$ and $x_i = \begin{cases} a_i & \text{if } i \in M \\ b_i & \text{otherwise} \end{cases}$.

Let $\mathcal{F}_M = \{L_M \mid M \subseteq \{1, 2, \dots, n\}\}$ play the role of $\mathcal{F}_2$ from the previous section.

**Example:**

With $n = 5$ and $M_1 = \{2, 3, 4\}$, $M_2 = \{1, 3, 5\}$ we have

$$\begin{pmatrix} 1 & 0 & 0 & 1 & 1 \\ 1 & 1 & 0 & 0 & 1 \end{pmatrix} \longmapsto \begin{pmatrix} L_{M_1}((1, 0, 0, 1, 1), (1, 1, 0, 0, 1)) \\ L_{M_2}((1, 0, 0, 1, 1), (1, 1, 0, 0, 1)) \end{pmatrix} = \begin{pmatrix} 1 & 1 & 0 & 0 & 1 \\ 1 & 1 & 0 & 0 & 1 \end{pmatrix}$$

The genetic crossover transformations are classified by the following property: If both parents have a 1 in the $i^{\text{th}}$ position then the offspring also has a 1 in the $i^{\text{th}}$ position. Likewise, if both parents have a 0 in the $i^{\text{th}}$ position then the offspring also has a 0 in the $i^{\text{th}}$ position. If, on the other hand, the alleles of the $i^{\text{th}}$ gene don't coincide, then the $i^{\text{th}}$ allele could be either a 0 or a 1.

It turns out, that if we relax the condition on the preservation of genes, so to speak, by half, meaning that If both parents have a 1 in the $i^{\text{th}}$ position then the offspring also has a 1 in the $i^{\text{th}}$ position, but, in any other case, there is no requirement on the $i^{\text{th}}$ gene: it could be either a 0 or a 1, then one obtains a very special evolutionary heuristic search algorithm described below. In section 5 we shall see that such an evolutionary algorithm is virtually universal, since it describes virtually any other reasonable heuristic search algorithm. (see theorem 5.5, corollary 5.6 and corollary 5.7.)

**Binary Semi-Genetic Algorithm:**

**Definition 4.1** Fix $m \geq 2$ and $\mathbf{u} = (u_1, u_2, \dots, u_n) \in S$. Define a semi-crossover transformation $F_{\mathbf{u}\,m} : S^m \to S$ as follows: For any given matrix

$$P = \begin{pmatrix} a_{11} & a_{12} & \dots & a_{1n} \\ a_{21} & a_{22} & \dots & a_{2n} \\ \vdots & \vdots & \ddots & \vdots \\ a_{m1} & a_{m2} & \dots & a_{mn} \end{pmatrix}$$

in $S^m$ $F_{\mathbf{u}\,m}(P) = \mathbf{x} = (x_1, x_2, \dots x_n) \in S$ where

$$x_i = \begin{cases} a_{1i} \text{ if } \forall\, 1 \leq j \leq k \leq m \;\, a_{ji} = a_{ki} = 1 \\ u_i \text{ otherwise} \end{cases}$$

In other words, $F_{\mathbf{u}\,m}$ preserves the $i^{\text{th}}$ gene if it is equal to 1 in all of the rows of $P$, and replaces it with $u_i$ otherwise. Denote by $\mathcal{F}_m = \{F_{\mathbf{u}\,m} \mid \mathbf{u} \in S\}$ the family of all semi-crossover transformations.

**Example:**
With $n = 5$ and $\mathbf{u}_1 = (0, 1, 1, 0, 1)$, $\mathbf{u}_2 = (0, 1, 0, 0, 1)$ we have

$$\begin{pmatrix} 1 & 0 & 0 & 1 & 1 \\ 1 & 1 & 0 & 0 & 1 \end{pmatrix} \longmapsto \begin{pmatrix} F_{\mathbf{u}_1\,2}((1, 0, 0, 1, 1),\ (1, 1, 0, 0, 1)) \\ F_{\mathbf{u}_2\,2}((1, 0, 0, 1, 1),\ (1, 1, 0, 0, 1)) \end{pmatrix} = \begin{pmatrix} 1 & 1 & 1 & 0 & 1 \\ 1 & 1 & 0 & 0 & 1 \end{pmatrix}$$

Notice, that if 1 is present in the $i^{th}$ position of both parents, then it remains in the $i^{\text{th}}$ position of both offsprings. There are absolutely no other restrictions, though.

# 5 the binary embedding theorem

**Question:** Under which conditions can a given heuristic search algorithm be encoded by a binary semi-genetic or, better yet, by a binary genetic algorithm?

The main idea behind answering the question above is to observe that the families of invariant subsets naturally determine the corresponding families of transformations fixing them. The rigorous machinery is fully developed in the appendix of [5], and is also available upon request from the author.

Let $\Gamma$ denote a family of transformations from $\Omega^m$ into $\Omega$.

Let $\Lambda_\Gamma = \{S \mid S \subseteq \Omega,\ T(S^m) \subseteq S\ \forall\, T \in \Gamma\}$ denote the family of all invariant subsets under the action of $\Gamma$.

Under certain slightly technical conditions on the family of transformations $\Gamma$ (these conditions are satisfied by both, the family of all crossover transformations and the family of all semi-crossover transformations. All of the rigorous details can be found in [5], and are also available upon request from the author.) the family of transformations

$$\widehat{\Gamma} = \{T \mid \forall\, \mathbf{x} \in \Omega^m\ \exists \text{ a transformation } T_{\mathbf{x}} \in \Gamma \text{ such that } T(\mathbf{x}) = T_{\mathbf{x}}(\mathbf{x})\}$$

is the largest family of transformations such that $\Lambda_{\widehat{\Gamma}} = \Lambda_\Gamma$.

As we have seen in the section 3, a given evolutionary heuristic search algorithm is entirely determined by the families of its “reproduction” transformations. This motivates the following definition:

**Definition 5.1** A heuristic $k$-tuple $\mathit{\Omega} = (\Omega, \Gamma_1, \Gamma_2, \ldots \Gamma_k)$ is a $k+1$-tuple where $\Omega$ denotes an arbitrary set while $\Gamma_i$ is just a family of transformations from $\Omega^{m_i}$ into $\Omega$ and $m_1 < m_2 < m_i < \ldots < m_k$. We say that the $k$-tuple of integers $(m_1, m_2, \ldots, m_k)$ is the arity of the heuristic $k$-tuple $(\Omega, \Gamma_1, \Gamma_2, \ldots \Gamma_k)$. We also say that the collection $\Lambda_{\mathit{\Omega}} = \bigcap_{1 \le i \le k} \Lambda_{\Gamma_i}$ is the collection of $\mathit{\Omega}$-invariant subsets.

For $x \in \Omega$, denote by $S_x^{\mathit{\Omega}}$ the smallest element of $\Lambda_{\mathit{\Omega}}$ containing $x$. (Notice that $\Lambda_{\mathit{\Omega}}$ is closed under arbitrary intersections so that $S_x^{\mathit{\Omega}} = \bigcap_{K \in \Lambda_{\Gamma_i},\, x \in K} K$.)

In section 4 we have described the binary semi-genetic algorithm by the following heuristic $k$-tuple:

**Definition 5.2** Let $S = \{0, 1\}$. We shall say that $(S, \widehat{\mathcal{F}_{m_1}}, \widehat{\mathcal{F}_{m_2}}, \ldots, \widehat{\mathcal{F}_{m_k}})$ is a semi-genetic heuristic $k$-tuple of dimension $n$, where $m_1 < m_2 < \ldots < m_k$.

The following definition provides the means for the comparison of the various heuristic $k$-tuples. An encoding of $\Omega$ by $\Phi$ is simply a mapping $\delta : \Omega \to \Phi$. $\forall\, w \in \Omega\ \delta(w)$ is just the code of $w$ in $\Phi$. If the mapping $\delta : \Omega \to \Phi$ is one-to-one, then one can completely recover any $w \in \Omega$ from its code $\delta(w)$. In other words, $\Omega$ is completely identified with the subset $\delta(\Omega) \subseteq \Phi$. If $\mathit{\Omega} = (\Omega,\ \mathit{\Gamma}_1,\ \mathit{\Gamma}_2, \ldots \mathit{\Gamma}_k)$ and $\mathit{\Phi} = (\Phi, \Theta_1,\ \Theta_2, \ldots, \mathit{\Gamma}_k)$ are two heuristic $k$-tuples of the same arity, a natural way to compare $\mathit{\Omega}$ with $\mathit{\Phi}$ is to construct an "encoding" mapping $\delta : \Omega \to \Phi$ which respects the "reproduction transformations". This motivates the following definition:

**Definition 5.3** Given two heuristic $k$-tuples $\mathit{\Omega} = (\Omega,\ \mathit{\Gamma}_1,\ \mathit{\Gamma}_2, \ldots, \mathit{\Gamma}_k)$ and $\mathit{\Phi} = (\Phi,\ \Theta_1, \Theta_2, \ldots, \Theta_k)$ of the same arity, ( see definition 5.1 ) a morphism $\delta : \mathit{\Omega} \to \mathit{\Phi}$ is just a function $\delta : \Omega \to \Phi$ which respects the reproduction transformations, meaning that $\forall\ 1 \le i \le k$ and $\forall\ T \in \mathit{\Gamma}_i\ \exists\ F \in \Theta_i$ such that $\forall\ w_1, w_2, \ldots, w_{m_i} \in \Omega$ we have $F(\delta(w_1), \delta(w_2), \ldots, \delta(w_{m_i})) = \delta(T(w_1, w_2, \ldots, w_{m_i}))$.

We say that a morphism $\delta : \mathit{\Omega} \hookrightarrow \mathit{\Phi}$ is an embedding if the underlying function $\delta : \Omega \to \Phi$ is one-to-one.

The binary embedding theorem establishes an explicit one-to-one correspondence between the set of all embeddings of a given heuristic $k$-tuple into binary semi-genetic algorithms and a certain collection of ordered $n$-tuples of $\Omega$-invariant subsets.

**Definition 5.4** Fix any heuristic $k$-tuple $\mathit{\Omega} = (\Omega, \mathit{\Gamma}_1, \mathit{\Gamma}_2, \ldots, \mathit{\Gamma}_k)$. We say that collection

$$\Upsilon_n = \{\mathcal{I} \mid \mathcal{I} = (I_1, I_2, \ldots, I_n)\ I_j \in \Lambda_{\mathit{\Omega}},\ \forall\, x,\, y \in \Omega \text{ with } x \ne y\ \exists\, 1 \le j \le n$$

$$\text{such that either } (x \in I_j \text{ and } y \notin I_j) \text{ or vise versa: } (y \in I_j \text{ and } x \notin I_j)\}$$

is a family of separating $n$-tuples.

**Theorem 5.5** *Fix a heuristic $k$-tuple $\mathit{\Omega} = (\Omega, \mathit{\Gamma}_1, \mathit{\Gamma}_2, \ldots, \mathit{\Gamma}_k)$. We now have the following bijection $\phi : \Upsilon_n \to F^n_{\mathit{\Omega}}$ which is defined explicitly as follows: Given an ordered $n$-tuple of sets from $\Lambda_{\mathit{\Omega}}$, call it $\mathcal{I} = (I_1, I_2, \ldots, I_n) \in \Upsilon_n$, (see definition 5.4) let $\phi(\mathcal{I}) = \delta_{\mathcal{I}}$ where $\delta_{\mathcal{I}}(x) = (x_1, x_2, \ldots, x_n) \in S = \{0,1\}^n$ with*

$$x_j = \begin{cases} 1 & \text{if } x \in I_j \\ 0 & \text{otherwise} \end{cases} \quad \forall x \in \Omega.$$

*Proof:* Due to space limitation, a detailed argument is available upon request from the author. □

It turns out that the conditions under which a given heuristic $k$-tuple can be embedded into a binary semi-genetic heuristic $k$-tuple are rather mild and naturally occurring as the following two corollaries demonstrate:

**Corollary 5.6** *Given a heuristic k-tuple $\Omega = (\Omega, \Gamma_1, \Gamma_2, \ldots, \Gamma_k)$, $\Omega$, the following are equivalent:*

1. *$\Omega$ can be embedded into an $n$-dimensional semi-genetic heuristic $k$-tuple for some $n$.*
2. *$\forall x, y \in \Omega$ with $x \neq y$ we have either $x \notin S_y^{\Omega}$ (see definition 5.1) or vise versa: $y \notin S_x^{\Omega}$.*
3. *$\forall x, y \in \Omega$ with $x \neq y$ we have $S_x^{\Omega} \neq S_y^{\Omega}$. (Another way to say this, is that the map sending $x$ to $S_x^{\Omega}$ is one-to-one.)*

*Moreover, if an embedding exists for some $n$, then there exists one for $n = |\Omega|$. We also must have $n \geq \lceil \log_2 |\Omega| \rceil$.*

*Proof:* One simply shows that $\forall x, y \in \Omega$ with $x \neq y$ we have either $x \notin S_y^{\Omega}$ or $y \notin S_x^{\Omega}$ if and only if $|\Omega|$-tuple $\mathcal{S} = (S_{x_1}^{\Omega}, S_{x_2}^{\Omega}, \ldots, S_{x_{|\Omega|}}^{\Omega})$ where $\{x_i\}_{i=1}^{n}$ is an enumeration of all the elements of $\Omega$ is separating ( i. e. $\mathcal{S} \in \Upsilon_n$, see definition 5.4 ) if and only if $\Upsilon_n \neq \emptyset$ which, in turn, according to theorem 5.5, happens if and only if $\Omega$ can be embedded into an $n$-dimensional semi-genetic heuristic $k$-tuple for some $n$. This establishes the equivalence of 1 and 2. Clearly , 2 implies 3. To see the converse, we show that "Not 2" implies "Not 3". Indeed, if $x \in S_y^{\Omega}$ and $y \notin S_x^{\Omega}$, then, by minimality, (see definition 5.1) we have $S_x^{\Omega} \subseteq S_y^{\Omega}$ and $S_y^{\Omega} \subseteq S_x^{\Omega}$, so that $S_x^{\Omega} = S_y^{\Omega}$. Due to space limitations, a detailed argument is available upon request from the author. □

**Corollary 5.7** *Given a heuristic k-tuple $\Omega = (\Omega, \Gamma_1, \Gamma_2, \ldots, \Gamma_k)$, if $\forall\ 1 \leq j \leq k$ and for every $T \in \Gamma_j$, $T$ is idempotent ( in other words, $\forall\ x \in \Omega\ T(x, x, \ldots, x) = x$ ) then $\Omega$ can be embedded into a binary semi-genetic heuristic $k$-tuple of dimension less than or equal to $|\Omega|$.*

*Proof:* The desired conclusion follows immediately from corollary 5.6 by observing that $\forall\ x, y \in \Omega$ with $x \neq y$ we have $S_x^{\Omega} = \{x\}$ so that $x \in \{x\} = S_x^{\Omega}$ while $y \notin \{x\} = S_x^{\Omega}$. □

## 6 Conclusions and Future Work

In a classical binary genetic algorithm crossover works by swapping the alleles, while in the generalized version it works by preserving only the good allele ( = 1 ) and may or may not preserve the 0 gene. (see Definition 4.1) It seems interesting to know that such an algorithm is almost universal in the sense of Corollary 5.6 and Corollary 5.7. Notice that the conditions of Corollary 5.7 are quite natural to assume. They basically say that two or more identical individuals produce the offspring which is identically the same as the parent individual. Corollary 5.6 also shows that the dimension of the embedding can always be made less than or equal to the size of the underlying set, $\Omega$. It can be shown that, in general, the dimension can not be reduced any further, but the author conjectures, that, due to the rigidity of the collection of $m$-fixable family of subsets (see Appendix A of [5] for the definitions and machinery. The material is also available upon request from the author.), under some mild conditions, the dimension may be reduced drastically. This provides at least one possible direction for the future research.

Another natural question to ask is the following: Under which conditions can a given heuristic search algorithm be encoded by a classical (not necessarily binary) genetic algorithm? It turns out that the conditions involve some basic Abstract Algebra: In fact, a given heuristic $k$-tuple $\Omega$ can be "encoded" by a genetic algorithm (not necessarily a binary one) if and only if there exists a way to enlarge a set $\Omega$ to a superset $\Psi$ so that there exists a ring structure on $\Psi$ with comaximal ideals $I_1, I_2, \ldots I_n$ for which $\bigcap_{j=1}^{n} I_j = 0$ and $\forall\ 1 \leq j \leq n$ any union of cosets of $I_j$ intersected with $\Omega$ is in $\Lambda_\Omega$. The proof of this fact involves Chinese Remainder Theorem (see, for instance, Dummitt and Foote [4]) together with a few other technical facts (due to space limitations, these are available upon request from the author) used in ways similar to their usage in the proof of Theorem 5.5. An alternative approach has been developed by Nicholas J. Radcliffe [6]. Notice, however that Radcliffe's work relies on the notion of a "formae" which is less general than Mitavskiy's notion of the $m$-fixable family of subsets described in detail in [5]. In particular there is no way to use Radcliiffe's formae to describe the family of semi-genetic crossover operators, while the family of $m$-fixable subsets describes absolutely any family of $m$-ary reproduction transformations on an arbitrary, representation independent search space. (see Appendix A of [5] for details. Also available upon request from the author.) This type of theorems will be studied in my future research.

## 7 Acknowledgements

I want to thank Professor John Holland for the helpful discussions and for the encouragement I've received from him to write this paper. I also want to thank my thesis advisor, Professor Andreas Blass for the numerous helpful advisor meetings which have stimulated some of the ideas for this and for my future work. Finally I would like to thank my fellow graduate student of mathematics, Ronald Walker for a few very helpful discussions, and the University of Michigan Complex Systems Group for the suggestions regarding the organization of this paper.

## Bibliography

[1] ANTONISSE, J., "A new interpretation of Schema Notation that Overturns the Binary Encoding Constraint", *Proceedings of the Third International Conference on Genetic Algorithms* (J. SCHAFFER ed.), Morgan Kaufmann (1989), 86–97.

[2] MICHALEWICZ, Z., *Genetic Algorithms + Data Structures = Evolution Programs*, Springer-Verlag (1996).

[3] VOSE, M., "Generalizing the Notion of a Schema in Genetic Algorithms", *Artificial Intelligence* **50(3)** (1991), 385–396.

[4] DUMMIT, D., FOOTE, R., *Abstract Algebra*, Prentice-Hall, Inc. (1991).

[5] MITAVSKIY, B., "Crossover Invariant Subsets of the Search Space for Evolutionary Algorithms", *Evolutionary Computation* **12(1)** (2004), 19–46.

[6] RADCLIFFE, N., "The Algebra of Genetic Algorithms", *Annals of Math and Artificial Intelligence* **10** (1994), 339–384.

[7] WATSON, R., "Recombination Without Respect: Schema Combination and Disruption in Genetic Algorithm Crossover", *Proceedings of the 2000 Genetic and Evolutionary Computation Conference* (D. WHITLY ed.), Morgan Kaufmann (2000), 112–119.

# Part II:

# Models

# Chapter 1

# Cell-like space charge configurations formed by self-organization in laboratory

**Erzilia Lozneanu** and **Mircea Sanduloviciu**
Department of Plasma Physics
Complexity Science Group
Al. I. Cuza University
6600 Iasi, Romania
msandu@uaic.ro

A phenomenological model of self-organization explaining the emergence of a complexity with features that apparently satisfy the specific criteria usually required for recognizing the appearance of life in laboratory is presented. The described phenomenology, justified by laboratory experiments, is essentially based on local self-enhancement and long-range inhibition. The complexity represents a primitive organism self-assembled in a gaseous medium revealing, immediately after its "birth", many of the prerequisite features that attribute them the quality to evolve, under suitable conditions, into a living cell.

## 1. Introduction

In this paper we would like to report on the possibility to create in laboratory a complex space charge configuration (CSCC) representing, in our opinion, the simplest possible system able to reveal operations usually attributed to a biological cell. It appears in a cold physical plasma, *i.e.* a medium presumable similar to those existent under prebiotic Earth's conditions, when an electrical spark creates a well-located nonequilibrium plasma. In spite of its gaseous nature, such a CSCC satisfies to a large extend the criteria usually required to recognize the creation of life in laboratory. Thus, similar to biological cells, the boundary of a self-assembled CSCC provides a selective confinement of an environment that qualitatively differs from the surrounding medium. The boundary appears as a spherical self-consistent electrical double layer (DL) able to sustain and control operations such as: (i) capture and transformation of energy, (ii) preferential and rhythmic exchange of matter across the system boundary and internal transformation of matter by the means of a continuous "synthesis" of all components of the system. After its formation, the CSCC is able to replicate, by division, and to emit and receive information.

As proved by experimental and simulation studies [1-4] self-organization occurs in an intermittent fashion (intermittent self-organization) or in a stepwise fashion (cascading self-organization). Intermittent self-organization occurs when the system is gradually driven away from equilibrium by continuous injection of matter and energy, while cascading self-organization occurs when matter and energy are suddenly injected into the system so that it relaxes stepwise towards a minimum energy state. Because the succession of physical processes involved in the intrinsically nonlinear mechanism at the origin of the cascading scenario of self-organization is very fast, its identifications was not yet possible. However, it was possible to obtain information on this mechanism starting from the experimentally proved fact that the CSCC, created by a cascading scenario of self-organization, strictly reveals the same features as those of a CSCC created by an intermittent scenario of self-organization [1,3,4]. Based on these experimental results, in the following we explain the creation of a CSCC by an electrical spark, considering the well-identified physical processes the successive development of which explains the generation of a CSCC by an intermittent scenario of self-organization.

## 2. Phenomenological model of a gaseous cell

Since life necessarily exists in the form of cells, the accumulation of charged particles in the form of a membranous boundary is the first structural requirement for the creation of a minimal prebiotic system. As mentioned, in a cold laboratory plasma in thermodynamic equilibrium, the premise for the formation of a gaseous membrane is the generation of a well localized non-equilibrium relatively high temperature plasma in a point where an electrical spark strikes the surface of a positively biased electrode [4]. Because of the differences in the mobility and thermal diffusivities of the electrons and the positive ions, the former are quickly collected by the positive electrode, so that a positive "nucleus" in the form of an ion-rich plasma appears. Acting as a gas anode, the potential of which depends on the positive electrode potential, the nucleus attracts, in the following phase of its evolution, electrons from the surrounding cold plasma. When the potential of the positive electrode, and implicitly of the gas anode, is so high that the accelerated electrons obtain kinetic energies sufficient to produce excitations and ionizations of the neutrals, the conditions for the self-assembly of a CSCC following an intermittent scenario of self-organization are realized [1,3-8]. It is well known that the excitation and ionization cross section functions depend on the kinetic energy of the electrons in such a way that the former suddenly increases for lower kinetic energies than the latter. Consequently, a net negative space charge populated with electrons that have lost their kinetic energy by excitation of neutrals at the specific energy levels is formed, in agreement with the equation $A + e_{fast} \rightarrow A^* + e_{slow}$. In this equation A* is an excited – but still neutral – atom, which, after about $10^{-8}$ s, returns into the ground state by emitting a photon with the energy $h\nu$. Therefore the region where the net negative charge is located appears as a luminous sheet that surrounds the positive nucleus [3]. Its extension depends on the scheme of the excitation levels of the respective gas atoms. The electrons, after having lost kinetic energy in excitation processes form a net negative space charge. This is dynamically maintained since the part of the accumulated electrons that disappear by recombination, diffusion etc. is continuously replaced by those electrons that have lost their momentum after excitations.

The appearance of a well-located net negative space charge represents the first phase in the pattern formation mechanism. Thus, the net negative space charge determines the location of the electric field in a relative small region at the border of the positive nucleus. This creates the premise for the following evolution sequence of the space charge into a CSCC. It is related to the electrons that have not produced excitations and, as a consequence of the acceleration in the electric field, they obtain sufficient kinetic energy to produce ionization in the nucleus according to the process $A + e_{fast} \rightarrow A^+ + 2e_{slow}$. Here $A^+$ is a single-ionized positive ion. Since the electrons that have produced ionizations and those resulting from these processes have low kinetic energies and are located in a relatively high electric field, formed between the net negative space charge and the positive electrode, the electrode quickly collects them. As a consequence, in this initial phase of the CSCC self-assembling process, there are two adjacent space charges of opposite sign in front of the anode: a positive one located in the nucleus and a negative one in the form of a sheet surrounding the nucleus. Between them the electrostatic forces act as long-range correlations, so that the adjacent space charge layers naturally associate in the form of a DL. So an electric field appears, within which the electrons obtain additional energies, so that the requirements necessary for the self-assembly mechanism of the DL are fulfilled at the boundary of the nucleus. The strength of this field depends on the densities of the two adjacent opposite net space charges, which in turn depend on the excitation and ionization rates and, implicitly, on the potential of the positive electrode. When the potential of the positive electrode is so high that the potential drop developed on the DL reaches the ionization potential of the gas, an instability will start to develop because a higher ion density in the nucleus causes an increase of the local electric field. As a consequence, the kinetic energy of the electrons accelerated towards the nucleus increases so that the ionization rate also grows. The result is an additional amount of positive ions that is added to the previous one. Thus, the local electric field and, implicitly, the ionization rate are further increased. In turn, the increase of the ionization rate produces an additional increase of the positive ion density and, consequently, a further grow of the electric field, and so on. As a result of this positive feedback mechanism the density of positive ions quickly grows in the region where the ionization cross section function suddenly increases (*i.e.* adjacent to the well-localized net negative space charge). In this way a *self-enhancement* mechanism for the production of positive ions working at the boundary of the nucleus governs the further evolution of the space charge configuration [8]. Once started at a given position, the sudden increase of the production rate of positive ions leads to an overall "*activation*" of this process. Therefore the self-enhancement of the production of positive ions alone is not sufficient to generate stable patterns in plasma. Their generation becomes, however, possible if the self-enhancement of the positive ion production is complemented by a mechanism able to act as a "*long-range inhibitor*" without impeding the incipient self-enhancement mechanism of the production of positive ions. The long-range inhibition mechanism is related to the creation of a negative space charge by neutral excitations acting as an "antagonist" to the positive one. Since also the excitation rates of neutrals depend on the kinetic energy of electrons, the increase of the electric field intensity, related to the self-enhancement of the production of positive ions, determines also the growth of the density of the adjacent negative space charge.

Simultaneously with the grow of the positive ion density, the local electric field increases, so that the region where the negative space charge forms by accumulation of those electrons that have lost their momentum by neutrals excitations is shifted away from the positive electrode. This "expansion" phase of the space charge configuration ceases because the production rate of positive ion cannot increase above a certain value when the neutral gas pressure is maintained constant. In the final phase of the CSCC evolution the negative space charge "balances" the positive space charge situated between it and the positive electrode. After this evolution the space charge configuration appears as a stable self-confined luminous, nearly spherical, gaseous body attached at the anode. Its self-assembling process governed by the electrostatic forces acting as correlations between the two adjacent net space charges of the DL, does not require additional energy since the transition occurs to a state characterized by a local minimum of the potential energy.

We note that in this stage of self-organization of the CSCC the DL from its border ensures its spatial stability by maintaining a local electric field that accelerates electrons at energies sufficient for producing within it the processes required for replacement of all of its components [1,8,9]. This becomes possible only when the transport of thermalized plasma electrons is ensured by the work done by the external dc power supply. A higher degree of self-organization of the CSCC appears when, after additional matter and energy injection, the CSCC transits into an open stationary state during which it undertakes a part of the work necessary for its self-existence. This is realized by a proper dynamics of the DL during which a rhythmic exchange of matter and energy with the surrounding environment takes place [6]. Both of these degrees of self-organization of the CSCC correspond to the "pre-natal" stages because their existence requires the presence of the external dc power supply.

The most interesting phenomenon observed in physical plasmas appears when the CSCC is created in low voltage thermionic arcs. In those plasma devices the CSCC emergence can also be explained considering an intermittent scenario of self-organization [11]. After its genesis the existence of the CSCC is possible in a free-floating steady state during which the DL at its border is subjected to a successive detachment and reformation processes. These phenomena reveal striking similarities with those observed when a CSCC is attached at the positive electrode of a plasma diode [6,12] or when it is created in an hf electric field [13]. Surprising is the fact that, during this free floating state, the mean potential of the nucleus of a CSCC exceeds the ionization potential of the gas also when the anode potential is much smaller than that. This proves that there exists a recharging mechanism of the nucleus of the CSCC by which its periodic discharging related to the transport of positive ions by the DLs, which periodically detach from its boundary, is compensated. This recharging mechanism, at present investigated in our laboratories, can be tentatively explained considering the experimental results that have proved that a moving DL is able to accelerate thermal electrons of the plasma, through which it propagates [14]. The recharging of the nucleus of the CSCC with positive ions becomes possible taking into account the Maxwellian energy distribution of the plasma electrons. Thus, obtaining kinetic energy by acceleration in the field of the moving (expanding) DL and also in the electric field of the net positive space charge (remaining in the nucleus after the DL detachment), the electrons reach the nucleus with energies sufficient to produce direct or stepwise ionizations. Since ionizations are accompanied by a heating process the temperature of the nucleus increases. This temperature increase is produced specially by the plasma electrons from the high energetic tail of the energy distribution. As a consequence a part of the electrons from the nucleus of the CSCC are ejected by thermal diffusion. After the ejection of electrons the potential of the nucleus attains again the value for which at its boundary a stable DL is self-assembled.

The periodicity of the process is related to the fact that after the detachment from the CSCC the spherical DL expands, so that the flux of electrons crossing it becomes too small to ensure its self-assembling process. Thus it decays. Thereafter the electrons, bounded at the DL space charge configuration, become free and are accelerated, as a bunch, towards the positive nucleus. Reaching the boundary of the nucleus, where a new DL is in a stable state, the flux of electrons increases the ionization rate so that the potential drop of the DL reaches the critical value for which its detachment process starts over. So, an internal working feedback mechanism ensures a rhythmic exchange of matter and energy between the CSCCC and the environment. Preliminary investigations performed in our laboratories seem to show that after the "birth" of the free-floating CSCC in a thermionic diode, its further existence does not require work from the external dc power supply. On the contrary, it is able to produce work by direct conversion of thermal energy in electric energy.

For high gas pressures, the described DL detachment and reformation processes take place in a relatively small region at the border of the free-floating CSCC. Conveniently, this region could be considered as playing the role of a "membrane" that protects the CSCC from the surrounding environment. Since the detachment of the DL involves the extraction and transport of positive ions from the nucleus of the CSCC to the surrounding plasma, a pressure difference appears between the former and the latter. Thus the periodic detachment of DLs from the boundary of the CSCC border implies a periodic "inhalation" of fresh neutrals into the nucleus. So, a CSCC also mimics the breathing process proper to all living systems.

By revealing the above described qualities, the free floating CSCC self-assembled in a thermionic diode is, to the best of our knowledge, the first autonomous complexity manufactured under controllable laboratory conditions able to ensure its survive by operations controlled by the DL at its boundary. Note that the excitation and ionization processes, the collective effects of which are at the basis of the operations performed by the DL are produced for relatively small values of electron kinetic energy.

For a biologist it could be of interest that the self-organization scenario, which explains the emergence of a CSCC, is essentially based on opposite space charge separation related to the symmetry breaking of specific quantum cross section functions. If a phenomenon, which is in principle similar to one acting in plasma, could also occur in chemical media, in which autocatalytic processes determine pattern formation, this could be a fascinating problem of further investigations. In this context we remember that, starting from the fact that symmetry breaking is a universally present phenomenon in biology, in his paper "Chemical Basis of Morphogenesis" Turing [15] proposed a mechanism for the generation of biological patterns. He considered a system of equations for chemical reactions, coupled by diffusion, which would deliver solutions that could break the symmetry of the initial state of a system. In the same context we remember that self-organization, related to symmetry breaking due to fluctuations in chemical systems with reaction and diffusion are regarded by Prigogine as a clue to the origin of life [16].

With respect to the cell models hitherto proposed, based on electronic circuits, constructed in order to simulate the electric activity of a biological system, the CSCC created by cascading self-organization, initiated by an electrical spark, reveals a electrical activity, potentially related to direct conversion of thermal energy in electric energy. A phenomenon that illustrates such a possibility is, in our opinion, the ball lightning, the occasional appearance of which proves the ability of Nature to create well located ordered space charge configurations [4]. As mentioned in the literature, the ball lightning deaggregation is accompanied by strong electrical oscillations. This seems to prove the presence of a steady state related to a dynamics of the DL at its boundary similar to that observed at a CSCC created in laboratory. The interpretation of the ball lightning as a "giant" cell seems to be justified if the described cascading scenario of self-organization actually stays at its origin [4]. Produced in an atmosphere that essentially differs from that presumably existent under pre-biotic Earth conditions, after its "birth" the evolution of a ball lightning ceases after a relatively short lifetime. This was, very probable, not the case when the CSCC self-assembly process was initiated under the prebiotic Earth conditions by a simple spark. Occurring in a medium, presumed to be a chemically reactive plasma, the possibility of a further evolution of a CSCC into the contemporary cell becomes a potentially possible alternative. In this context we remember that experiments performed in chemically reactive gases, simulating the prebiotic Earth conditions, have proved that micro-spheres (vesicles) protected from the environment by membranes able to support a potential drop, are self-assembled in electrical discharges [17].

As revealed by Nature, the creation of a living cell requires the self-assembly of a framework in the form of the cell membrane mainly constituted of lipids and proteins. The most important parts of this framework are the channels that, by a specific electric activity, control the matter and energy exchange between the nucleus of the cell and the surrounding medium. The force that maintains the ionic current through a channel has its origin in the electric potential produced by the gradients of the concentration of the different ion species inside and outside the cell. In the steady state of the cell, the local ionic influx compensates the ionic efflux. Why this gradient appears and acts, is today a challenging problem of Biology. In this context new information offered by biological observations have proved the presence of *pH* modulations. These are accompanied by the appearance of spatio-temporal patterns originating from a hypothetical self-organization scenario presumably related to a symmetry breaking mechanism [18]. For explaining the periodic pattern in the diffusion currents, induced by concentration gradients, a theory of electrodynamical instabilities was proposed linked to the specific properties of the membrane conductance [18].

An alternative explanation for the presence of spatio-temporal patterns in biological observations could be based on a self-organization mechanism as that one described in this paper. Such a mechanism becomes possible if a biological cell is the result of the evolution of a gaseous cell, formed by a cascading self-organization scenario in a chemical reactive medium, as that presumably present under primeval

Earth conditions. In that case the membrane of the cell must contain, in order to ensure its viability, channels able to maintain a local gradient of different ion species. This could be possible if at the ends of the channels "micro"-DLs are situated with qualities remembering their recent history. This means that the micro-DL preserves its initial ability to sustain and control an anomalous transport of matter and energy through the channel, by the described dynamics. It has obtained this ability during its creation under prebiotic Earth conditions. In this way the living state of a cell can be related to a mechanism able to explain the presence of periodic current patterns observed in the channels of its membrane. This mechanism can tentatively explain also the manner by which the pumping process is sustained in the channels of the cell membrane.

As mentioned, the CSCCs, created in plasma by self-organization, also reveal other interesting phenomena such as self-multiplication by division and exchange of information [19]. This latter behavior is realized by the emission of electromagnetic energy with an appropriate frequency by a CSCC during its steady (viable) state and its resonant absorption by another CSCC.

## 3. Conclusions

The identification of the physical causes of pattern formation in plasmas reveals the presence of a mechanism of self-organization that substantially advances the knowledge concerning the creation of complex systems in general. Based on phenomena such as local self-enhancement and long-range inhibition this scenario of self-organization shows striking similarities to those considered to lie at the heart of biological pattern formation [20]. Therefore for biologists the described self-organization scenario could be interesting accepting that this can explain the emergence of a primitive organism that could be a prerequisite condition for making possible the more complex chemical reactions. The space charges arrangement and the intrinsic nonlinear mechanism that ensure the "viability" of the CSCC are premises very probable necessary for a further chemical evolution into a organism revealing features as those proper to a contemporary cell. Additionally the described self-organization scenario suggest a new insight concerning the actual origin of various biological events as the morphogenesis, the polarization, the acquisition of nutriments and the mechanism by which electrical signals control the rhythm of a biochemical system. Clearly a major task of experimental studies are necessary to verify whether or not a spark produced at a positive electrode immersed in a chemical reactive plasma could initiate an "universal" self-organization process whose final product can eventually explain the origin of the life. Such experiments are at present performed in our laboratories.

**Acknowledgments.** The authors sincerely thank Professor Dr. Roman Schrittwieser from the University of Innsbruck, Austria for reading the paper and for many useful suggestions. This research was financially supported by the World Bank and by the Romanian Ministry of Education and Research under the Grant No.28199, CNCSIS code 47.

## References

[1] LOZNEANU, Erzilia, Sebastian POPESCU and Mircea SANDULOVICIU (2001) "Plasma experiments with relevance for complexity science", *InterJournal of Complex Systems*, article, 357, http: www.interjournal. org. and the references there in.

[2]. SATO, Tetsuya (1998) "Simulation Science", *J. Plasma Fusion Res.* SERIES **2**, 3-7.

[3] LOZNEANU, Erzilia, Virginia POPESCU, Sebastian POPESCU and Mircea SANDULOVICIU. (2002) "Spatial and spatiotemporal patterns formed after self-organization in plasma", *IEEE Transaction on Plasma Science* **30** No.1 February.

[4] SANDULOVICIU, Mircea. and Erzilia LOZNEANU (2000) "The ball lightning as a self-organization phenomenon", *J. Geophys. Res. Atmospheres*, **105**, 4719-4727.

[5] SANDULOVICIU, Mircea 1993 "Quantum processes at the origin of self-organization and chaos in current carrying gaseous conductors", XXI Int. Conf. Phen. Ionized Gases, Bochum. Proceedings, vol III Invited Papers pp. 279- 286.

[6] SANDULOVICIU, Mircea, Catalin BORCIA and Gabriela LEU (1995) "Self-organization phenomena in current carrying plasma related to non-linearity of the current versus voltage characteristic", *Phys. Lett.* A **208**, 136-142.

[7] SANDULOVICIU, Mircea, Viorel MELNIG and Catalin BORCIA(1997) "Spontaneously generated temporal patterns correlated with the dynamics of self-organized coherent space charge configuration formed in plasma", *Phys. Lett.* A **229**, 354-361.

[8] LOZNEANU, Erzilia and Mircea SANDULOVICIU (2001) "On the mechanism of patterns formation in gaseous conductors", XXV Int. Conf. Phenom. Ionized Gases. Nagoya (Ed. Toshio Goto Nagoya University Nagoya, Japan). Proc. Vol. 3 165-166.

[9] LOZNEANU, Erzilia, Virginia POPESCU and Mircea SANDULOVICIU (2002) "Negative differential resistance related to self-organization phenomena in a dc gas discharge", *J. Appl. Phys.* in print. (Aug.1. 2002).

[10] ALEXANDROAEI, Dumitru and Mircea SANDULOVICIU (1987) "Dynamics of double layer formed in the transition region between two negative plasma glows" *Phys. Lett.* A **122**, 173-177.

[11] SANDULOVICU, Mircea, Erzilia LOZNEANU, and Gabriela LEU (1997) "Spontaneous formation of ordered spatio-temporal structures in laboratory and nature", *Double Layers* World Scientific (edited by Senday "Plasma Forum"Tohoku University) 401-406.

[12] LOZNEANU, Erzilia and Mircea SANDULOVICIU (2001) "On the current limiting phenomenon in a "collisionless" plasma diode", XXV Int. Conf. Phenom. Ionized Gases. Nagoya (Ed. Toshio Goto Nagoya University Nagoya, Japan). Proc. Vol. 3 185-186.

[13] SANDULOVICIU, Mircea (1987) "Pulsatory plasmoids formed in high frequency electric fields", *Plasma Phys. Control. Fusion* **29**, 1687-1694.

[14] SANDULOVICIU, Mircea and Erzilia LOZNEANU (1996) "Experiments proving particles acceleration by electrical double layers", Int. Conf. Plasma Phys. Nagoya 1996 (Editors H. Sugay and T. Hayashi) Proceedings volume 2 pp. 1618-1621.

[15] TURING, Alan Mathison (1952) "The chemical basis of morphogenesis" *Phil. Trans. R. Soc. London* Ser. B **237**, 37-72.

[16] PRIGOGINE, Ilya and Gregoire NICOLIS (1967) "On symmetry-breaking instabilities in dissipative systems", *J. Chem. Phys.* **46**, 3542-3550.

[17] SIMIONESCU, Cristofor *et al.* (1995) "Cold plasma synthesis and the complexity of chemical evolution systems", *Models in Chem.* **132**, 367-378 and references therein.

[18] LEONETTI, Marc and Elisabeth DUBOIS-VIOLETTE (1998) "Theory of Electrodynamic Instabilities in Biological Cells", *Phys. Rev. Lett.* **81**, 1977-1980.

[19] SANDULOVICIU, Mircea (1997) "On the microscopic basis of self-organization", *J. Tech. Phys.* (Warshawa) **38**, 265-267.

[20] KOCH, A. J. and H. MEINHARDT (1994) "Biological pattern formation: from basic mechanism to complex structures", *Rev. Mod. Phys.* **66**, 1481-1506.

# Chapter 2

# Local Complexity and Global Nonlinear Modes in Large Arrays of Elastic Rods in an Air Cross-Flow

**Masaharu Kuroda**
Applied Complexity Engineering Group,
National Institute of Advanced Industrial Science and Technology,
Japan
m-kuroda@aist.go.jp

**Francis C. Moon**
Sibley School of Mechanical and Aerospace Engineering,
Cornell University,
USA
fcm3@cornell.edu

## 1. Introduction

Through the ages in engineering fields, many researchers have been interested in flow of a fluid around an object and vibration of the object. This report experimentally investigates a periodic structure in a cross-flow, especially, an elastic structure such as a rod array in an air cross-flow. Complex dynamics caused by coupling between the fluid and periodic structure is famous as an outstanding problem in the research field of fluid-related vibration or fluid-induced vibration [1-6]. Not only as a scientifically prolific research target, but also as an experimental model such as a pipe array of a heat exchanger system, it is very useful in engineering to investigate its dynamic characteristics. From the viewpoint of a pattern-formation problem in spatio-temporal dynamics of a complex system, we attempt to newly consider this kind of well-known and persistent fluid-engineering problem.

In this study, we discuss experimental nonlinear dynamics of a large array of up to 1000 elements in a cross flow. In this paper we extend the results reported in the letter of Moon and Kuroda [7] for a 1000 rod array. The experimental data presented here is related to recent theoretical work of Homer and Hogan [8] who have presented an impact dynamics model for a large number of pipes vibrating in a heat exchanger.

## 2. Experimental Set-Up and Conditions

Figure 1 shows the experimental set-up. A rectangular array structure of cantilevered rods placed in a wind tunnel is shown schematically. As shown in this figure, rod-like structures are cantilevered at the base and are free to vibrate at the top. Coupling between rods consists of fluid forces and contact when vibration amplitude becomes too large. Fluid forces are of two types: fluid-elastic nearest-neighbor forces equivalent to springs and dampers; and non-nearest neighbor forces produced by vortices leaving the forward rows of cylinders and affecting dynamics of rearward cylinder rows [9].

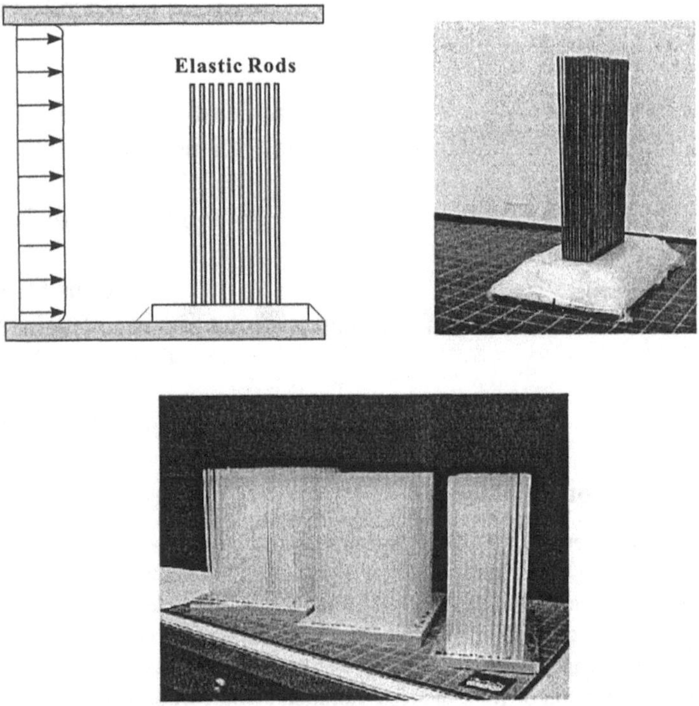


**Figure. 1.** Schematic diagram of the experimental set-up

The wind tunnel is a low turbulence system with a 25.6 cm x 25.6-cm cross section. Wind speeds ranged from 0.0-12.0 m/s. The Reynolds number based on rod diameter ranges from 200-900. Photos show the entire view of an array structure in the wind-tunnel observatory section. Equipped with small accelerometer probes, a test rig is arranged inside the observation part and flood-lights, which are necessary for videotaping, are located under the observation part. Also, photographic images with shutter speed on the order of the vibration period of 10 Hz were taken. They also show various rod arrays for differing materials, gap ratios, and numbers of elements used in experiments.

## 3. Steel-Rod Array Experiments

Each array used here is composed of steel rods with 1.59 mm diameter and 17.1cm length. The first eigen frequency of the steel rod is 26 Hz.

### 3.1. Local Complexity in an Array of 90 Steel-Rods

Observations of tip vibration dynamics in the 90 rod case reveal complex rod-motion patterns. Some rods appear nearly stationary, others vibrate in a straight-line motion at an angle to upstream flow, and the rest vibrate in elliptical patterns sometimes associated with rod to rod contact. Figure 2 shows the ratio of rods in straight-line motion and elliptical orbits as a function of wind speed. Clearly, this illustrates onset of motion at a critical wind speed and growth of ratio of rods in linear motions after a full turbulence regime. This may be related to greater incidence of rod impact at higher wind velocities.

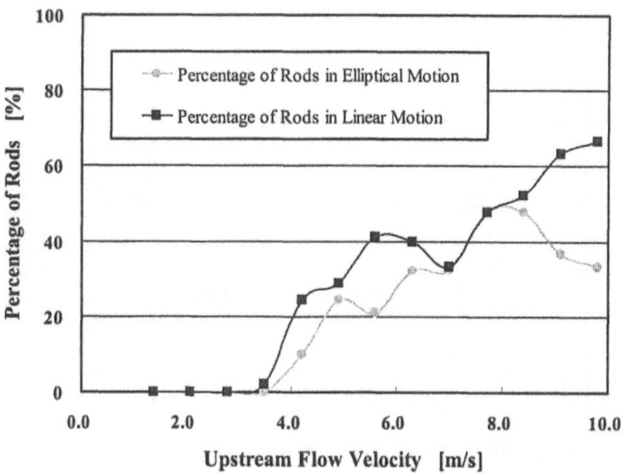


**Figure. 2.** Percentage of rods in straight-line and those in elliptical motion

While rod frequencies lie close to their eigen frequencies, phases and types of motion, i.e., stationary, straight-line, and elliptical orbit, show no regular modal pattern and change over time. It is important to emphasize that linear theory would predict 2 x N x M eigen modes for an N x M array of rods. However, no clear pattern emerges as wind velocity in the wind-tunnel increases. Thus, we will apply an entropy measure of complexity to describe the dynamics [10].

### *3.1.1. Entropy Measure for Complexity*

Entropy measures employed here are based on the information entropy

$$S = \Sigma\, p_n \log (1/ p_n), \qquad (1)$$

where $p_n$ is a probability measure for occurrence of a certain spatial pattern [11]. We explored the possibility of defining entropy using previously obtained data expressed by discrete variables. Two photographs, Figs. 3(a) and 3(b), show examples of calculated entropies. Figures 4(a) and 4(b), which are matched with Figs. 3(a) and 3(b), respectively, are used for entropy calculation.

(a) A Photo at 7.7m/s flow velocity by 2.0 sec. shutter speed

(b) A Photo at 5.6m/s flow velocity by 2.0 sec. shutter speed

**Figure. 3.** Photo examples of the array of 90 vibrating rods

| Photo Number 55 | No Motion = 0 | Linear Motion = 1 | Elliptical Motion = −1 |
|---|---|---|---|

| Column Number | Front Row | Middle Row | Back Row |
|---|---|---|---|
| 1 | −1 | 1 | 1 |
| 2 | −1 | 1 | 1 |
| 3 | 1 | 1 | 1 |
| 4 | 1 | 1 | 1 |
| 5 | 1 | 1 | 0 |
| 6 | −1 | 1 | 1 |
| 7 | 1 | 1 | 1 |
| 8 | 1 | 1 | 1 |
| 9 | −1 | 1 | 1 |
| 10 | −1 | 1 | 1 |
| 11 | −1 | −1 | −1 |
| 12 | 1 | 1 | 1 |
| 13 | 1 | 1 | 1 |
| 14 | −1 | −1 | −1 |
| 15 | 1 | 1 | 1 |
| 16 | −1 | 1 | 1 |
| 17 | 1 | 1 | 0 |
| 18 | 1 | −1 | 1 |
| 19 | 1 | 1 | 1 |
| 20 | 1 | 0 | 1 |
| 21 | 1 | 1 | 1 |
| 22 | 1 | 1 | 1 |
| 23 | 1 | 1 | −1 |
| 24 | 1 | 1 | 1 |
| 25 | 1 | 1 | −1 |
| 26 | −1 | 1 | 1 |
| 27 | 1 | 1 | 1 |
| 28 | 1 | 1 | 1 |
| 29 | 1 | 1 | 1 |
| 30 | 1 | −1 | 1 |

(a) The spatial Pattern for Fig. 3(a)

| Photo Number 67 | No Motion = 0 | Linear Motion = 1 | Elliptical Motion = −1 |
|---|---|---|---|

| Column Number | Front Row | Middle Row | Back Row |
|---|---|---|---|
| 1 | −1 | −1 | 0 |
| 2 | 1 | −1 | 0 |
| 3 | 1 | 1 | 1 |
| 4 | 0 | 1 | 0 |
| 5 | 0 | 1 | 1 |
| 6 | 1 | −1 | 1 |
| 7 | 1 | 1 | 1 |
| 8 | 1 | −1 | 0 |
| 9 | 1 | −1 | 0 |
| 10 | 1 | −1 | 1 |
| 11 | −1 | 1 | −1 |
| 12 | −1 | 1 | 1 |
| 13 | 1 | 1 | 1 |
| 14 | 1 | −1 | 1 |
| 15 | 1 | −1 | −1 |
| 16 | 1 | −1 | 1 |
| 17 | 1 | 1 | 1 |
| 18 | 1 | 1 | 1 |
| 19 | 1 | 1 | 1 |
| 20 | 1 | −1 | −1 |
| 21 | 1 | 1 | −1 |
| 22 | 1 | 1 | 1 |
| 23 | 1 | 1 | 1 |
| 24 | 1 | 1 | 1 |
| 25 | 1 | −1 | 1 |
| 26 | 0 | 1 | 0 |
| 27 | 1 | 1 | 0 |
| 28 | 0 | 1 | 0 |
| 29 | 1 | 1 | 1 |
| 30 | 1 | 1 | 1 |

(b) The spatial Pattern for Fig. 3(b)

**Figure. 4.** Spatial pattern examples of the array of 90 vibrating rods

Entropy is calculated here for two-dimensional spatial patterns. We call this cluster pattern entropy (CPE). From Fig. 5, it is understood that cluster pattern entropy took the maximum value, but not in the full turbulence case.

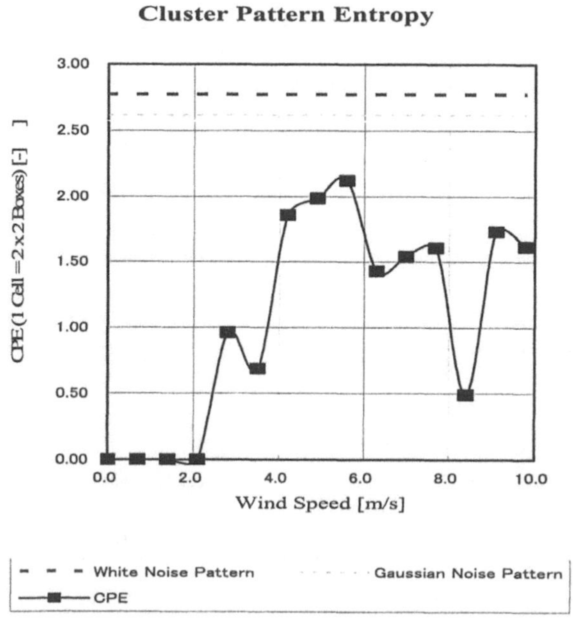


**Figure. 5.** Cluster Pattern Entropy (CPE) as a function of wind speed

### 3.2. Spatio-Temporal Patterns Appearing in an Array of 300 Steel-Rods

The array has 30 columns and 10 rows with the gap ratio 1.0. Observed from the top in the low wind-speed regime, rod tops seem to oscillate individually and do not show significant difference in density distribution over time. If rod alignment were more precise, we could have observed some kind of organized movement even in this wind velocity regime.

Figure 6 exhibits videotape frames of rod-array behavior from above and corresponding contour maps of rod-density distribution in the middle wind-speed regime. The white area on the contour maps is manifestly larger than before; the blue area appears as though it has torn itself and dense and scattered areas alternate remarkably over time. Finally, clusters, i.e., collective movements of some rods, emerge in the rod-density distribution. We confirmed that these clusters do not move along a specific pattern at this stage, but rather just move and freely collide with one another.

Figure 7 displays videotape frames of rod-array behavior from above and corresponding contour maps of rod-density distribution in the high wind-speed regime. The white area on contour maps continues to widen; rifts and gaps appear on the blue area and the number of clusters displays continued increase. Furthermore, it is most characteristic and important that those clusters are linked together as a chain and that the chain moves in a diagonal direction to the center part of the backmost row repeatedly, as a wave does. From the videotape data, it is clear that wave-like motions originating at both sides collide at the center part on the back row.

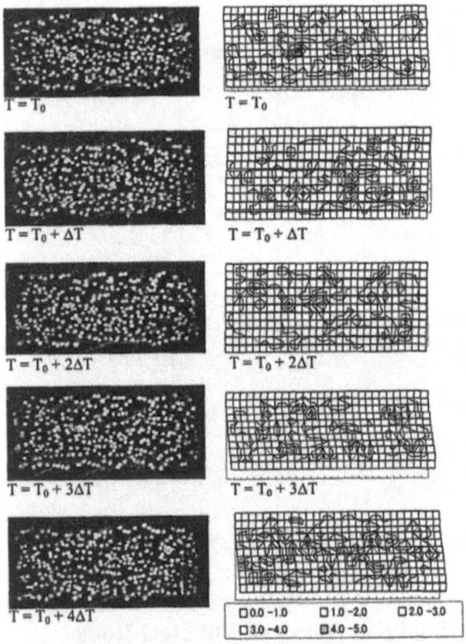


**Figure. 6.** A sequence of VTR frames and rod-density distribution at 7.0 m/s wind velocity

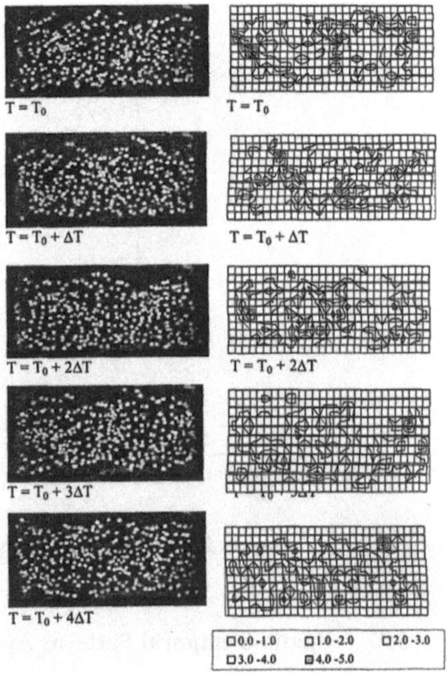


**Figure. 7.** A sequence of VTR frames and rod-density distribution at 10.5 m/s wind velocity

*3.2.1. Discussion*

Behavior of this wave-like motion looks like a soliton. This fact is the reason why formulation of a mathematical model like the Toda-lattice, which is well-known in the field of nonlinear lattice dynamics, prospectively leads to clarification of these complex phenomena appearing in an array of fluid-elastic oscillators. Unfortunately, it is not certain whether waves collapse or pass through each other without soliton-like effect, or if they reflect each other at that point; the number of rows is insufficient to make that determination.

Furthermore, complex behavior patterns of oscillating rods at each wind-velocity regime seem to present an analog of cellular automata. Especially in the high-wind velocity regime, the pattern seems to resemble Wolfram's Class 4 in cellular automata dynamics [12].

### 3.3. Time Series Data from Accelerometers

On the other hand, time-series data on one rod, shown in Figs. 8 and 9, suggests that bursting phenomena are probably related to impacts between rods as wind speed is increased. These time-series data were obtained from a small accelerometer attached 1 cm up from the foot of the rod in the middle of the front row. In this location the accelerometer was not sensitive to low frequency first mode vibrations of the rod but was very sensitive to the higher modes in the rods produced by impacts. This meant that the sensor did not respond to pre-impact flow induced vibrations.

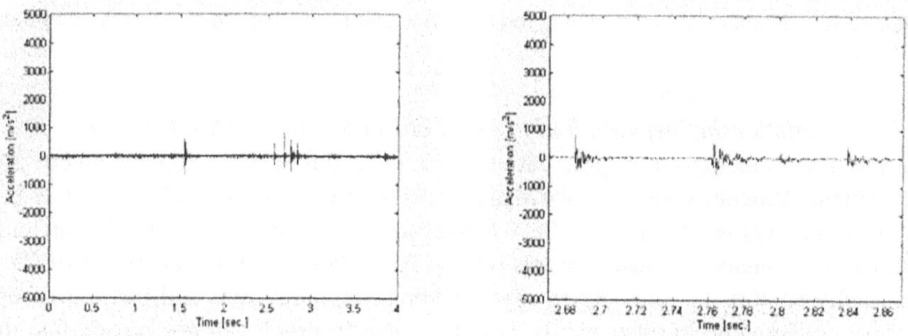


(Magnification)

**Figure. 8.** Acceleration at the front-center rod at 7.0 m/s wind speed

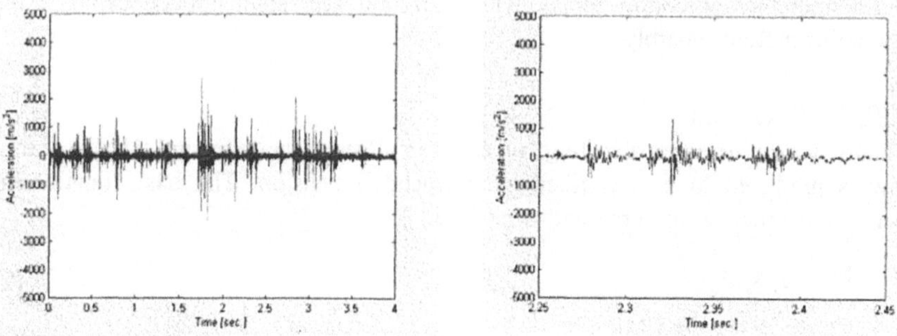


(Magnification)

**Figure. 9.** Acceleration at the front-center rod at 11.2 m/s wind speed

In addition, Fig. 10 shows a plot of the number of bursting waveforms for 24 sec. as a function of wind speed. Bursting phenomena take place due to rod collisions. Bursting peaks were counted by the program coded on MATLAB.

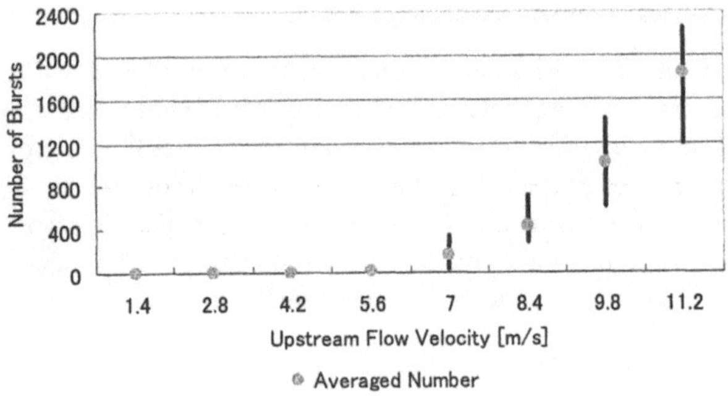


**Figure. 10.** Number of bursts among rods as a function of wind speed at a threshold of 5.0 x $10^3$ m/s$^2$

*3.3.1. Relationship between Rod-to-Rod Collisions and Self-Organization*

In the low wind-speed regime around 3.5 m/s, fluid-elastic forces govern rod movement. Videotape records show that rods are free to move individually in this wind speed regime. In the middle wind-speed regime of about 7.0 m/s, not only fluid-elastic forces, but also impacts among rods start affecting rod behavior. It is especially notable that boundaries appear among oscillating rods and that rod groups behave collectively. In other words, rod clustering occurs. It can not be said that rod clusters move with a clear temporal pattern, but they appear to be just pushing each other. In the high wind speed regime of approximately 10.5 m/s, dominant forces determining rod behavior shift to impact from fluid-elastic forces. This is understood from the fact that bursting phenomena frequency at 10.5 m/s is 10 times higher than at 7.0 m/s, as shown in Fig. 10. As a result, global wave-like motion takes place. Furthermore, slow-motion replay with a 1/2000 sec. shutter-speed captures this wave-like motion superbly.

*3.3.2. Power Law*

Finally, we confirm that a scaling law exists in cluster generation and that the scaling law is predicted to be described by a fractal dimension. The basic relationship between the number of bursts and wind speed is

$$V^2 = c\,N^{\alpha}, \tag{2}$$

where V represents wind velocity, N shows the number of bursts and c is a constant. The term $V^2$ is proportional to input energy into the rod array. By taking the logarithm of both sides of Eq. (2), we obtain a log N-log V plot,

$$Y = \text{const.} + (\alpha/2)\,X, \tag{3}$$

where $X = \text{Log } N$, $Y = \text{Log } V$ and $\alpha$ is the fractal dimension. We obtain the graph slope; Figure 11 indicates that $\alpha$ is almost 0.25. Therefore, energy input into the rod array by wind and the number of bursts follow the power law with a fractal dimension of 0.25.

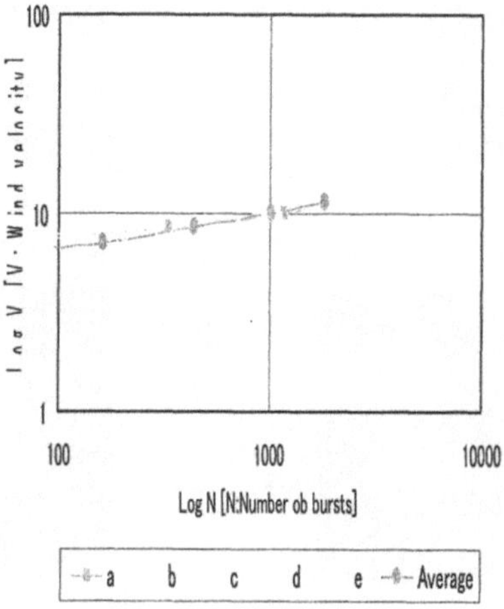


$V^2 = cN^{\alpha}$, $\alpha = 0.242$

**Figure. 11.** Power law between the number of bursting signals and wind velocity at an acceleration threshold of $5.0 \times 10^3$ m/s$^2$

## 4. Polycarbonate-Rod Array Experiments

Each array used here is composed of polycarbonate rods with 3.18 mm diameter and 20.0 cm length. The first eigen frequency of the polycarbonate rod is 18 Hz.

In these experiments, specially manufactured polycarbonate rods were made to insure straight cylinders. These rods allowed greater frontal cross section without a large increase in stiffness. With the transparent rods, we could shine light from underneath the rod array so that the camera could see points of light moving as the array vibrated, instead of using reflected light from the tops of the steel rods.

### 4.1. Global Nonlinear Modes observed in an Array of 300 Polycarbonate-Rods

Global nonlinear modes in large arrays of elastic rod oscillators were discovered while local complex behavior was revealed through videotaped experimental results. Figures 12 and 13 show videotape frames capturing the behavior of a 300 polycarbonate rod array from the above at 11.20 m/s wind speed. The array has 25 columns and 12 rows with the gap ratio 1.5. Each two-picture set shows successive images; the rod-array spatial pattern oscillates from that in the upper picture to that of the lower picture.

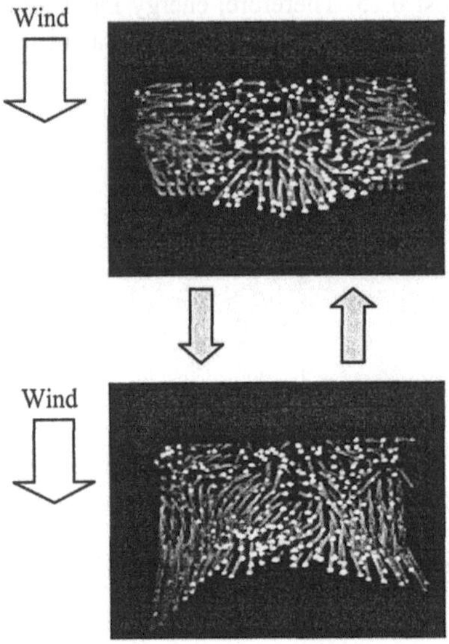


**Figure. 12.** Symmetrical mode of 25 column x 12 row array of 300 rods

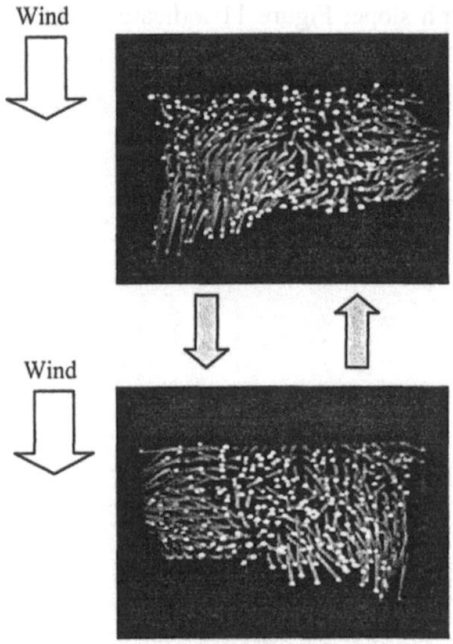


**Figure. 13.** Asymmetrical mode of 25 column x 12 row array of 300 rods

Furthermore, it was clarified that observed nonlinear global modes are limited to two types: a symmetrical mode and asymmetrical mode. Also, the overall vibrating rod-array shape switches from one mode to another repeatedly over time [13].

## 4.2. Global Nonlinear Modes observed in an Array of 1000 Polycarbonate-Rods

Figures 14 and 15 show videotape frames capturing behavior of a 1000-polycarbonate-rod array from above at 8.40 m/s wind speed. The array has 25 columns and 40 rows with the gap ratio 1.5. Each two-picture set shows successive images, and the rod-array spatial pattern oscillates from that in the upper picture to that of the lower picture. By examining videotaped rod motions, we found that several characteristic frequencies of motion exist: the frequency of the switch between symmetrical mode and asymmetrical mode, those of waves proceeding on boundary edges of the array, and those of waves passing through the inner area of the array.

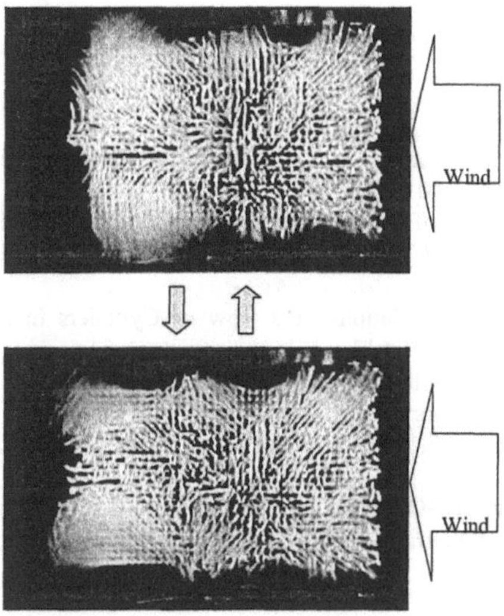


**Figure. 14.** Symmetrical mode of 25 column x 40 row array of 1000 rods

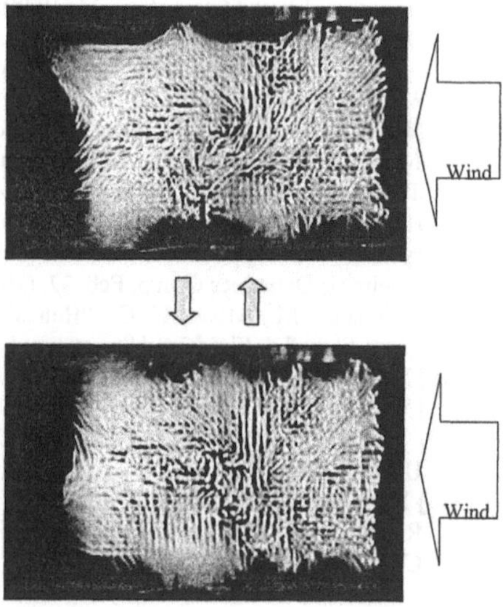


**Figure. 15.** Asymmetrical mode of 25 column x 40 row array of 1000 rods

## 5. Conclusions

Non-stationary complex phenomena occurring in large arrays of up to 1000 vibrating-rods in the wind tunnel were investigated by experiment.

As the intensity of interaction between neighboring elements (frequency of collisions among rods in this case) increases, a set of the elements (a rod-array in this case) achieves globally better-organized behavior. Also, the organized behavior produces a transfiguration in quality in a staircase pattern when it crosses over a threshold as the phase transition of matter does. In this manner, specifically, individual rods, clusters of rods, and a wave of a chain of clusters comprise the central players of dynamic order-formation shifts at each stage. Finally, conformation of the rod-array collective behaviors leads to two types of global nonlinear modes: symmetrical mode and asymmetrical mode.

## References

[1] Coutts, M. P., Grace, J. (eds.), *Wind and Trees*, Cambridge University Press, Cambridge, UK, (1995).

[2] Finnigan, J. J., "Turbulence in Waving Wheat," *Boundary-Layer Meteorology*, 16 (1979), 181-211.

[3] Finnigan, J. J., Mulhearn, P. J., "Modeling Waving Crops in a Wind Tunnel," *Boundary-Layer Meteorology*, 14 (1978), 253-277.

[4] Grace, J., *Plant Response to Wind*, Academic Press, London, (1977).

[5] Inoue, E., "Studies of Phenomena of Waving Plants ("HONAMI") Caused by Wind," *Agric. Meterol.* (Japan), 11 (1955), 18-22.

[6] Niklas, K. J., *Plant Biomechanics*, Chapters 7 & 9, University of Chicago Press, (1922).

[7] Moon, F. C., Kuroda, M., "Spatio-Temporal Dynamics in Large Arrays of Fluid-Elastic Toda-Type Oscillators," *Physics Letters A*, 287 (2001), 379-384.

[8] Homer, M., Hogan, S. J. (2001) "A simple model of impact dynamics in many dimensional systems, with applications to heat exchangers" *University of Bristol Preprint*, Applied Nonlinear Dynamics Group, Feb. 27, (2001).

[9] Thothadri, M., Moon, F. C., "Helical Wave Oscillations in a Row of Cylinders in a Cross-Flow," *J. Fluids and Structures*, 12 (1998), 591-613.

[10] Moon, F. C., *Chaotic and Fractal Dynamics*, J. Wiley & Sons, NY, (1992).

[11] Williams, G. P., *Chaos Theory Tamed*, Joseph Henry Press, Washington D.C., (1999).

[12] Wolfram, S., *Cellular Automata and Complexity*, Addison-Wesley Publishing Company, USA, (1994).

[13] Kuroda, M., Moon, F. C., "Complexity and Self-Organization in Large Arrays of Elastic Rods in an Air Cross-Flow," *EXPERIMENTAL CHAOS: 6th Experimental Chaos Conference*, CP622 (2002), 365-372.

# Chapter 3

# Self-organization in Navier-Stokes turbulence

**Jacques Lewalle**
Syracuse University
jlewalle@syr.edu

Fluid turbulence is undoubtedly a complex phenomenon, but the relation between the fundamental laws and the emergence of large-scale coherent eddies from differential interactions in a random background has remained elusive. Such a connection is attempted in this paper. The key analytical step is to express the Navier-Stokes equations in the Hermitian wavelet representation, which adds one independent spectral variable to the spatial description at any instant. In this expanded representation, the equations reduce naturally to a form consistent with pairwise nonlinear interactions governed by known kernels. The emergence of strong eddies from a random background, and their eventual breakdown, are illustrated by a 2-dimensional simulation.

## 1 Introduction

Fluid turbulence is, with various attributions (see e.g. [10], p.3), '... *the last great unsolved problem of classical physics'*. One might accept as likely that it belongs in the new field of dynamics of complex systems, which generate 'complicated' behavior from simple inputs. Nicolis & Prigogine [14] focused on chaotic dynamics and generic instabilities as relevant to complexity in the mathematical sciences, but no theory of turbulence has yet evolved on these premises. A complex system usually includes many interacting parts, resulting in cooperative (rather than merely collective) behavior (Bar-Yam [2]). Also significant is the difference between fine- and coarse-grained descriptions, which introduces the concept of scale. Scaling properties are usually statistical, but the ergodic assumption is not expected to hold for organized structures. Badii & Politi [1] emphasize a *hierarchy* of parts, interactions and scaling as relevant to complex physical systems. For generic parts $Y(x)$, the (arguably) simplest

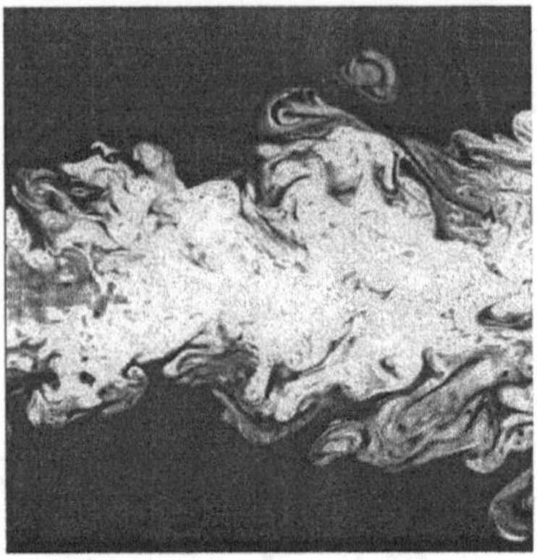

**Figure 1**: longitudinal section of a turbulent jet, obtained by laser induced fluorescence, courtesy of C. Fukushima and J. Westerveel (Technical University of Delft, the Netherlands) and of efluids.com. The dye concentration reflects the cumulative history of mixing as the jet develops from left ot right.

evolution equation associated with complex behavior is

$$\frac{dY}{dt} = A * (YY) \tag{1}$$

The local kernel $A(x)$ is convolved (asterisk) with nonlinear expressions, resulting in interactions with nearby parts only. Based on such a lattice model, Hansen & Tabeling [9] illustrated the emergence of eddies (see also [3]), paving the way for a more fundamental study based on exact equations.

This highlights the need to identify relevant 'parts' in the turbulence problem. Among the contenders, Fourier modes are unsatisfactory in spite of significant successes, because they conceal the localized nature of eddies. With the Karhunen-Loeve eigenmodes, success is most definite in systems where few modes are active. Finally, vortices have been effective in two-dimensional turbulence [4, 7], but less so in three-dimensional flows. Each of these fundamental 'parts' captures some important feature of turbulence while others remain elusive. Badii & Politi [1] point out that '*...The concept of complexity is closely related to that of understanding...*'. In the turbulence problem, our lack of understanding is epitomized by the poor match between established descriptive concepts (e.g. eddies) and analytical formulations.

The example shown on Fig. 1 illustrates the presence of large and small scale eddies in a section of a jet. As instabilities break up the large eddies into smaller ones, energy cascades from large to smaller scale in an intermittent process. The spatial and spectral distribution and variable intensity of eddies, and their embedding patterns, is the core of the turbulence problem.

# 2 Basic equations

## 2.1 Fundamental laws

Fluid turbulence falls within the scope of classical dynamics (Newton). The extension to continua by Euler introduced the nonlinearity essential to the existence of turbulence. Then, Cauchy showed that tangential (viscous) stresses must be added to Euler's isotropic pressure $p$. The result for the simplest case (Newtonian incompressible fluids) is due to Navier and Stokes [1] :

$$\partial_t u_i + u_j \partial_j u_i = -\frac{1}{\rho}\partial_i p + \nu \partial^2_{jj} u_i. \tag{2}$$

Here $\rho$ is the fluid's density and $\nu$ its viscosity. Eq.(2) constitutes the inescapable mathematical and physical basis for turbulence dynamics. Since it has three components, and four unknowns $u_i$ and $p$, it must be supplemented by mass conservation $\partial_i u_i = 0$. Hence, we have a system of four field equations for velocity and pressure at each point. The elimination of pressure has been done traditionally in Fourier space (see e.g. [13]) or through the use of vorticity $\underline{\omega} = \nabla \times \underline{U}$. This paper is based on a new alternative.

Let us define flexion as the Laplacian of velocity $\alpha_i = \partial^2_{jj} u_i$. The application of the Laplacian to Eq.(2) and simple substitutions give

$$\partial_t \alpha_i - \nu \partial^2_{jj} \alpha_i = \partial^3_{ijk}(u_j u_k) - \partial^3_{jkk}(u_i u_j) \tag{3}$$

The left side of the Eq.(3) shows the familiar diffusion equation; the cost of the elimination of pressure is seen on the right side, with triple derivatives of the nonlinear terms somehow responsible for the phenomena of turbulence. Assuming a solution $\alpha_i$ for the flexion vector, the velocity field can be reconstructed using the Biot-Savart equation (3-D version shown here)

$$u_i = -\frac{1}{4\pi}\int \frac{\alpha_i(x')}{\mid x - x' \mid} dV'. \tag{4}$$

## 2.2 Wavelet transforms and reformulation

Eq.(3) is more familiar in its Fourier representation [13]. Then, the spectral decomposition obscures the spatial intermittence (i.e. the presence of organized motion at each scale), whereas in the spatial representation, eddies of all scales are superposed. In the wavelet representation [5, 6], we can select the point of compromise between spectral and spatial accuracy to suit our own purposes.

[1] For notations, the components of the velocity vector $\underline{U}$ will be indexed as $u_i$, with $i = 1...N$ in $N$ dimensions. $N$ will have the default value 3 in this paper; use of $N = 2$ will be clear from the context. The total or material time derivative $\frac{d}{dt}$, the local or apparent time derivative $\frac{\partial}{\partial t}$, and the spatial derivatives $\frac{\partial}{\partial x_i}$ will be abbreviated respectively as $d_t$, $\partial_t$ and $\partial_i$. Second and higher-order derivatives will follow the notational pattern $\partial_i \partial_j = \partial^2_{ij}$. Summation over repeated indices is assumed. Depending on context, $x$ will stand either for one Cartesian coordinate, as in $(x, y)$, or for the generic coordinate vector $\underline{x}$.

Wavelets [5, 6] can be discrete (orthogonal) or continuous (redundant). Either way, when used in integral transforms, they meet the requirements of the existence of an inverse transform (no loss of information) and of Parseval's theorem ($L^2$ normalization). A variant of the Mexican hat continuous wavelet is used here.

Based on the normalized $N$-dimensional Gaussian filter $F_s(x) = e^{-x^2/4s} / (2\sqrt{\pi s})^N$ of scale $s$ (note that $s$ has dimensions $L^2$), the wavelet transform of a velocity component $u_i(x)$ at any time $t$ is identical to filtered flexion

$$\tilde{u}_i(x, s, t) = \nabla^2 F_s * u_i(x, t) = \psi_s(x) * u_i(x, t) = F_s * \alpha_i(x, t) = F_s * \tilde{u}_i(x, 0, t), \quad (5)$$

where the asterisk denotes spatial convolution[2]. While, in general [5, 6], the inverse wavelet transform includes a convolution, in this case it takes the particularly simple form [11]:

$$u_i(x, t) = -\int_0^\infty (s\tilde{u}_i(x, s, t))\frac{ds}{s} = -\int_0^\infty F_s * \alpha_i \, ds. \quad (6)$$

The mapping from flexion to velocity is equivalent to Eq.(4), but Eq.(6) is valid for any space dimension $N$ [12]. Thus the quantity $s\tilde{u}_i$ appears as a natural building block, with superposition over all (logarithmic) scales to reconstruct the field $u_i$ or (when squared) its energy density. For a 1-dimensional signal, the distribution of energy is shown on Fig. 2. The lumping of the energy density into quasi-discrete objects in the $(x - s)$-space suggests that the property $s\tilde{u}_i$ is one possible analytical tool to capture eddies. In contrast to the lumpy $s\tilde{u}_i$, the wavelet coefficients $\tilde{u}_i$ show a gradual relaxation from $s = 0$ toward larger scales, and increasing the wavelet scale corresponds to coarse-graining of the flexion field.

The equation governing the evolution of the wavelet coefficients is obtained by filtering of Eq.(3), yielding

$$(\partial_t - \nu\partial_{jj}^2)\tilde{u}_i = \partial_{ijk}^3 F_s * (u_j u_k) - \partial_{jkk}^3 F_s * (u_i u_j) = F_{ijk} * (u_j u_k) - F_{jkk} * (u_i u_j) \quad (7)$$

The definition of $F_{ijk}$ is self-explanatory. The nonlinear terms (right side) include the convolution with a known localized interaction kernel $F_{ijk}$, as in the generic model Eq.(1). However, the kernel acts not on the wavelet coefficient $\tilde{u}_i$, but, through the inverse transform formula, on wavelet coefficients at all scales, obtained by filtering of the fine-grained coefficient. It is sufficient to solve a fine-grained version (small $s$) of this equation, with all larger scale (coarser grained) solutions implied by filtering.

## 3 A 2-D periodic simulation

The spectral flexion equation (7) describes three distinct phenomena that combine into turbulence: diffusion, spatial organization and spectral exchanges.

[2]The numerical coefficients and specific scaling reflect the particular definition of the wavelet transform adopted here; alternatives can be found e.g. in [5, 15], and it is important to use an internally consistent set of relations. The present selection was derived in [11].

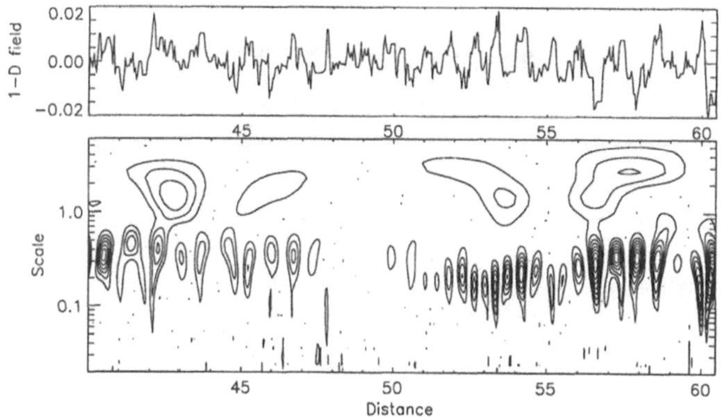


**Figure 2**: Local spectral energy density contours for the signal shown in the upper box; units of space and scale match, but are otherwise arbitrary.

Simplifying the problem to two dimensions on a $32^2$-grid, we can use $u$ and $v$ components in the cartesian directions $x$ and $y$, respectively, rather than the generic indices. This gives

$$(\partial_t - \nu\partial_{jj}^2)\tilde{u} = [F_{xxy} * (uv) + F_{xyy} * (vv) - F_{xyy} * (uu) - F_{yyy} * (uv)] \quad (8)$$

and

$$(\partial_t - \nu\partial_{jj}^2)\tilde{v} = [F_{xxy} * (uu) + F_{xyy} * (uv) - F_{xxx} * (uv) - F_{xxy} * (vv)]. \quad (9)$$

The interaction kernels $F_{ijk}$ can be constructed as products of factors in the cartesian directions; each factor is an approximation of derivatives of a bell-shaped curve. The fourth-order central-difference coefficients were adopted as a model. The resulting discrete and continuous interaction kernels are in excellent qualitative and quantitative agreement. Note that, in 2 dimensions, the 3 derivatives must arrange themselves so that an even-ordered derivative applies in one direction (and the kernel is symmetric in that direction) and an odd-ordered derivative applies in the other direction (with anti-symmetry of the kernel). Because each kernel contains derivatives in at least one direction, its average is zero, so that the dynamics conserve the total amount of $\tilde{u}$ and $\tilde{v}$ in the system, and merely rearrange them spatially.

## 3.1 2-D velocity field reconstruction

Velocity field reconstruction from Eq.(6) is necessary to evaluate the nonlinear terms. In addition, vorticity is a useful property for the graphical interpretation

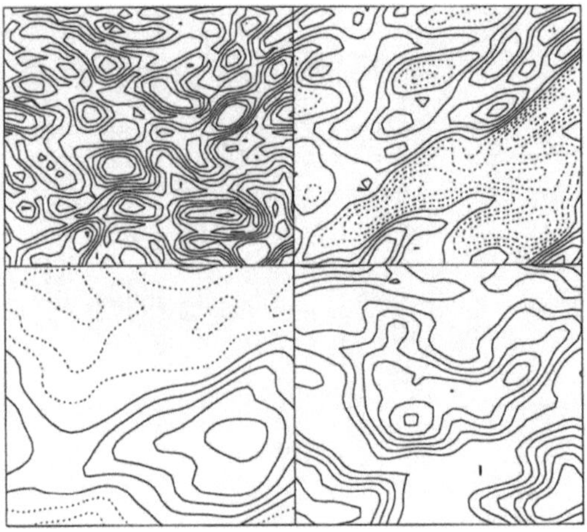

**Figure 3**: Clockwise from top left: flexion, vorticity, velocity and streamfunction magnitudes obtained from the initial (incompressible) flexion components. Each map has unit variance, contour lines from -3 to +3 by intervals of 0.5, dotted lines for negative values.

of 2-D flows, because only the component normal to the plane of the flow is non-zero. The relations between the streamfunction (defined below), velocity, vorticity and flexion also provide accuracy diagnostics, and are summarized next.

The streamfunction $\psi(x, y)$ is a function such that the velociy components $u$ and $v$ can be obtained from the relations $u = -\partial_y \psi$ and $v = \partial_x \psi$. For any sufficiently smooth such function $\psi$, the continuity equation is automatically satisfied. In 2-D flows, the vorticity vector has only one component $\omega = \partial_x v - \partial_y u$ normal to the plane of the flow. Then, the flexion components are given by $\alpha_x = \nabla^2 u = -\partial_y \omega$ and $\alpha_y = \nabla^2 v = \partial_x \omega$. It is a simple matter to verify that $\omega = \nabla^2 \psi$ and $\alpha_i = \nabla^2 u_i$. This classical sequence can be inverted from flexion to velocity (Eq.6) to vorticity to streamfunction, the latter resulting from the inversion of the Laplacian $\psi = -\int_0^\infty F_s * \omega ds$ similar to Eq.(6).

The initial conditions to be used below served as an example of the reconstruction technique and accuracy check. They were generated by assigning random numbers to the grid values of the streamfunction and filtering. The velocity, vorticity and flexion fields were then calculated by central differences. Then, starting with the flexion fields, the other properties were reconstructed as described above, and the values compared to the originals. Errors of the order of 2% were observed. After normalization of the variance for each map, the results are plotted on Fig. 3. The vorticity map will be used as we track the evolution of the field under Navier-Stokes dynamics. Two additional quantities provide useful diagnostics. Given a wavelet coefficient $\tilde{u}$, a scale-dependent local Reynolds number, or eddy Reynolds number, can be defined as $Re_s = \frac{s^{3/2}\tilde{u}}{\nu}$. Similarly, the eddy inverse time scale (turnover rate) is given by $f = s^{1/2}\tilde{u} = \frac{Re_s \nu}{s}$.

**Figure 4**: The evolution of normalized vorticity shown on Fig. 3 at dimensionless times 0.25, 0.4 and 0.5. Contour levels range from -2 to +2.

# 4 Simulation results and discussion

The evolution in Eqs. (8) and (9) was tracked with a second-order Runge-Kutta scheme. The results are independent of the time increment. The supremum of the inverse time scale is used to adjust the time-step to current conditions. Overall time is non-dimensionalized by the initial time scale. An initial eddy Reynolds number (supremum) of 150 was used. Results at times 0.25, 0.4 and 0.5 are shown on Fig. 4. In the first phase of evolution, the field organizes itself into a large-scale vorticity distribution. Little convection is observed in the poorly organized field. The maximum eddy $Re_s$, distinct from the traditional mean $Re$, increases with the dominant scale to approximately 2000, a twelve-fold increase. In the second phase of evolution, the large eddies break down into smaller eddies again. Simulations on larger grids, and in 3-D, will be necessary to diagnose the differences in correlations and scaling relations between the initial (unstructured) eddy distribution and the later outcome of a cascade process. This is the subject of current work by several groups (e.g. Farge, private communication, 2002).

A century-and-a-quarter of efforts have yielded remarkably few exact results and no physical theory of Navier-Stokes turbulence. The source of difficulty is that the basic empirical concepts of turbulence (scaling, intermittence, eddies, cascade) are poorly expressed by the traditional analytical tools. The wavelet representation captures the local spectral behavior and translates it into a lumpy distribution of magnitude and energy. The Mexican hat transform is identical to filtered flexion and thus related to vorticity as well. Filtering, of course, expresses the fine- vs. coarse-graining of renormalization approaches. With these tools, the Navier-Stokes dynamics for the wavelet coefficients are local and non-differential in space, and the nonlinear interactions are governed by fixed known localized kernels and involve local wavelet coefficients at all scales. On this basis, the wavelet representation of the Navier-Stokes equations is unusually promising.

# Bibliography

[1] BADII, Remo and A. POLITI, *Complexity: Hierarchical structures and scaling in physics*, Cambridge Univ. Press (1997).

[2] BAR-YAM, Yaneer, *Dynamics of complex systems*, Perseus Books (1997).

[3] BOHR, Tomas, M.H. JENSEN, G. PALADIN and A. VULPIANI, *Dynamical Systems Approach to Turbulence*, Cambridge Univ. Press (1998).

[4] CHORIN, Alexander J., *Vorticity and turbulence*, Springer Verlag (1994).

[5] DAUBECHIES, Ingrid, *Ten Lectures on Wavelets*, S.I.A.M. (1992).

[6] FARGE, Marie, "Wavelet transforms and their applications to turbulence", Ann. Rev. Fluid Mech. **24** (1992), 395-457.

[7] FARGE, Marie, K. SCHNEIDER and N. KEVLAHAN, "Non-gaussianity and coherent vortex simulation for two-dimensional turbulence using an adpative orthogonal wavelet basis", Phys. Fluids **11** (1999), 2187–2201.

[8] FRISCH, Uriel, *Turbulence: the legacy of A.N. Kolmogorv*, Cambridge U. Press (1995).

[9] HANSEN, A.E., and P. TABELING, "Coherent structures in two-dimensional decaying turbulence", Nonlinearity **13** (2000), C1-C3.

[10] HOLMES, Philip, J.L. LUMLEY and G. BERKOOZ, *Turbulence, coherent structures, dynamical systems and symmetry*, Cambridge U. Press (1996).

[11] LEWALLE, Jacques, "Formal improvements in the solution of the wavelet-transformed Poisson and diffusion equations", J. Math. Phys. **39** (1998), 4119–4128.

[12] LEWALLE, Jacques, "A filtering and wavelet formulation for incompressible turbulence", J. Turbulence **1** (2000), 004, 1–16.

[13] McCOMB, W.D., *The Physics of Fluid Turbulence*, Clarendon Press (1990).

[14] NICOLIS, Gregoire, and I. PRIGOGINE, *Exploring complexity*, W.H. Freeman and Co. (New York) (1989).

[15] PERRIER, Valérie, and C. BASDEVANT, "Besov norms in terms of the continuous wavelet transform. Application to structure functions", *Math. Models and Methods in Appl. Sciences* **6** (1996), 649–664.

# Chapter 4

# *The* Game (Introduction to Digital Physics)*

**Plamen Petrov**
ppetrov@digitalphysics.org

In the present brief article we introduce the main idea of Digital Physics in the form of an abstract game.

## 1 Introduction

Let us imagine two computer programmers playing the following strange game:

Each of the programmers has his/her own computer, and they both sit in front of each other in such a way that they are able to see each other faces but not what is going on on the screen of the "opponent's" computer.

They both program something and their computers are powerful enough (i.e. have sufficient memory) to permit them to complete their tasks.

The first programmer, let us call him "God", has the right to program whatever He likes, as far as it can be represented on the screen as a cellular automaton (CA)[1]. He also has the right to "choose" some initial configuration (IC) for His CA, whatever He likes, as far as the configuration is a finite one.

---

* This article is a part of [Petrov 2002].

[1] Probably the best known cellular automaton is Conway's "Game of Life" [Gardner 1970, Berlekamp 1982]; for a good introduction to cellular automata in general, check out [Wolfram 1984].

(And this is everything that "God" must do, i.e. He makes no further "moves" in this Game. He simply "invents" some CA, some finite initial configuration within that CA, and then sits comfortably in His chair enjoying what is going on His screen, waiting for the other programmer to complete his "turn"...)

Similarly, the other programmer, let us call him "scientist" (or even: "digital physics scientist", for a reason that shall become clear later) makes his "move" by programming his computer to show some CA on the screen starting from some finite initial configuration.

The "scientist" tries to "choose" such a CA and finite IC within it that is either the same as (i.e. "isomorphic" to) the CA and IC programmed by his opponent ("God") or, at least, his CA and IC are programmed in such a way that when the time passes, "scientist's" CA is able to *emulate in real time*[2] what is going on the "opponent's" ("God's") screen.

(Actually, we can even think that both programmers play the game "in parallel", i.e. there is no need to assume that "God" makes His "move" first, He "informs" the other programmer He is ready with His "almighty CA", and only *then* the "scientist" makes his "turn". No! We can safely assume that *both* players are simply programming their "CA plus IC" in parallel, and when both are ready we can check out what is the 'result' of the whole Game...)

We shall say that "God has won the Game", if the God's "CA+IC" are either different or, at least, "non-emulatible" by the scientist's "CA+IC".

Otherwise, of course, we say that "the scientist has won the Game" against God, and what is on the scientist's screen is actually a TOE, a "Theory of Everything".

## 2 Historical notes

This popular explanation (in the form of a game) is probably the shortest possible way we can introduce the main idea of a relatively new scientific field, a crossing between theoretical computer science and theoretical physics, known as "Digital Physics".

(That's why we call our "scientist" also a "digital physics scientist"...)

A little bit of history:

---

[2] A CA is said to emulate another CA "in real time" if there is a constant $c$ such that the following holds: for every configuration of the first CA at time steps $t_0 < t_1 < \ldots < t_i$, there exists an isomorphic configuration of the second CA at time steps $t'_0 < t'_1 < \ldots < t'_i$ where $t'_j - t'_{j-1} = c(t_j - t_{j-1})$ for $1 \leq j \leq i$. Emulation in real time is a term proposed by Shönhage [Shönhage 1980]. For brevity, later in this article and in all articles from this series [Petrov 2002] we shall talk merely about "emulation" instead of "emulation in real time".

Sometime around 1960, Edward Fredkin, a professor from MIT, proposed the spectacular idea that our *real*(!) Universe is actually a giant cellular automaton [Wright 1988].

Originally, Fredkin's idea has been experimented with the facts known from contemporary physics. The basic method has been to "model" laws of physics using CAs as mathematical models, as well as to check out what kind of conjectures about our real world we can make assuming the possibility that both space and time are *discrete*, but not continuous, and that the laws of physics are *deterministic*.

This line of research is known as "Digital Mechanics" [Fredkin 1990] and has been developed by Fredkin and others for the past 40 years. It has led to a number of important results (for both computer science and physics), the most well-known of which is probably the so-called "Billiard Ball Model" (BBM) [Fredkin 1982] and the corresponding BBM CA [Margolus 1984].

The BBM CA is a theoretical model that shows that (in principle) there is no problem in constructing a universal computer that dissipates no energy while making computations.

In more general terms, BBM CA and the (larger) class of so-called "Reversible Universal Cellular Automata" (RUCA) have been proposed as a discrete mathematical model of the second law of thermodynamics, as well as a possible clue to look for the solution of the more general problem of the Universe as a CA.

Fredkin's approach, i.e. "Digital Mechanics" (DM), is largely based on physics and what it suggests for our abstract CA-based mathematical models.

In 1993 the author of these lines proposed another idea, aimed directly at the point of finding the solution to the main problem of the "Universe as a CA" suggestion.

This idea, known as "Petrov's hypothesis", is a new way to look at the whole problem and what we usually know as "physics".

The idea gave birth to another interesting line of thought and research, known as "Digital Physics" (DP).[3]

In DP, we assume the suggestion made by Fredkin is correct, and we call it "Fredkin's thesis".

However, instead of trying to use facts from contemporary physics, Petrov's hypothesis suggests it might be possible to re-construct "The CA" that is supposed to be an exact informational model of our Universe *without* looking into... *any*(!) physics.

DP uses a *completely* new idea to look at the whole of physics, as we know it; namely, DP is a *non-experimental* physics.

DP is a "computer scientist's" look at physics, a strange "physics" without... *any* physics at all!

---

[3] In his earlier works Fredkin also speaks about "Digital Physics"; we have no pretensions on the "mental copyright" of the term used. In the present text, we merely use "DM" and "DP" abbreviations in order to distinguish between the two *methods* (or "viewpoints") to look at the whole problem, where DM denotes Fredkin's approach, and DP – the approach proposed by the author.

The task of DP, as we see it, is to find at least one possible combination of "CA plus IC" (actually, an ordered pair <CA, IC>) that "solves" the Game described above...

# 3 Conclusion

But let us get back to the Game.

Someone may say:

"This Game is a meaningless one! The 'scientist' can always win the Game, if he merely chooses some computationally universal CA, i.e. UCA. For example, the scientist can choose the Game of Life (GOL) as a CA, and since GOL is known to be an UCA[4], he can be sure he wins the Game!"

No, it is not that easy!

The Game has some subtle element within it that most people usually do not pay enough attention to when they first start thinking about it... But that component of the Game is really a *very* important one!

Please note that we want "God" and "scientist" not only to "choose" some CAs; we also want them to choose some finite ICs within their CAs, too!

And what they "choose" is their final choice, i.e. they have no right to "change their mind" and decide to "choose something else" in the "future".

CAs are *fixed*, ICs are *fixed*, too!

(So to speak, "the dice has been rolled", and there is no turning back!)

So, please note now that it is not sufficient for the "scientist" to choose some UCA, like GOL, for example. Since if he chooses the GOL as a CA, but then pays no attention to his IC, he may choose something as simple as a "still configuration", "traffic lights", or a lonely "glider" traveling in a void space full of "zeros"...

(For simplicity, let us concentrate on the last proposed choice: a lonely glider)

Let us assume also, for the moment, that "God" Has decided to choose GOL as well.

But clever God may decide to choose *two* (or more!) gliders (as an initial configuration) that collide at some point in the CA space, giving birth to something as chaotic as one is able to think about!

So, who will be the winner? God or the scientist?

Of course, God wins the Game in this case: His CA might be identical to the CA of the scientist, but still, what is going on His screen is in no way compatible with the simple computational process taking place on the scientist's screen. To be more

[4] The classical reference for Conway's "Game of Life" (GOL) is [Gardner 1970]; the proof that GOL is a universal cellular automaton (UCA) is in [Berlekamp 1982].

precise: what is going on on God's screen, in this case, *cannot* be *emulated* on the scientists screen, and that's for sure!

The lonely glider, traveling in the void space of the GOL, is certainly *not* something even getting close to what one usually imagines as a "complex computational process". No doubt, it is not!

That is why we can talk about a "(computationally) universal IC" (UIC) that lies within some (fixed) UCA. The task of DP, in this sense, is to find at least one UCA and some UIC within it. A hard task, indeed!

Still, not everything looks so "impossible" when one starts to think seriously about it, and, especially, when he/she starts thinking about the so-called "Petrov's hypothesis", proposed by the author of these lines.

Let us call the ordered pair <UCA, UIC> and the following computational process (that starts from UIC and takes part in that particular UCA) a "universal algorithm". Without looking too deeply into the technical details, what Petrov's hypothesis says, in short, is that there exists a *single*(!) "universal algorithm", and *that* algorithm is actually *The* Algorithm of our Universe!

That is the whole point; that is what DP *is*.

Actually, it is unimportant whether we shall write "universal algorithm" or "Universal Algorithm" (i.e. *The* Algorithm that runs our *real* Universe), since they both turn out to be one and the same thing...

## Bibliography


[1] Berlekamp, E., Conway, J. & Guy, R., 1982, Winning ways for your mathematical plays, Academic Press (New York), see chapter 25.

[2] Fredkin, E., & Toffoli, T., 1982, Conservative Logic, International Journal of Theoretical Physics 21, pp. 219-253. Online version: http://digitalphilosophy.org/download_documents/ConservativeLogic.pdf

[3] Fredkin, E., 1990, Digital Mechanics: An Informational Process Based on Reversible Universal CA, Physica D 45, pp. 254-270. Online version: http://digitalphilosophy.org/dm_paper.htm

[4] Gardner, M., 1970, The Fantastic Combinations of John Conway's New Solitaire Game of 'Life', Scientific American 223, pp. 120-123. Online version: http://hensel.lifepatterns.net/october1970.html

[5] Margolus, N., 1984, Physics-like models of computation, Physica D 10, pp. 81-95. Online version: http://kh.bu.edu/qcl/pdf/margolun198461680918.pdf

[6] Petrov, P., 2002, Series of Brief Articles about Digital Physics (DP), in preparation. For the latest update, check out: http://digitalphysics.org/Publications/Petrov/DPSeries/

[7] Schönhage, A., 1980, Storage modification machines, SIAM Journal on Computing 9, pp. 490–508.

[8] Wolfram, S., 1984, Computer Software in Science and Mathematics, Scientific American 251, pp. 188-203. Translated into Russian as: Nauchnye

issledovaniya, in: Sovremennyj Kompyuter, Mir (1986) pp. 158-173. Online version: http://www.stephenwolfram.com/publications/articles/general/84-computer/

[9] Wright, R., 1989, Did the Universe just happen?, Atlantic Monthly, pp. 29-44. Translated in Bulgarian as: Kompjutar li e Vselenata? (Is the Universe a Computer?), Spectar 66 (1989) pp. 42-51. Online version (in English): http://digitalphysics.org/Publications/Wri88a/html/ Online version (in Bulgarian): http://digitalphysics.org/Publications/Wri88a-Bul/html/

# Chapter 5

# Epsilon-Pseudo-Orbits and Applications

**Douglas E. Norton**
Department of Mathematical Sciences
Villanova University
douglas.norton@villanova.edu

## 1. Introduction

We consider dynamical systems consisting of the iteration of continuous functions on compact metric spaces. Basic definitions and results on chain recurrence and the Conley Decomposition Theorem in this setting are presented. An $\varepsilon$-pseudo-orbit approximation for the dynamics, with $\varepsilon$ of fixed size, is presented as a mathematical representation of a computer model of such a discrete dynamical system. We see that the Conley decomposition of a space can be approximated by an $\varepsilon$-coarse Conley decomposition in this setting.

This paper was presented at the New England Complex Systems Institute International Conference on Complex Systems, June 9-14, 2002, at Nashua, New Hampshire. The author thanks the organizers for the opportunity to participate in a very interesting meeting.

## 2. The Setting

Charles Conley presents in [1] a fundamental result now called the Conley Decomposition Theorem. It provides for the decomposition of every flow on a compact metric space into a part that exhibits a particular type of recurrence and a part in which the dynamics are essentially one-way. The theorem and its implications have been investigated in several settings, from semi-flows to homeomorphisms; see, for example, [2-7]. Here, we consider a model for corresponding results in one of the standard contemporary introductions to the study of dynamical systems: the iteration of continuous functions on computers.

Let $(X, d)$ denote a compact metric space and $f : X \to X$ denote a continuous function mapping $X$ into itself. (Notice that $f$ is not necessarily invertible.) The *forward orbit* of $x_0 \in X$ is the set of all forward iterates of $x_0$. The point $x_0$ is a *fixed point* for $f$ if $f(x_0) = x_0$. The point $x_0$ is a *periodic point* for $f$ if $f^n(x_0) = x_0$ for some $n > 0$, and its forward orbit is called a *periodic forward orbit.* If $x_n = x_0$ but $x_k \neq x_0$ for $0 < k < n$, then $n$ is called the *period* of the forward orbit. Finally, an *orbit* is a bi-infinite sequence of points $( \cdots, x_{-2}, x_{-1}, x_0, x_1, \cdots )$ such that $f(x_n) = x_{n+1}$ for every integer $n$.

## 3. Basic Definitions and Results

To represent the output of a computer performing the iteration of the function, we consider the idea of *ε-pseudo-orbit.* This idea goes back to Anosov, Bowen, and Conley in [8-11, 1]. A sequence of points $(x_0, x_1, \cdots, x_n)$ in $X$ which satisfies the condition $d(f(x_{k-1}), x_k) < \varepsilon$ for $k = 1, 2, \cdots, n$ is called an *ε-pseudo-orbit of length n*.

For $x$ and $y$ in $X$, we will write $x > y$ to mean that for every $\varepsilon > 0$, there is an ε-pseudo-orbit from $x$ to $y$; that is, for every $\varepsilon > 0$ there is an $n > 0$ so that $(x_0, x_1, \cdots, x_n)$ is an ε-pseudo-orbit with $x_0 = x$ and $x_n = y$. Notice that $x > y$ does not simply mean that there is an ε-pseudo-orbit from $x$ to $y$; it is a statement for *all* $\varepsilon > 0$.

A point $x \in X$ is called *chain recurrent* if $x > x$. That is, $x$ is chain recurrent if there is an ε-pseudo-orbit from $x$ back to itself for any choice of $\varepsilon > 0$. $R(f)$ = *the chain recurrent set of f* = $\{ x \in X: x$ is chain recurrent$\}$. Since any true orbit is an ε-pseudo-orbit for any ε, any periodic point is chain recurrent. Some of the basic results on chain recurrence can be found in [1-3, 5-6].

For $x$ and $y$ in $X$, we will write $x \sim y$ to mean that $x > y$ and $y > x$. That is, for every $\varepsilon > 0$ there exist ε-pseudo-orbits $(x_0, x_1, \cdots, x_m)$ and $(y_0, y_1, \cdots, y_n)$ with $x_0 = x = y_n$ and $x_m = y = y_0$. It can be shown that $R(f)$ is closed and that the relation "~" is an equivalence relation on $R(f)$. These and other results are proved in this setting in [6].

The relation "~" partitions $R(f)$ into equivalence classes, sometimes called *chain components* or *chain classes*. Each chain class is *chain transitive*: that is, within each chain class, any two points have an ε-pseudo-orbit connecting them, for any $\varepsilon > 0$. Given any two points within a chain class and any $\varepsilon > 0$, there is a periodic ε-pseudo-orbit containing both points. Chain classes are by definition the maximal chain transitive sets in the sense that no chain class lies within a chain transitive set that is strictly larger than itself.

A key result in [1, 4-5], proved in this setting in [6], is that the chain recurrent set can be characterized in terms of the attractors of the space. If $A^*$ is the repeller complementary to attractor $A$, we have the following.

**Theorem.** $R(f) = \bigcap\{\, A \cup A^* : A \text{ is an attractor for } f \text{ on } X \,\}$.

Chain classes of the chain recurrent set are found in intersections of attractors and repellers of the system. A point not in the chain recurrent set lies in the domain(s) of attraction for some attractor(s) and under the action of the function heads toward its $\omega$-limit set in a unique chain class. Such points outside the chain recurrent set exhibit behavior that is said to be gradient-like.

The idea of a gradient-like portion of the space is an extension from gradient flows of the idea of functions that decrease on solutions, called *Lyapunov functions*. A space is called *gradient-like* if there is some continuous real-valued function that is strictly decreasing on nonconstant solutions.

**Definition.** A *complete Lyapunov function* for the space $X$ with respect to a continuous function $f$ is a continuous, real-valued function $g$ on $X$ satisfying:

(a) $g(f(x)) < g(x)$ for $x \in R(f)$;

(b) $g(R(f))$ is a compact nowhere dense subset of $\mathbf{R}$;

(c) if $x, y \in R(f)$, then $g(x) = g(y)$ if and only if $x \sim y$; that is, for any $c \in g(R(f))$, $g^{-1}(c)$ is a chain class of $R(f)$.

**Theorem.** *If $f$ is a continuous function on a compact metric space $X$, then there is a complete Lyapunov function $g: X \to \mathbf{R}$ for $f$.*

The structure and the Lyapunov function combine to characterize the basic dynamical composition of the space in what is sometimes called the Fundamental Theorem of Dynamical Systems. See [12].

**Theorem: The Conley Decomposition Theorem.** *Let $(X, d)$ denote a compact metric space and $f: X \to X$ denote a continuous function mapping $X$ into itself. Then the dynamical system consisting of the iteration of $f$ on $X$ decomposes the space $X$ into a chain recurrent part and a gradient-like part.*

The partial order ">" on the points of the space $X$ generates a partial order on the chain classes of $X$ which is then reflected by the complete Lyapunov function on $X$. If $\{C_k\}$ represents the chain classes of $R(f)$, then if there is an orbit from $C_i$ to $C_j$, then $g(C_i) > g(C_j)$, and the components of $R(f)$ can be ordered respecting the decreasing requirement of $g$. In general, there are many orderings of the components of $R(f)$ by different complete Lyapunov functions, all of which respect the order imposed by the dynamics. Finally, it is the chain classes and appropriate unions with connecting orbits from the gradient-like portion of the space which form the isolated invariant sets of the space. See [1] for the original theory in the setting of flows and [6] for the results in this setting.

## 4. New Definitions and Results

Computer models of dynamical systems do not represent the more theoretical "for every $\varepsilon > 0$" but rather reflect a fixed finite bound on the deviation size at each iteration. The focus of the ongoing research described here is to fix $\varepsilon > 0$, make the definitions that correspond to the standard definitions that let $\varepsilon$ go to zero, and compare results to those in the standard setting. For proofs and related results, see [7]. For $x$ and $y$ in $X$, we will write $x >^{\varepsilon} y$ to mean that for the fixed $\varepsilon > 0$, there is an $\varepsilon$-pseudo-orbit from $x$ to $y$. That is, there is an $n > 0$ so that $(x_0, x_1, \cdots, x_n)$ is an $\varepsilon$-pseudo-orbit with $x_0 = x$ and $x_n = y$. There is no requirement beyond the existence of one $\varepsilon$-pseudo-orbit for the given $\varepsilon > 0$. Then a point $x \in X$ is called *$\varepsilon$-chain recurrent* if $x >^{\varepsilon} x$. That is, $x$ is $\varepsilon$-chain recurrent if there is an $\varepsilon$-pseudo-orbit from $x$ back to itself for our given fixed $\varepsilon > 0$. We define the *$\varepsilon$-chain recurrent set of $f$* to be $R_\varepsilon(f)$ = $\{ x \in X : x \text{ is } \varepsilon\text{-chain recurrent} \}$. We have the following results.

**Proposition.** *For any $\varepsilon > 0$, $R(f) \subseteq R_\varepsilon(f)$.*

**Proposition.** $R(f) = \cap R_\varepsilon(f)$.

For $x$ and $y$ in $X$, we will write $x \sim^{\varepsilon} y$ to mean that $x >^{\varepsilon} y$ and $y >^{\varepsilon} x$. That is, for our fixed $\varepsilon$, there exist $\varepsilon$-pseudo-orbits $(x_0, x_1, \cdots, x_m)$ and $(y_0, y_1, \cdots, y_n)$ with $x_0 = x = y_n$ and $x_m = y = y_0$. Then "$\sim^{\varepsilon}$" is defined on the $\varepsilon$-chain recurrent set and partitions $R_\varepsilon(f)$ into equivalence classes, so that the relation "$\sim^{\varepsilon}$" is an equivalence relation on $R_\varepsilon(f)$.

The Conley Decomposition Theorem has an analogue in the case of fixed $\varepsilon$. See [7].

**Theorem: The Conley Decomposition Theorem for Fixed $\varepsilon$.** Let $(X, \text{d})$ denote a compact metric space and $f: X \to X$ denote a continuous function mapping $X$ into itself. Let $\varepsilon > 0$ be fixed. Then the dynamical system consisting of the iteration of $f$ on $X$ decomposes the space $X$ into an $\varepsilon$-chain recurrent part and a gradient-like part. That is, the one-way nature of the orbits of points outside the $\varepsilon$-chain recurrent can be described by that part of the Lyapunov function inherited from the chain recurrent decomposition which decreases on those orbits.

The main result is the following.

**Theorem.** $R_\varepsilon(f) \to R(f)$ as $\varepsilon \to 0+$, where the set convergence is with respect to the Hausdorff metric.

The collection of $\varepsilon$-chain recurrent sets converges in the Hausdorff metric to the chain recurrent set of the system. That is, as we take $\varepsilon$ smaller and smaller, the $\varepsilon$-chain recurrent set is a better and better approximation of the chain recurrent set in a very well-defined way. This result suggests that the $\varepsilon$-chain recurrent set offers an appropriate representation of the chain recurrent set if $\varepsilon$ is the level of greatest possible accuracy.

For example, in computer representations of dynamical systems of iterated maps, smaller and smaller values of $\varepsilon$ give closer and closer views of the chain recurrent set as seen through the $\varepsilon$-chain recurrent sets. We can study the $\varepsilon$-decomposition of the space and consider an $\varepsilon$-graph of the decomposition corresponding to the decomposition graph of the transitivity relation as described in the standard setting; see [7], for example.

Of course, there are phenomena that one could expect to miss in computer experiments for any finite value of $\varepsilon$. On the other hand, it is possible for $R_\varepsilon(f)$ to have infinitely-many chain classes; see [13] for an example. With either a finite or infinite number of chain classes, $R_\varepsilon(f)$ both shows some of the significant dynamics and fails to capture all the dynamical structure that the standard chain recurrent set reflects. However, if the chain recurrent set has only finitely many chain classes, then by the convergence in the Hausdorff metric, there is an $\varepsilon > 0$ so that $R_\varepsilon(f)$ has the same number of classes.

**Proposition.** *If the number of chain classes in the chain recurrent set is finite, then the ε-graph of the system and the standard decomposition graph of the system are the same for some* $\varepsilon > 0$.

The ε-decomposition of the space, then, provides a good tool for the analysis of the dynamics on the space. If the actual decomposition of the space has finitely many chain classes, then for a sufficiently small $\varepsilon$, the ε-decomposition reflects the dynamics of the space. If the actual decomposition of the space has infinitely many chain classes, then smaller and smaller values of $\varepsilon$ provide increasingly accurate reflections of the dynamics.

## 5. Applications

One recent application of ε-pseudo-orbits is in the field of neurosciences. A spike in action potential reflecting neural excitability can be described as a subthreshold oscillation having a large amplitude periodic pseudo-orbit nearby. See [14] for details.

Recent applications of ε-pseudo-orbits within mathematics are varied, from shadowing and approximation to the structure matrix, attractors, and computation. See [15-20] for some current uses of ε-pseudo-orbits across a spectrum of dynamical systems contexts.

## References


[1]C. Conley, *Isolated Invariant Sets and the Morse Index*, CBMS Regional Conf. Ser. in Math., Vol. 38 , Amer. Math. Soc., Providence, R.I. (1978).

[2]L. Block and J. E. Franke, "The Chain Recurrent Set for Maps of the Interval", *Proc. Amer. Math. Soc.* **87**, 723-727 (1983).

[3]L. Block and J. E. Franke, "The Chain Recurrent Set, Attractors, and Explosions", *Ergod. Th. & Dynam. Sys.* **5**, 321-327 (1985).

[4]J. Franks, "A Variation on the Poincaré-Birkhoff Theorem", in *Hamiltonian Dynamical Systems,* (Edited by K.R. Meyer and D.G. Saari), Amer. Math. Soc., Providence, R.I., 111-117 (1988).

[5]M. Hurley, "Bifurcation and Chain Recurrence", *Ergod. Th. & Dynam. Sys.* **3**, 231-240 (1983).

[6]D. E. Norton, "The Conley Decomposition Theorem for Maps: A Metric Approach", *Commentarii Mathematici Universitatis Sancti Pauli* **44**, 151-173 (1995).

[7]D. E. Norton, "Coarse-Grain Dynamics and the Conley Decomposition Theorem", to appear in *Mathematical and Computer Modelling.*

[8]D. V. Anosov, *Geodesic Flows on Closed Riemannian Manifolds of Negative Curvature*, Proc. Steklov Inst. Math. **90** (1967), translation by American Math. Soc. (1969).

[9]R. Bowen, *Equilibrium States and the Ergodic Theory of Axiom A Diffeomorphisms*, Lecture Notes in Mathematics, No. 470, Springer-Verlag, New York (1975).

[10]R. Bowen, *On Axiom A Diffeomorphisms*, CBMS Regional Conf. Ser. in Math., No. 35, Amer. Math. Soc., Providence, R.I. (1978).

[11]C. Conley, *The Gradient Structure of a Flow: I*, IBM RC 3932 (#17806), July 17, 1972. Reprinted in *Ergod. Th. & Dynam. Sys.* **8***, 11-26 (1988).

[12]D. E. Norton, "The Fundamental Theorem of Dynamical Systems", *Commentationes Mathematicae Universitatis Carolinae* **36**, 585-597 (1995).

[13]D. E. Norton, "An Example of Infinitely Many Chain Recurrence Classes For Fixed Epsilon in a Compact Space", to appear in *Questions and Answers in General Topology.*

[14]E. M. Izhikevich, "Neural Excitability, Spiking, and Bursting", *International Journal of Bifurcation and Chaos* **10**, 1171-1266 (2000).

[15]G. Osipenko, "Symbolic Analysis of the Chain Recurrent Trajectories of Dynamical Systems", *Differential Equations and Control Processes* **4**, electronic journal: http://www.neva.ru/journal (1998).

[16]Y. Zhang, "On the Average Shadowing Property", preprint, http://www.pku.edu.cn/academic/xb/2001/_01e510.html (2001).

[17]S. Aytar, "The Structure Matrix of a Dynamical System", International Conference on Mathematical Modeling and Scientific Computing, Middle East Technical University and Selcuk University, Ankara and Konya, Turkey, April 2001.

[18]F. J. A. Jacobs and J. A. J. Metz, "On the Concept of Attractor in Community-Dynamical Processes", preprint.

[19]M. Dellnitz and O. Junge, "Set Oriented Numerical Methods for Dynamical Systems", preprint, http://www.upb.de/math/~agdellnitz.

[20]K. Mischaikow, "Topological Techniques for Efficient Rigorous Computations in Dynamics", preprint.

# Chapter 6

# On the "mystery" of differential negative resistance

**Sebastian Popescu, Erzilia Lozneanu** and **Mircea Sanduloviciu**
Department of Plasma Physics
Complexity Science Group
Al. I. Cuza University
6600 Iasi, Romania
seba@uaic.ro

Investigating the causes of the nonlinear behavior of a gaseous conductor we identified the presence of two states of the complex space charge configuration self-assembled in front of the anode. These states correspond to two levels of self-organization from which the first one is related to spatial pattern whereas the second one to spatiotemporal pattern. Their emergence through two kinds of instabilities produced for two critical distances from thermodynamic equilibrium is emphasized in the current voltage characteristic as an S-shaped, respectively Z-shaped bistability. Their presence attributes to the gaseous conductor the ability to work as an S-shaped, respectively an N-shaped negative differential resistance.

## 1. Introduction

Many physical, chemical or biological systems driven away from thermodynamical equilibrium may give birth to a rich variety of patterns displaying strong structural and dynamical resemblances. The study of these nonlinear structures reveals many interesting features as multiplicities of states, hysteresis, oscillations and eventually chaos. In a characteristic in which a response parameter of the analyzed system is represented as a function of a control parameter, the multiplicity of states (*e.g.* bistabilities) is intimately related to the hysteresis phenomenon proving that the system has memory [1].

Informative for the genuine origin of the memory effect are experimental investigations recently performed on gaseous conductors [2,3]. Thus, when the gaseous conductor (plasma) is gradually driven away from the thermodynamic equilibrium the system reveals two different bistable regions (bifurcations) that successively appear in the form of an S-shaped, respectively an N-shaped **n**egative **d**ifferential **r**esistance (NDR). Their appearance emphasizes the development of two successive levels of self-organization [2,3].

The aim of the present paper is to show that the S-shaped DNR is related to a first level of self-organization, the final product of which is a stable (static) complex space charge configuration (CSCC). The stability of CSCC is ensured by an electrical double layer (DL) at its boundary. The sensitive dependence on the current of the emergence and deaggregation of the stable CSCC attributes to the gaseous conductor, in which such a self-organization structure appears, the quality of an S-shaped NDR. When the distance from the equilibrium is further increased by increasing the inflow of matter and energy, the CSCC transits into a new, higher level of self-organization. This is a steady state of the CSCC during which the DL from its border sustains and controls a rhythmic exchange of matter and energy between the complexity and the surrounding environment.

The presented experimental results offer, in our opinion, a new insight into a phenomenology essentially based on a self-organization scenario, the consideration of which could substantially advance the theoretical basis of a class of phenomena frequently classified as mysteries of Physics.

## 2. Levels of self-organization at the origin of S-, respectively Z-shaped bistabilities

The bistability type of a nonlinear system is usually related to different electrical transport mechanisms illustrated in the shape of the current-voltage characteristics. Thus the S-type characteristic specific to pnpn-diodes, quantum-dot structures etc. are related to a transport mechanism that involves impact-ionization breakdown, whereas the Z-shape characteristic is currently associated to moving domains, a mechanism observed in Gunn diodes, p-germanium etc. The fact that distinct shapes of the current-voltage characteristics were observed when the nonlinearity of different systems was investigated has led to the opinion that the physical processes at their origin are related to different basic phenomena, two of them mentioned above. That this is not so will be illustrated on the example of a gaseous conductor, the current-voltage characteristic of which reveals, when the anode voltage is gradually increased, the successive appearance of the S-shaped, respectively of the Z-shaped bistabilities. The cause of their appearance

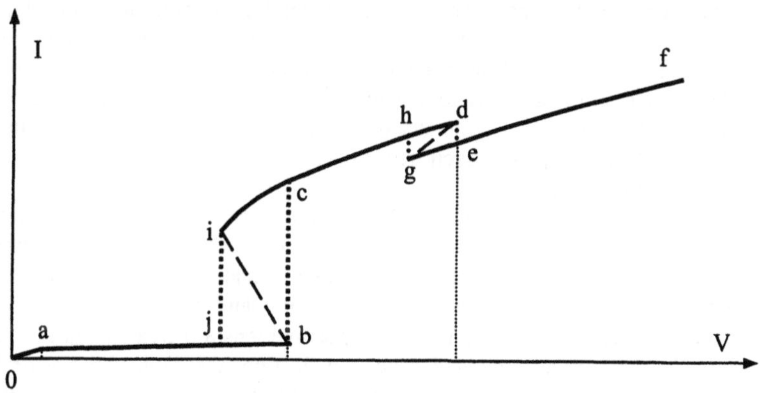


**Figure 1.** Typical static current *versus* voltage characteristic for a diode-like plasma system. The solid and dashed lines denote the stable, respectively unstable branches, and the dotted ones mark jumps.

is an evolution mechanism during which a spatial pattern (stable state of the CSCC) emerges first in the gaseous conductor and, when the voltage of the dc power supply is gradually increased, spatiotemporal patterns appears (steady state of the CSCC). These levels of self-organization evolve through two instabilities by a mechanism whereby the first level favors further evolution by increasing the nonlinearity and the distance from equilibrium.

The aim of our paper is to demonstrate that these bistabilities are related to a hierarchy of self-organization phenomena, the genesis of which involve key processes as symmetry breaking, bifurcations and long-range order *i.e.,* concepts at present considered as basic for the science of complexities.

The static current-voltage characteristic for a plasma diode is presented in Fig. 1, the nonlinear conductor being is plasma, *i.e.* a gaseous medium that contains free electrons, ions and atoms in the ground state. Its shape reveals that the gaseous conductor in a plasma diode evolves, when the voltage of the dc power supply connected to it is gradually increased and decreased, through a succession of abrupt variations of its current transport behavior. Such abrupt transitions can only arise in nonlinear systems that are driven away from equilibrium. These appear beyond a critical threshold when the system becomes unstable transiting into a new conduction state.

Applying a positive potential on an electrode playing the role of the anode in a device working as a plasma diode an electric field appears in front of it, which penetrates the gas on a certain distance. This distance depends on the voltage of the anode and on the plasma parameters. The electrons that "feel" the presence of the electric field are accelerated towards the anode. For values of the anode potential for which the kinetic energy of the electrons is not sufficient to produce excitations of the gas atoms the current collected by it increase linearly in agreement with Ohm's law (branch **0**-a). If the voltage of the dc power supply is gradually increased so that the electric field created by the anode accelerates the electrons at kinetic energies for which the atom excitation cross section suddenly increases, a part of electrons will lose their momentum. As a consequence, at a certain distance from the anode, which depends on its potential, a net negative space charge is formed. Acting as a barrier for the current its development simultaneously with the increase of the anode potential determines the appearance, in the current-voltage characteristic, of the branch **a-b** in which the current is nearly constant although the voltage on the gaseous conductor increases. The accumulation of electrons in a region where the translation symmetry of the excitation cross section, as a function of electron kinetic energy, is broken represents the first phase of spatial pattern formation in a plasma diode.

When the voltage of the dc power supply is additionally increased the electrons that have not lost their momentum after atom excitations will obtain the kinetic energy for which the ionization cross section suddenly increases. Because of the small differences of electrons' kinetic energy for which the excitation and ionization cross-sections suddenly increase, a relatively great number of positive ions are created adjacent to the region where electrons are accumulated. Since the electrons that ionized the neutrals, as well as those resulting from such a process, are quickly collected by the anode, a plasma enriched in positive ions appears between the net negative space charge and the anode. As a result of the electrostatic forces acting between the well-located negative space charge and the adjacent net positive space charge, the space charge configuration naturally evolves into an ordered spatial structure known as DL. This is located at a certain distance from the anode that depends in its potential. Its potential drop depends on the excitation and ionization rates sustained by the electrons accelerated towards the anode. The cause that produces the abrupt increase of the electrical conductance of the gaseous conductor emphasized in Fig. 1 by the sudden increase of the current when its density reaches a critical value (marked by **b** in Fig. 1) is the appearance of a first level of self-organization. This level of self-organization is revealed by the spontaneously self-assembly in front of the anode of a nearly spherical CSCC, the spatial stability of which being ensured by a DL. It appears when the correlation forces that act between the two adjacent opposite space charges located in front of the anode reach a critical value for which any small increase of the voltage of the dc power supply make the current to grow above the threshold value (marked by **b** in Fig.1) for which a first instability of the gaseous conductor starts. As already shown [2,3], this instability is related to a self-enhancement process of the production of positive ions that appears when the potential drop on the DL reaches a critical value. This critical value corresponds to the state of the DL for which the electrons accelerated within it produce by ionization, in a relatively small region, adjacent to the net negative space charge, a net positive charge able to balance it. The amount of positive ions created in this way in the vicinity of the net negative space charge increases the local potential drop in which the thermalized plasma electrons are accelerated. Because of electrons' kinetic energy dependence of the production rate of positive ions, a new amount of positive ions are added to the previously existent one. At its turn this determines a new increase of the local electric field in which the thermalized plasma electrons are accelerated and so on. After this positive feedback mechanism the region where the electrons are accumulated after excitation is shifted away from the anode. So the region where the concentration of positive ions grows suffers a displacement simultaneously with the region in which the net negative space charge is located. Owing to the fact that the production rate of positive ions is limited by the number of atoms present in the gaseous conductor when the pressure is maintained constant, the departure of the DL from the anode ceases at a certain distance from it. During this unstable phase the DL expands to the state for which the net negative space charge equilibrates the adjacent net positive space charge placed in its next vicinity but also that located in the nucleus of the CSCC. Taking into account

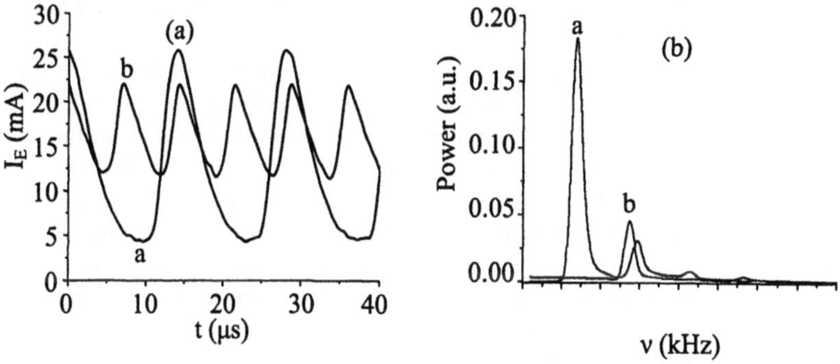


**Figure 2.** Oscillations generated by the S-shaped (corresponding to point **c** in Fig. 1) and Z-shaped bistabilities (corresponding to point **e** in Fig. 1); Power spectra for the oscillations shown in Fig. 2(a).

that the expanding DL reveals a behavior similar to surfaces that separates two different media, *i.e.* surface tension behavior, the final shape of the DL corresponds to a state of local minimum of the free energy. Therefore the DL tends to evolve to a spherical fireball shape. However because the self-assembling mechanism of the DL requires a constant supply with electrons transported by the current, the CSCC must maintain the electrical contact with the anode. As a consequence, the CSCC appears in a plasma diode attached to the anode. Under such conditions the electrons that produce and result after ionizations in the nucleus of the CSCC are constantly collected by the anode, so that the flux of thermal electrons that pass through the DL from ensures the DL continuous self-assembly.

The abrupt increase of the current simultaneously observed with the emergence of the CSCC proves that the DL at its boundary accelerates electrons at kinetic energies for which positive ions are created. This signifies that the DL acts as an internal source of charged particles, the appearance of which increases the conductance of the gaseous conductor so that, for the same voltage of the anode, the current delivered by the dc power supply becomes greater. Since a nonlinear conductor is usually connected to the dc power supply through a load resistor the abrupt increase of the current is accompanied by a similar decrease of the potential supported by the conductor. This decrease of the potential drop on the conductor depends on the value of the load resistor.

The spatially ordered distribution of the opposite net electrical space charges in front of the anode displays all the characteristics of a spatial pattern formed by self-organization. Thus its emerge has the following characteristics. (i) It is self-assembled in a nonlinear system driven away from thermodynamical equilibrium when the external constraint (in our case the potential applied on the electrode) reaches a critical value. (ii) Its self-assembly process is initiated by spatial separation of net opposite space charges related to the symmetry breaking and spatial separation of the characteristic functions of the system (*i.e.* the neutral excitation and ionization cross sections as functions of the electron kinetic energy). (iii) The ordered structure self-assemblies as a result of collective interactions between large groups of structural elements of the system (electrostatic attraction between opposite space charges). (iv) The groups of structural elements spontaneously emerge into a DL when the electrostatic forces, acting between them as long-range correlation, attain a critical value for which the space charge configuration naturally evolves into a state characterized by a local minimum of the free energy.

Although the self-assembly mechanism of the CSCC in front of the anode involves all key processes of a self-organized system this level of self-organization refers only to the emergence of a stationary spatial pattern. Another question related to this level of self-organization is the fact that the maintenance of the ordered spatial structure requires work performed by the dc power supply connected to the conductor. This work is done for ensuring the transport of thermalized plasma electrons towards the DL at the boundary of the stable CSCC. Only under such conditions the DL can ensure, by local acceleration of the thermalized plasma electrons, the physical phenomena required for its continuously self-assembling process. Therefore this state of self-organization is a metastable one because it requires work for maintaining stability.

In agreement with the above said the S-shaped bistability and its ability to work as a NDR has its origin in the described self-organization process during which matter and energy originated from the external dc power supply are stored in the CSCC. Since the self-assembling and de-aggregation processes of the CSCC sensitively depends on the current, fact demonstrated by the hysteresis loop **b-c-i-j-b**, it become possible to trigger these processes by oscillations that naturally appear when the gaseous conductor contains (or is connected to) a system able to perform natural oscillations. In such a case a sufficiently strong random variation of the current can stimulate oscillations when the voltage of the anode is placed in the voltage range where the gaseous conductor emphasizes an S-shaped bistability. Such current oscillations actually appear in a plasma device the current-voltage characteristic of which has the shape showed in Fig. 1. The shape of such oscillations obtained from a double plasma machine that works as a plasma diode are presented in Fig. 2. Their appearance was explained [4] revealing by experimental measurements that the CSCC has the ability to work as an S-shaped NDR and also as an oscillatory circuit. The reactive elements of this oscillatory circuit are the DL from its boundary, working as a capacitor, and an inductor, the physical origin of which is related to the differences in the masses of electrons and positive ions. The inductive behavior of the CSCC appears during its dynamics related to the periodic assembling and total (or partial) de-aggregation of the CSCC.

The S-shaped NDR displayed by the characteristic in Fig. 1 by the dashed line is particularly called current controlled NDR. As already showed it is associated to the self-assembling, in front of the anode, of a stable CSCC. Its genesis involves accumulation of charged particles and electric field energy, both of them provided by the external power supply. Since the self-assembling process of the CSCC sensitively depends on the current, it becomes possible to drive its formation and de-aggregation by changing the current. Therefore under conditions for which the CSCC reveals the behavior of a system able to perform natural oscillations the plasma device acts as a plasma oscillator. The appearance and maintenance of these oscillations is related to an aleatory process by which the potential of the anode is suddenly varied with a value for which the S-shaped bistability can work as an S-shaped NDR. These oscillations appear abruptly without revealing an amplitude growing process. Their appearance is an illustration of the way by which a saddle-node bifurcation appears in a plasma device. Their maintenance is ensured by an internal feedback mechanism by which the oscillations themselves drive the accumulation and the release of matter and energy related to the modification of the internal space charge configuration of the CSCC. Note that the CSCC actually reveals an internal "fine structure" [5] that makes it able to release only a part of the matter and energy accumulated during its self-assembling process. Therefore the S-shaped NDR can sustain oscillations by transiting from an unstable point placed on the line that represents the S-shaped NDR and the corresponding maximal value of the current. As known, the area of the hysteresis loop corresponds to the power extracted from the dc power supply during the CSCC self-assembling process.

As revealed by the I(V)-characteristic presented in Fig. 1, simultaneously with the emergence of the stable CSCC, the gaseous conductor transits into a state for which the same current is transported through the plasma for a smaller voltage applied on it. This reveals that the system "locked" in a state for which the power required to maintain it is ensured by a minimal "effort" from the external dc power supply. This minimal value of the power extracted from the dc power supply is related to the emergence of the stable CSCC acting as a new source of charged particles. It is located close to another minimum that appear when the voltage of the anode and, consequently, the current reaches the critical value marked by **d** in Fig. 1. For this value of the anode voltage the static current voltage characteristic shows the Z-shaped bistability. In that case the current decreases abruptly for the same value of the anode voltage. This new conduction state of the gaseous conductor for which the power required from the dc power supply becomes again minimal is related to the transition of the CSCC into the above-mentioned steady state. As shown in Fig. 2, simultaneously with this transition the oscillations generated by the plasma device have twice the frequency and half of the amplitude of the oscillations stimulated by the S-shaped NDR. Simultaneously with the sudden transition into this new conduction state the current collected by the anode becomes periodically limited. This phenomenon appreciated as the most celebrated diodic event [6] was explained [7] considering a new kind of instabilty that appears when the potential of the anode reaches the critical value for which the gaseous conductor reveals Z-shaped bistability. This bistability develops when the excitation and ionization rates at the two "ends" of the DL at the boundary of the stable CSCC become so high that the equilibrium between the two adjacent opposite space charges can be realized after ionization in a small region placed in the vicinity of the negative net space charge. Under such conditions the DL from the boundary of the CSCC becomes unstable because a small increase of the anode voltage that determines its departure from the anode initiates a new mechanism by which the DL is able to ensure its own existence. Thus, transiting into a moving state through a medium that contains thermalized electrons, namely the plasma, the DL becomes able to self-adjust its velocity at the value for which the additional flux of electrons transiting it is equal with the flux of electrons related to the decreasing of the current. Transiting into a moving phase the DL undertakes a work that diminishes the work required from the external dc power supply to maintain the new conduction state of the gaseous conductor.

As the genuine cause of the Z-shaped bistability it was identified the periodic limitation of the current related to the transition of the CSCC into a steady state during which the DL at its boundary is periodically detached and re-assembled [7]. Since after the departure of the DL from the boundary of the CSCC the conditions for the self-assembling of a new DL are present, an internal triggering mechanism appears, ensuring the periodicity of the consecutive formation, detachment and disruption of DLs from the CSCC boundary. Thus, after the departure of the DL from the anode the conditions for the self-assembling of a new DL in the same region where the first one was self-assembled, appear. The development of the negative space charge of the new DL acts as a barrier for the current, so that this is diminished at the critical value for which the existence conditions for the moving DL disappear. As a result, the moving DL disrupts. During the disruption process the electrons from the negative side of the DL becomes free moving as a bunch towards the anode. Reaching the region where the new DL is in the self-assembling phase the flux of electrons traversing it, and implicitly the ionization rate, suddenly increases at the value for which the new DL starts its moving phase. In this way an internal triggering mechanism ensures the successively self-assembly, detachment and de-aggregation of DLs from the boundary of the CSCC. The described dynamics of the DL related to the Z-shaped bistability of the gaseous conductor is experimentally proved by the appearance of oscillations with twice the frequency and half the amplitude of the oscillations sustained by the S-shaped bistability. This demonstrates that the dynamics of the successively self-assembled and de-aggregated DLs is produced in the time span corresponding to the period of the oscillations sustained by the S-shaped NDR.

We note that periodic limitation of the current related to the successive self-assembly and de-aggregation of DLs from the border of the CSCC requires a relatively small amount of matter and energy extracted from the dc power supply. Additionally, this periodical variation of the current does not represent genuine oscillations as those sustained in a resonant circuit by the S-shaped NDR. Therefore connecting an oscillatory circuit with a high quality factor to such kind of NDR, the oscillations reveal an amplitude growing process. The phenomenology related to the periodic limitation of the current is an illustrative model for the Hopf bifurcation.

## 3. Conclusions

The presented new information offered by plasma experiments concerning the actual physical basis of the S-shaped, respectively Z-shaped bistability and implicitly of the S-shaped and N-shaped NDRs reveal that two levels of self-organization are at their origin. The first level of self-organization is emphasized in the gaseous conductor by the emergence of a CSCC, the self-assembly and de-aggregation of which sensitively depend on the current collected by the anode. In that case the oscillations appear abruptly when to the S-shaped NDR is coupled a system able to perform natural oscillations.

The Z-shaped bistability and the current limiting phenomenon related to the N-shaped NDR are related to a higher level of self-organization emphasized by the presence of a CSCC able to ensure its existence in a steady state. During this steady state matter and energy are periodically exchanged between the CSCC and the environment. In this way the steady CSCC acts as the "vital" part of an oscillator, *i.e.* it stimulates and sustains oscillations in a system working as a resonator. In the last case the oscillations appear after an amplitude growing process.

### References

[1] NICOLIS, Gregoire and Ilya PRIGOGINE, *Exploring Complexity- an Introduction*, Freeman & Co. (1989).

[2] LOZNEANU, Erzilia, Sebastian POPESCU and Mircea SANDULOVICIU, "Plasma experiments with relevance for complexity science" 3rd ICCS, Nashua, NH, USA 2000, *InterJournal of Complex Systems*, article 357 (2001) http://www.interjournal.org

[3] LOZNEANU, Erzilia and Mircea SANDULOVICIU, "On the mechanism of patterns formation in gaseous conductors" XXV International Conference on Phenomena in Ionized Gases 2001 Nagoya, Japan, Ed. Toshio Goto, Proceedings vol. 3, p 165-166.

[4] POHOATA, Valentin *et al.* 2002 "Nonlinear phenomena related to self-organization observed in a double plasma machine", *Phys. Rev.* E (submitted).

[5] LOZNEANU, Erzilia *et al.* (2001) "On the origin of flicker noise in various plasmas", *J. Plasma Fusion Res.* SERIES **4**, 331-334.

[6] ENDER, A. Ya., H. KOLINSKYN, V. I. KUZNETSOV and H. SCHAMEL (2000) *Phys. Rep.* **328**, 1-72.

[7] LOZNEANU Erzilia and Mircea SANDULOVICIU, "On the current limiting phenomena in a "collisionless" plasma diode" XXV International Conference on Phenomena in Ionized Gases 2001 Nagoya, Japan, Ed. Toshio Goto, Proceedings vol. 3, p 185-186.

# Chapter 7

# Dissipative Structures and the Origins of Life

**Robert Melamede, Ph.D.**
University of Colorado
1420 Austin Bluffs Parkway
PO Box 7150
Colorado Springs, CO 80933-7150
rmelamed@uccs.edu

## 1. Introduction

Modern open system thermodynamics, as pioneered by Prigogine, provides a new framework for examining living and associated systems [Ji 1987] [Schorr and Schroeder 1989] [Loye and Eisler 1987]. This paper will show how the behavior of molecules can lead to biological events. Novel hypotheses are proposed that integrate open system, far from equilibrium, physical principles with the many layers of biological complexity. An abstract conceptual development provides a unifying perspective of prebiotic evolution, the origins of the genetic code, and the origins of life itself. An understanding of the nature of health and disease arises as a natural extension of physical phenomena and includes thermodynamic interpretations of the basic living processes of cell division and cell death. The reiteration of underlying physical principles provides new definitions of speciation and individuality. Consideration of man's place in an evolving biosphere is also examined.

### 1.1 Background

The guiding principle applied to living systems in this paper is that the entropy of the universe must always increase (dS/dT>0) even if an open system within it decreases, albeit to a lesser extent [Prigogine 1980]. While at first glance this statement might appear to be counter-intuitive and in violation of the second law of thermodynamics, it is not.

**dS/dt=dSi/dt + dSe/dt** [Prigogine 1980]

**dS** = total entropy of a system
**dSi** = entropy produced by the system (internal)
**dSe** = net entropy flowing into the system (exchange)
**dS** can only be negative if dSe is sufficiently negative to overcome dSi, dSi is always positive.

Creativity, the production of new forms of order (new ideas, information + new physical entities = negative entropy) will occur as long as the entropy of the universe increases to a greater extent than if the entropy of a system within it had not decreased.

## 2 Sources of Creativity

Paradoxically, new forms of order can be created as a system tends towards equilibrium as well as when a system is pushed away from equilibrium.. A collection of molecules at equilibrium will have assumed its most probable distribution and typically, entropy would be maximized. Yet, if the system were not at equilibrium, it would move towards it. In doing so, creative events could occur. For example, if non-equilibrium were in the form of reduced conformational isomers or chemical species compared to what would exist at equilibrium, the move towards equilibrium would require creative events to generate these species [Wicken 1976]. Thus, the creation of order and the destruction of order exist in a time independent balance for any system at equilibrium. Free energy will be at a minimum and/or entropy will be maximized. Intimately linked to both of these processes is the very meaning of time.

Creative events also occur when a system far from equilibrium is pushed beyond a critical point such that a flow dependant phase change occurs and a dissipative structure is formed. The creation of a dissipative structure, and its subsequent maintenance under constant conditions, leaves the system with a time independent entropy value. Entropy, however, is not maximized. Simultaneously, the steady state production of entropy to the universe continues at an enhanced rate for as long as the difference in the generating potential exists and the structure is maintained.

## 3 Time

Recent work from Prigogine's group provides a resolution to the "time paradox", the problem of an irreversible world that is typically described by physics in which time is reversible [Prigogine 1997] Time is manifest in the movement of a system towards equilibrium, at which point it is lost. A system at equilibrium has no events that may be used to measure time. Both the equilibrium "steady state" of generation and destruction of chemical species, and the stable, dynamic, steady state that exists far from equilibrium, are characterized by time independent molecular distributions. The meaning of time in the two systems cannot be the same since a continuous degradation of potential due to the presence of an organized system is required only in the latter case. Thus, a dissipative structure requires that a portion of the energy flowing through it be used to maintain the far from equilibrium, stable, steady state, while the entropy of the universe increases at a more rapid rate than would occur if the dissipative structure did not exist. It seems that the degree of organization within either system, equilibrium or dissipative, is timeless. Essentially, time is dependant on a system's distance from equilibrium. Hence, time is made when a system moves away from equilibrium and time is lost when a system move towards equilibrium. A system may move from equilibrium in a linear (near equilibrium) or nonlinear manner (dissipative structure formation). A dissipative structure existing in a stable steady state may be seen as storing time.

## 4 Evolution

Biologically speaking, evolution is the process by which species develop from earlier forms of life. More generally, evolution is the gradual development of something into a more complex form. In the following sections, physical processes will be described that result in the increasingly complex forms that become biological species.

### 4.1 Evolution Phase I: The Generation of Chemical Diversity

Both nonequilibrium (but not dissipative) and equilibrium processes generate chemical diversity by conventional chemical mechanisms that are driven by photodynamic, geothermal and electrical potentials.

### 4.2 Evolution Phase II: Dissipative Structures First Appear

The chemical diversity created in Phase I is sufficient to add chemical potential as a gradient producing source that, in addition to the energetic sources mentioned above, lead to the formation of dissipative structures.

### 4.3 Evolution Phase III: Simple Interactions between Dissipative Structures Form

Relationships (source and sink) developed between dissipative structures composed of prebiotic chemical pathways such that they became dependant on each other. The prebiotic precursor pathways are what will evolve into to carbohydrate, lipid, amino acid and nucleic acid pathways. The primodial genetic code need not have been between proteins and nucleic acids, but could have been between prebiotic dissipative structures that would evolve into the biochemical pathways of carbohydrate, nucleic acid and amino acid synthesis. This line of thought provides an abstract mechanism that allows for the evolution of the genetic code to occur in the absence of cells, membranes, triplet codons, or life.

## 4.4 Evolution Phase IV: Complex Interactions between Dissipative Structures Form

Just as large collections of molecules existing far from equilibrium can spontaneously undergo rearrangements to form organized flow dependent structures, so can collections of interacting dissipative structures. This mechanism can account for the formation of a precellular environment that could have many of the attributes of a living cell, but need not have all of them. In turn, these systems could continue to evolve until livings system resulted. With this model, some interesting possibilities naturally arise. Did different versions of prelife/life first occur from which a founding organism was selected that became the tree of life? Alternatively, did many prelife/life forms appear out of the prebiotic incubator that were very similar and essentially represented a dynamical system that was an attractor, and this form became the founding organism?

The meaning of time again arises when considering the concepts presented in this section. Is there a relativity of time that corresponds to the dynamic hierarchies?

## 4.5 Evolution Phase V: the Cell

In general, a dissipative structure can be communicating with its' environment in one of three modes: maintaining, growing, or collapsing. Similarly, if a cell does not have sufficient negentropic flow ($dSe/dt>dSi/dt$) it must die. It is important to recognize that the flow of energy is controlled by gate keeping functions such as surface receptors that ultimately modulate gene expression and consequently catalytic activity. A cell, with sufficient flow only to maintain a low stable steady state, would be in the G0 stage of the cell cycle. In contrast, if a cell were taking in an excess quantity of negentropic potential the system could destabilize, with the fluctuations of intensive thermodynamic variables going to infinity. The thermodynamic system would bifurcate thus, the cell would have to grow and ultimately divide.

This line of thinking suggests that the cell cycle is a series of phase changes. With sufficient potential, the flow at each transition point destabilizes the system and forces the phase change to the next stage of the cell cycle. In order for a cell in G0 to enter G1, a stimulus must perturb the system in a manner that reconfigures its' interface with its supporting environment in a manner that promotes the transition to the higher metabolic activity associated with G1. This change can only occur successfully when the cell is in an appropriate environment of receptors and their ligands coupled with sufficient energetic potential. These two components represent the negentropic potential.

The G1/S transition represents a phase change in which the high level of metabolic flow is manifested in DNA replication. In other words, the condensation of nucleotide monomers into DNA, and the condensation of amino acids into proteins (histones) serve as sinks for the growing metabolic potential. This sink drives the high degree of flow dependant, cytosolic organization [Melamede and Wallace 1985] that is required for the coordinated replication of the huge lengths of DNA that in addition to replicating, are no longer compacted into relatively static chromosomal structures. The condensation of biochemical monomers into polymers that can serve as a driving force for organization provides a novel insight into what a possible role for "junk" DNA might be. The directional flow of metabolic potential into DNA acts a sink that is increased by "junk" DNA.

The G2/M transition can similarly be viewed as condensation driven. At this stage, the energy dependant mechanics of mitosis and cell division may be partially thermodynamically driven by the condensation of non-chromosomal DNA into a chromosomal structure. A latent heat of organization is released. Again, a potential is generated by energy released as flow dependant structures condense into more stable, less flow dependant, polymers. These phase changes may release energy that can be used by other cellular energy dependant processes. For example, if a cell were in an environment that was promoting cell division it could trigger a new round of cell division. In contrast, if a cell's environment was not sufficiently nurturing, the energy throughput and degree of flow dependant organization of the open system might stabilize at a new lower level.

## 4.6. Evolution Phase VI: Speciation

In this section, conventional evolution is examined from an open system thermodynamic perspective. Thermodynamically, a species is interpreted as a heterogeneous chemical distribution in the dynamic molecular collection of the biosphere. From this perspective, the individuals that compose a species can be viewed as chaotic oscillations of the chemical distribution that characterizes a particular species. Hence, an individual is a loop of a limit cycle. The dissipative structure-based mechanism for the origin of life can be reiterated to provide an explanation for the formation of new species that characterize evolution. As described by Kondepudi and Prigogine, excess potential and flow can destabilize a dissipative structure and lead to the formation of a new, more complex one. A species could, therefore, be destabilized by its' own success. A successful species must extract ample negentropic potential from its' environment, hence, food supply is a measure of this success. When a sufficient number of individuals in a species are subject to excess thermodynamic flow and its' associated destabilizing affects, the molecular distribution that characterized the species undergoes a non-linear rearrangement that results in the formation of a new species.

If thermodynamic principles are to be extended to the evolution of biological systems, a question that must be answered is "What are the intensive variables of a biological system that would be expected to have fluctuations approaching infinity prior to the phase change"? Fluctuations that occur in the ratio free radical production to negentropic flow is hypothesized to be an intensive variable in biological systems that is of primary importance. Ultimately, all living systems are based on electron flow. Respiring organisms pass electrons from reduced hydrocarbons to oxygen. Not all the energy that is released by this process is used for the maintenance and growth of an organism. In a healthy species, existing in dynamic equilibrium with its' environment, the entropic term is predominantly composed of biological waste and heat production. When species has an excess of potential (too much food), energy flow is not optimized. The net capacity to dissipate entropy is insufficient. The result of entropy flow imbalance is that entropy is trapped in the system. The trapped entropy manifests itself as a generalized increase in the steady state level of entropy found in the individuals that compose the species. An important manifestation of this increased internal entropy might be metabolically generated free radicals. A lower degree of functional organization is precisely the characteristic that results from excess free radical production. The general quality of a living system, both metabolically and genetically is destabilized by excess internal entropy production. A thermodynamic entropic view of life provides an understanding of why caloric restriction reduces age related diseases such as autoimmune, neuronal degenerative, cardiovascular disease, cancer and increases longevity. All of these imbalances are thought to have free radicals as a component of their etiology. Hence, despite many failures, successful new species emerge from the increased variability found when a biological system is destabilized by excess negentropic flow.

## 4.7 The Evolution of Body Systems

The evolution of species is characterized by the appearance of organized biological systems such as the digestive, excretory, nervous, immune, endocrine, and reproductive systems. A thermodynamic perspective on the evolution of these systems shows them to be a natural outgrowth of energy flow. The increased entropy production in the universe that resulted in the simplest life forms could again be realized by increasing the negentropic throughput resulting from the evolution of new more complex organisms.

*A priori,* a first step to increasing negentropic flow into a biological system would be to increase the size of the system. Limits defined by surface area to volume ratios restrain the size of cells and naturally lead to multicellularity. Once single celled organisms existed, they had the capacity to take energy and mass into the system, extract sufficient negative entropy to maintain the system, and to bifurcate when thermodynamically appropriate. Minimal changes could result in simple multi-cellular organisms such as found in the porifera. The next logical means for enhancing flow would be to increase the capacity for acquiring food and getting rid of waste, hence the early development of digestive and excretory systems. Essentially, dSe/dT drives a system's variability and the selection of more sophisticated input/output systems that included feedback loops. In order to optimize this flow term, feedback mechanisms are needed so that the ever-changing environment can be best adapted to. dSe/dt is in reality, the sum of two terms that represent the flow of energy (dSe(energy)/dt and mass dSe(mass)/dt. The nervous system can be viewed as the feedback mechanism that monitors the energy flow into and out of a system. Similarly, the immune system may be viewed as a feedback mechanism that monitors a system's mass flow. The endocrine system may be the feedback system that attempts to monitor the interface between a system's entropy production (dSi/dT) and flow (dSe/dt). The reproductive system integrates the flows of the complex system for bifurcation.

## 5 Aging, Health and Disease

The fundamental properties of life, aging, health, disease, and death may be simply characterized thermodynamically. Initially, aging is the time dependant movement of a system further from equilibrium when the system is young and expanding. The phase of physical growth and differentiation is followed by the return of a system to equilibrium as it ages and dies. The more rapidly a system returns to equilibrium, the more rapidly it ages. Thus, health is the ability of a system to maintain, or increase its' distance from equilibrium. Since living systems, as described in this article, are "simply" a hierarchy of interacting dissipative structures, disease may be viewed as a disproportionate movement of an organized sub-system towards equilibrium. Since all flow dependant structures composing an organism are dynamically interacting with each other, if one collapses, it can have far-reaching consequences. The thermodynamics that underlies a particular set of systemic circumstances may lead to apoptosis of a cell. The death of a cell could lead to the collapse of a previously thermodynamically stable tissue. The underlying loss of organized flow could reiterate and amplify itself through out the hierarchal organized system leading to organ failure and death of an individual that could expand to the loss of a species.

## 6 Conclusions

If, as proposed, dissipative structures are the foundations on which life is built, the knowledge of their formation and stability provides fundamental feedback to systems that they compose. Biological, political, social and economic systems are all natural outcomes of basic underlying physical principles. Our attempts to maintain the health of these systems must be guided by our understanding of the physical processes that created and maintain them [Lipton et al. 1998].

## References

Ji, S., 1987, A general theory of chemical cytotoxicity based on a molecular model of the living cell, the Bhopalator, *Arch Toxicol* **60**, 95-102.

Lipton, L., and R. Melamede 1998, in *Organizational Learning: The Essential Journey,* edited by

Loye, D., and R. Eisler, 1987, Chaos and transformation: implications of nonequilibrium theory for social science and society., *Behav Sci* **32**, 53-65.

Melamede, R. J., and S. S. Wallace 1985, in *A possible secondary role of thymine-containing DNA precursors,* edited by Serres, F. d., Plenum Press, (N.Y.) 67.
Prigogine, I. (ed.), 1980, *From Being to Becoming.* W.H. Freeman (San Fransisco).
Prigogine, I. (ed.), 1997, *The End of Certainty: Time, Chaos and the New Laws of Physics*. The Free Press (
Schorr, J. A., and C. A. Schroeder, 1989, Consciousness as a dissipative structure: an extension of the Newman model. *Nurs Sci Q* **2**, 183-193.
Wicken, J. S., 1976, Chemically organizing effects of entropy maximization., *J. Chem. Educ.* **53**, 623-627.

# Chapter 8

# Reverse Engineering Neurological Disease

**Christopher A. Shaw[1,2,3,4], Jason M.B. Wilson[4], and Iraj Khabazian[1]**

Departments of Ophthalmology[1], Physiology[2], Experimental Medicine[3], and Neuroscience[4]
University of British Columbia
828 W. 10th Ave, V5Z 1L8, Vancouver BC, Canada
phone : 604-875-4111 ext 68375
fax : 604-875-4376

cshaw@interchange.ubc.ca

## 1. Abstract

The key problem for preventing the onset of the age-dependent neurological diseases (Alzheimer's, Parkinson's, and ALS) lies in knowing what factor(s) triggers the cascade of events leading to neuron cell death. Similarly, halting such cascades before they have done irreparable harm to the nervous system requires knowing the stages in the cascade in the correct temporal sequence. Without addressing and solving these problems, only palliative efforts are possible. A number of features of each of these diseases, and a consideration of the unusual features of an unusual variant, ALS-parkisonism dementia complex (ALS-PDC) provide some clues to etiological factors that might be crucial. Epidemiological evidence from ALS-PDC suggests an environmental dietary neurotoxin as an early trigger for neurodegeneration. A murine model of ALS-PDC based on these findings has succeeded in duplicating all of the essential behavioral and pathological features of the human disease and provided insight into many stages in the neurodegenerative cascade. The insights gained from this model system will allow us to understand the sequence of events leading to initiation of disease end state, thereby providing a basis for future treatment strategies.

## 2. The Key Problems in Neurological Disease

The age-related neurological diseases, including Alzheimer's disease, Parkinson's disease and amyotrophic lateral sclerosis (ALS), are diagnosed only once significant behavioral deficits have been observed clinically. Alzheimer's disease involves the death of neurons of various regions of the cerebral cortex and the hippocampus and results in loss of cognitive functions such as memory and learning. In Parkinson's disease a part of the nigral-striatal system dies and the loss of dopamine containing neurons in the substantia nigra leads to loss of dopaminergic terminals that terminate in the striatum. This loss, in turn, impacts motor control leading to tremor and gait disturbances. ALS primarily involves the loss of spinal and cortical motor neurons, leading to increasing paralysis and death.

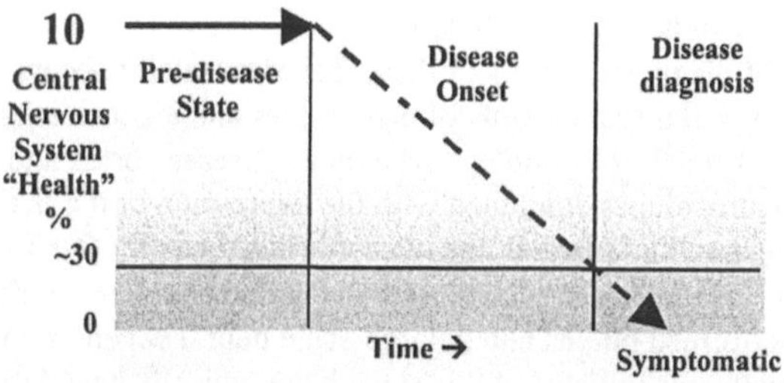


**Figure 1:** Schematic timeline of putative stages in sporadic forms of neurological disease. The schematic represents an idealized timeline from a condition of an early intact nervous system, particularly the neural subsets affected in the age-dependent neurodegenerative diseases Alzheimer's, Parkinson's, and ALS. The middle panel suggests that the diseases follow a linear decline, but the actual rate remains unknown (see Clarke et al., 2000). Clinical diagnosis occurs in most cases once behavioral symptoms have become overt. Between this stage and the death of the individual, it is believed that the majority of the neurons in the particular neural subset have or will die. The horizontal line represents the presumed 'threshold' level of neuron loss leading to clinical symptoms.

In each case, by the time clinical diagnosis is achieved, major damage has been done to the specific regions of the nervous system affected. Estimates of neuron loss vary, but may approach 60-70%, suggesting that neural compensation of surviving neurons occurs over long periods until some final threshold of functional neuron number is reached. Figure 1 is a schematic diagram illustrating the presumed stages of the various diseases, including the late stage of clinical diagnosis, some earlier stage involving the gradual loss of neurons, and some pre-disease state. Note for the purposes of this schematic that the rate of decline in the middle phase is drawn as a linear function, but other functions are actually more likely [Clarke et al., 2000]. Determining the actual rate of decline is an important experimental goal. The realization of this goal will have profound implications for understanding the mechanisms/processes underlying neuron loss, enhancing the prospects for early intervention thereby preventing the final stages of each disease from being reached. We will discuss this point in more detail below.

It is important to note that in spite of conventional views that each of these diseases are totally distinct entities, considerable overlap occurs as a combination of symptoms, biochemical features, and in regions of the central nervous system showing neural degeneration. Such overlap has suggested to various investigators the possibility that each of these age-related diseases may share common mechanism/pathways leading to neural degeneration [Calne & Eisen, 1990], if not common etiologies. Further support for this notion is provided by observations concerning ALS-parkinsonism dementia complex (ALS-PDC), a disease complex first described for the islands of Guam and Rota in the Marianas [see Kurland 1988 for review] and similar to disorders described for the Kii Peninsula of Japan and for parts of New Guinea. Details about ALS-PDC and its implications for understanding the factors responsible for all age-dependent neurological disorders will be discussed below.

The confirmation of diagnosis for any of these diseases usually occurs only postmortem by histological measurements of neuron loss and identification of specific molecular markers. For example, Alzheimer's disease cortex and hippocampus show neuronal loss combined with the expression of the 'hallmark' features of the disease, amyloid plaques and neurofibrillary tangles (NFT) of abnormal tau protein. Treatment for such diseases is undertaken once preliminary diagnosis has been performed but, as noted, large-scale neural degeneration has already occurred. Treatments initiated at this stage have typically been largely palliative, offering only limited amelioration of symptoms and life extension. For Parkinson's disease, intense current interest focuses on stem cell injections as potential means to reverse the neural damage. However, such measures raise theoretical, practical, and economic problems that are difficult, if not impossible, to overcome. Briefly, chaos theory and the dynamic nature of the developing nervous system, the myriad problems associated with recreating neural circuits outside of normal developmental periods, and the costs associated with stem cell transplantation, make this technology as the 'cure' for Parkinson's disease at best a faint hope. Other prospective treatments of end stage neurological disease, e.g., estrogen, anti-oxidants, etc. all face the same ultimate problem of attempting to restore function to already destroyed neural circuits.

Given such problems, a conventional triage approach would focus attention on early stages in the disease: An ideal solution would be prophylactic, but would require that the cause of the disease be known. Next best would be to halt disease progression before irreversible damage had been done to regions of the nervous system, but this would require that the various stages of disease progression be known, so that targeted, rather than random, pharmacological therapeutics could be applied.

It is obvious that we are not in the position to be able to do either of the latter since we do not know what etiological factors cause these diseases nor do we have any clear idea of the early stages in disease progression. While many molecules and abnormal cellular processes can be identified postmortem in humans and in some animal models, it is equally possible that such events represent not causal factors, but rather molecules that are altered 'co-incidentally' but not pathologically, or molecules or processes that may actually be compensatory (successful or failed). Postmortem examination alone cannot distinguish between these possibilities, nor can it put the myriad affected molecules into any sort of temporal sequence from early stages of the disease to final stages resulting in cell death. This fact alone leaves animal models as the most likely means to unravel the temporal sequence of events, a timeline that may prove crucial to successful future therapy. An effective model would have to mimic the essential behavioral and pathological stages in the disease including its progressive nature. It would be predictive, revealing features not found in previous human studies. In addition, the model would have to be induced in a

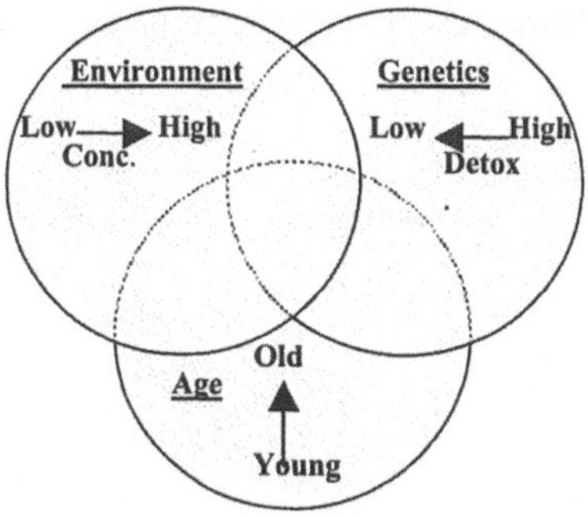


**Figure 2:** Potential synergies of causal and risk factors in sporadic neurological disease. The key causal factor(s) involved in such diseases are thought to reflect exposure to some toxin(s) that can arise from the environment or within the body and which may be synthetic or naturally occurring. This is represented by the set on the left side of the schematic. Range of toxicity effects run from left to right as low to high. Intersecting this is a set consisting of a propensity that could arise due to genetic polymorphism in efficiency of detoxification mechanisms (from right to left expressed as high to low). Genes with other features could also be involved (e.g., APOE alleles). The intersection of these two sets describes the individuals who may be at risk of developing the neurological disorder. Note that the intersecting region can increase or decrease depending on strength of either variable. Intersecting these two sets is the variable of age with risk factor increasing from young to old (bottom to top).

manner similar to that presumed to underlie the disease being studied. The model would also have to allow for the testing of potential therapeutics. Finally, the model would have to allow the stages of disease progression to be 'template matched' to presumably similar, but still unidentified, stages in the human disease.

The identification of the classes of potential causal factors is key to the discussion. In brief, these include 'gain of function' mutations, deletions, environmental toxins or a combination of epigenetic and genetic factors. While an early onset familial form of Parkinson's disease has been identified to be linked to abnormal α-synuclein [Sharma et. al., 2001], it now seems obvious that the vast majority of cases are late onset and non-genetic [Tanner et al., 1999]. Alzheimer's disease has early onset genetic factors involving abnormal tau proteins [McKeith et. al., 1996], but the incidence of this mutation in relation to the total Alzheimer's population is small. Similarly, the ALS population has a familial form, a small fraction of which involves a heritable mutation of the gene coding for superoxide dismutase (mSOD1) [Gaudette et. al., 2000]. For ALS, only 2 to 3% of all cases involve this mutation leaving the sporadic form accounting for virtually all others. The absence of a clear causal genetic component for the sporadic form of any of these diseases focuses our attention on potential etiological features in the environment, notably neurotoxins. Various examples of synthetic toxins have been documented [eg. Ballard et al., 1985] and numerous natural neurotoxins exist [Betarbet et al., 2000].

Although environmental neurotoxins, synthetic or natural, seem the most

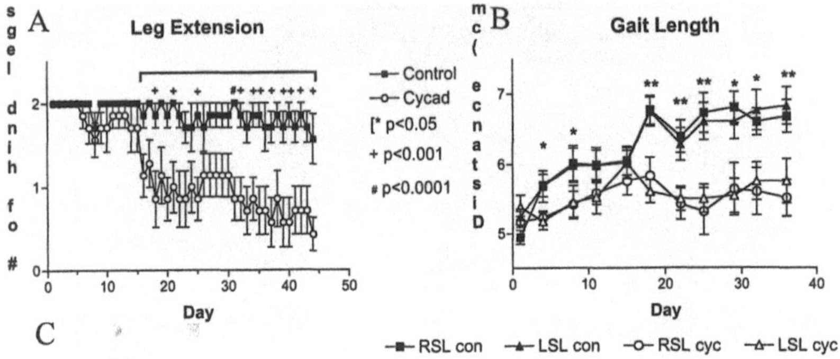


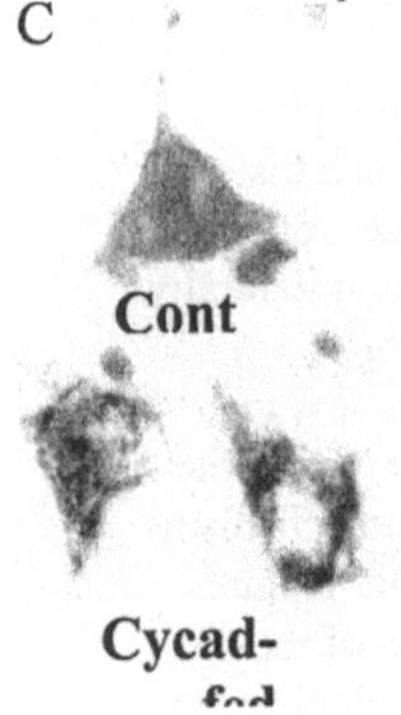


**Figure 3:** Effects of cycad feeding on motor behavior and CNS. See text for details and Wilson et al. (2002). A. Leg extension deficits in cycad-fed vs. control mice. B. Gait length in the same animals. C. Top panel: motor neuron in cervical spinal cord of a control mouse. Bottom panels: motor neurons showing pathological features in cervical spinal cord of cycad-fed animals (1000x mag).

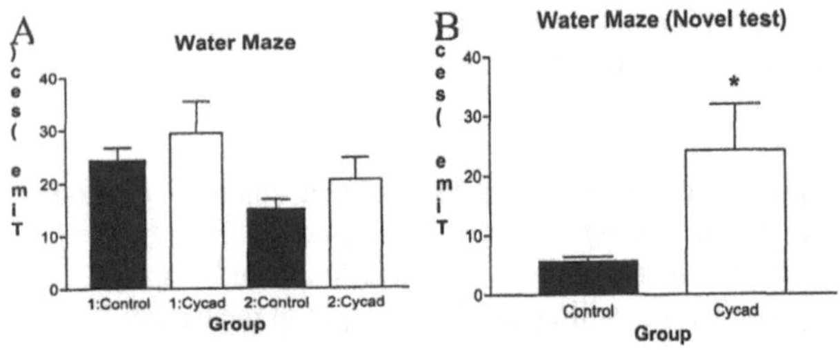


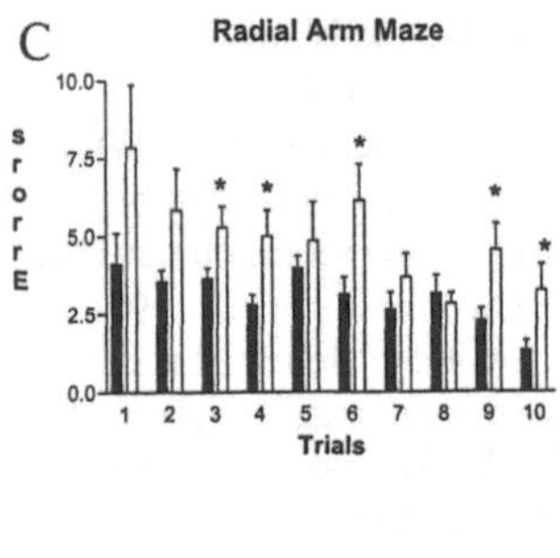


**Figure 4:** Effects of cycad feeding on cognitive behavior in CNS. Details as in Figure 3. A. Morris water maze showed longer latency to find the hidden platform in cycad-fed than control mice. These data did not reach significance, but highly significant effects were noted with a later "probe test" in which the platform was moved to a new location (B). C. Radial arm data showing significantly increased learning and memory deficits in cycad-fed

likely causal factors leading to neurodegenerative changes in the nervous system, it will be equally clear that if any given toxin were the sole cause, larger fractions of the population would have to be afflicted. As this is not the case (although disease incidence may be on the increase [see Martyn, 1994]), it seem most likely that the age-related neurodegenerative diseases arise due to an intersection of exposure to an environmental toxin and a genetic susceptibility (see Figure 2). This susceptibility may take a number of forms, including genetic polymorphisms. For example, it might involve toxin handling (e.g., detoxifying enzyme expression), absorption (e.g., transporter proteins), or an interaction with the handling of related molecules (e.g., APOE alleles). In this view, the intersection of the sets comprised of environmental factor and genetic susceptibility can expand or contract depending on toxin concentration and/or duration and the relative levels of expression of the genetic factor. Age is also a critical variable, with increasing age involved in each of the disorders discussed here.

Various animal models of environmental toxicity leading to parkinsonism-like features have been described [Ballard et. al., 1985]. Recently, Betarbet et al. [2000] demonstrated that rotenone, a natural pesticide, causes complex I mitochondrial damage leading to behavioral and pathophysiological outcomes in 50% of rats injected with the molecule.

Accepting that an animal model is the most likely means of identifying the stages of disease progression or for testing therapeutic approaches, it should be an important effort to identify animal models that could satisfy the above model criteria. In the following, we will describe a model that we believe satisfies these criteria and the data that support this view. The implications of this model for early treatment of pre-symptomatic neurodegenerative disease will also be discussed.

## 3. ALS-PDC and an Animal Model of This Neurological Disease

In the years after World War II, L.T. Kurland and various other investigators described in detail an unusual neurological disease complex, ALS-parkinsonism dementia complex (ALS-PDC). The disease could be expressed as a rather conventional form of ALS (termed 'lytico' or 'paralytico' by the Chamorro population of Guam), or as a form of Alzheimer's disease with strong parkinsonian features (parkinsonism-dementia complex or PDC, locally termed 'bodig'). The history of this disorder was believed by Kurland and other early investigators [for review, see Kurland, 1988] to be so unique that the disease could serve as a type of neurological "Rosetta Stone", the decipherment of which would unlock crucial clues to neurological disorders worldwide. Kurland and other neurologists identified a number of unusual features of ALS-PDC, including its (then) high level of incidence, the occasional overlap of symptoms of the neurological subsets, and often early age of onset. Epidemiology failed to find a genetic cause [Morris et. al., 1999] and the main clue was the notion that a toxin contained in the seed of the cycad palm was the crucial factor [Kurland, 1988]. Kurland and others noted that the disease incidence peaked within several years of massive cycad consumption by the indigenous Chamorro population and dramatically declined when cycad consumption declined. In spite of this, the early enthusiasm that neurotoxins contained in cycad seed were causal to ALS-PDC stalled due to results showing that toxins identified early in the studies did

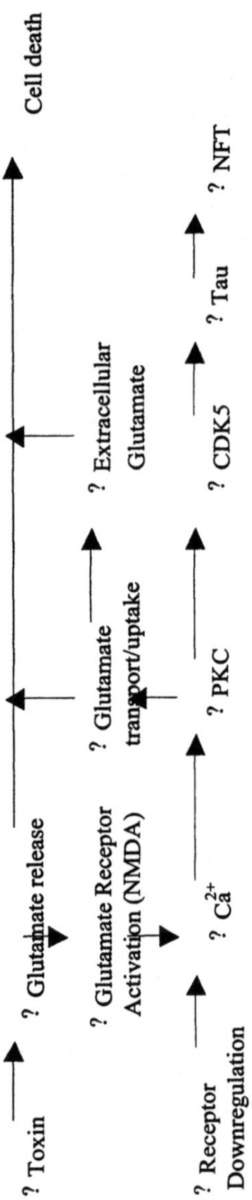


**Figure 5:** Preliminary outline of stages in the biochemical cascade leading to neuronal degeneration. See text. Small arrows indicate increases (? ) or decreases (? ) in amount or function. Big arrows indicate stages or pathways.

not remain in the flour of the cycad processed for consumption [Duncan et. al., 1990].

We recently reexamined the issue using quantitative isolation procedures combined with bioassays for neural activity and cell death [see Khabazian et al., 2002]. These studies identified the most toxic molecule contained in cycad as a sterol glucoside whose actions *in vitro* included the excitotoxic release of glutamate and increases in protein kinase activity. *In vivo* studies of mice fed washed cycad seed flour containing this sterol glucoside employed a battery of motor, cognitive, and olfactory behavioral measures that demonstrated a temporal sequence of behavioral deficits in all three systems [see Wilson et al., 2002]. In regard to motor neuron disorders (see Figure 3 and Wilson et al., 2002), the cycad-fed mice showed significant losses of the leg extension reflex (indicative of motor neuron dysfunction), had pronounced gait disturbances as well as losses of muscle strength and balance. Upon sacrifice, mice fed cycad showed TUNEL and caspase-3 positive cells indicative of apoptosis in spinal cord, cortex, hippocampus, substantia nigra and olfactory bulb. Motor neuron numbers were decreased significantly in ventral cord and various regions of cortex showed significant thinning. These regions of neural degeneration were consistent with the observed behavioral deficits. Cognitive deficits were observed in spatial learning tasks (Morris water maze) and reference memory (radial arm maze) with corresponding neurodegeneration seen in regions of cerebral cortex and hippocampus [see Figure 4 and Wilson et al., 2002]. In addition, the olfactory system showed a significant loss of function accompanied by disrupted structures and cell loss in the olfactory glomeruli. In various regions, key molecules associated with neuronal degeneration were altered. These included elevation in tau protein and various protein kinases (notably PKCs and CDK5) [see Khabazian et al., 2002]; in addition, elements of the glutamatergic system were severely affected including a dramatic decrease in two variants of the GLT1 glutamate transporter (GLT-1$\alpha$ and GLT-1B; also called EAAT2) [Reye et al., 2002] on astrocytes and a decrease in NMDA and AMPA receptor binding [Wilson et al., 2002b; Khabazian et al., 2002b]. Both effects were noted in regions of the CNS showing neural degeneration. The decrease of the GLT-1B transporter was quantified by Western blots and showed a primary loss of the 30 kDa cleavage product that could be partially restored by pre-treating the tested protein fraction with alkaline phosphatase. All of the features described above in our murine model are consistent with features of ALS-PDC, as well as key aspects of Alzheimer's, Parkinson's, and ALS, including a progressive development of neuronal dysfunction [see Wilson et al., 2002].

## 4. Reverse Engineering Neurological Disease

The data cited above have several key implications. First, the correspondence between the changes in behavioral outcome and histological indices of neurodegeneration in the animal model compared to ALS-PDC validates the hypothesis that ALS-PDC may be due to consumption of cycad toxins. Second, the overall similarities to many behavioral, biochemical, and pathophysiological outcomes in age-dependent neurodegenerative diseases suggests that similar mechanisms based on exposure to environmental neurotoxins may be common features of each. In regard to potential therapeutic treatment for early stage neurodegenerative diseases, the fact that we can mimic many of the essential characteristics of these diseases in a reproducible manner suggests that we can use this model system to create a timeline of neurodegenerative events. These events span the period from initiation (onset of exposure to an identified neurotoxin; see Khabazian et al., 2002] through the various stages of neural dysfunction culminating in neural cell death. By rerunning the experiment and sacrificing some animals at set time points, we should be able to create a detailed temporal sequence of molecular events leading to neural degeneration such that causal, co-incident, and compensatory molecules and events can be put into the correct sequence. The implications of the latter are that specific, targeted therapeutics could, in principle, be directed at particular abnormal biochemical events, ideally at early enough stages to prevent the final loss of neurons and neuronal function. Based on our data and studies in the literature, we have been able to begin this analysis as shown in Fig. 5. This schematic shows various putative stages of the neurodegenerative disease process, including the onset of pathological processes induced by the identified cycad toxin. In so doing, it predicts the temporal order of some crucial events. For example, conventional notions have suggested that NFT are an early causal feature of neurodegeneration in Alzheimer's disease and ALS-PDC. In contrast, our model suggests that NFT are 'down stream' of events such as the down-regulation of the glutamate transporter.

Being able to place the molecular events into correct temporal sequence is a key feature of our emerging model system. In addition, as details of the sequence emerge, we will be able to plot the function of neural cell loss with respect to time and answer the question of the quantitative relationship between the loss of particular neurons and the decline in the functions they control. We will also be able to determine if the loss of neurons and the underlying biochemical modifications over time describe linear, sigmoidal, exponential, or other functions. The type of function defined has huge implications for potential therapies. For example, exponential declines in cell number may imply single pathological 'one hit' events in which cell loss is constant over time [Clarke et al., 2000]; in contrast, a sigmoidal decline would suggest that cell death occurs following cumulative long-term damage. The former is consistent with an excitotoxic mechanism of cell death and with the data cited above, but not with mechanisms based on oxidative stress as proposed by various researchers. As a preliminary attempt to address such issues, we have plotted leg extension data (measuring motor neuron function) from mice fed washed cycad vs. time and found an exponential decline (Fig. 6) that closely resembles data generated in various models of Parkinson's disease [see example in Clarke et al., 2000].

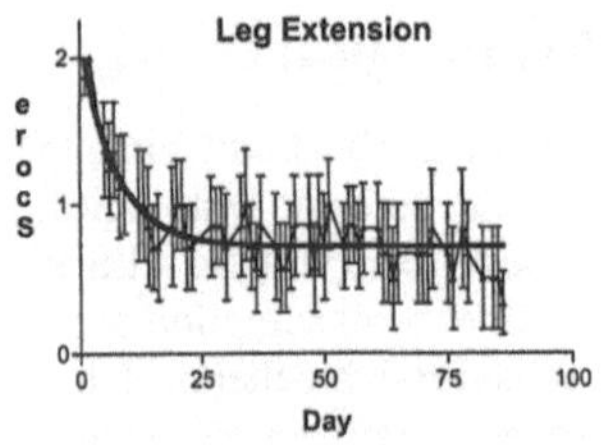


**Figure 6**: Loss of motor neuron function as measured by leg extension. The data represent the averaged results of 7 cycad-fed animals from the onset of cycad feeding to sacrifice. These data are from a different batch of cycad-fed animals than those shown in Figure 3. Note that the resulting curve is initially exponential but appears to reach a steady level for a period of time before showing a further decline (Curve fitting provides the equation $y=1.444*\exp^{(-0.138x)}+0.714$). These data are similar to data derived in a model of chemically-induced Parkinson's disease (Clarke et al., 2000).

## 5. Future Directions

The attempt to reverse engineer neurological disease using our model system is still in very preliminary stages. However, the success of this effort in establishing the correct temporal sequence of molecules involved in neurodegeneration cascades, the feedback and feed-forward loops involved, and the dynamical characteristics of what is seemingly an increasingly chaotic system seems promising. Once the above issues have been resolved, we will have a means to attempt the next crucial task of matching the unseen features of evolving human neurological diseases to known features in the model system. This approach may offer the best hope of successfully treating neurological disease processes before they have done irreparable harm to the nervous system.

## Acknowlegments

We thank M. Wong, J. Lai, H. Bavinton, J. Schulz, J. Avent, and E. Hawkes for discussions. C. Melder provided formatting of the ms. We are also grateful to U. Craig, D. Pow, R. Smith, and the late L.T. Kurland for insights and discussion. This work was supported by grants from the ALS Association, the Scottish Rite Charitable Foundation of Canada, and the Natural Engineering and Research Council of Canada.

## References

[1] Ballard, P.A., Tetrud, J.W., & Langston, JW., 1985 Permanent human parkinsonism due to 1-methyl-4-phenyl-1,2,3,6-tetrahydropyridine (MPTP): seven cases. Neurology 35(7): 949-956.

[2] Betarbet, R., Sherer, T.B., Mackenzie, G., Garcia-Osuna, M., Panov, A.V., & Grennamyre, J.T., 2000. Chronic systemic pesticide exposure reproduces features of Parkinson's disease. Nature Neuroscience 3(2): 1301-1306.

[3] Calne, D.B., & Eisen, A., 1990. Related Parkinson's disease, motoneuron disease and Alzheimer's disease: origins and interrelationship. Adv Neurol. 53: 355-360.

[4] Clarke, G, Collins, RA, Leavitt, BR, Andrews, DF, Hayden, MR, Lumsden, CJ, & Mcinnes, R.R., 2000. A one-hit model of cell death in inherited neuronal degenerations. Nature 406(6792): 195-199.

[5] Duncan, M.W., Steele, J.C., Kopin, I.J., & Markey, S.P., 1990. 2-amino-3-(methylamino)-propanoic adcid (BMAA) in cycad flour: an unlikely cause of amyotrophic lateral sclerosis and parkinsonism-dementia of Guam. Neurology 40: 767-772.

[6] Gaudette, M., Hirano, M., & Siddique, T., 2000. Current status of SOD1 mutations in familial amyotrophic lateral sclerosis. Amyotroph Lateral Scler Other Motor Neuron Disord. 1: 83-89.

[7] Khabazian, I., Bains, J.S., Williams, D.E., Cheung, J., Wilson, J.M.B., Pasqualotto, B.A., Pelech, S.L., Andersen, R.J., Wang, Y-T., Liu, L., Nagai, A., Kim, S.U., Craig, U-K, & Shaw, C.A., 2002. Isolation of various forms of sterol β-d-glucoside from the seed of cycas circinalis: neurotoxicity and implications for ALS-PDC. J. Neurochem. 82: 1-13.

[8] Khabazian, I., Pow, D.V., Krieger, C., & Shaw, C.A., 2002b. Effects of cycad toxins on glutamate transporters in a murine model of ALS-PDC.Society for Neuroscience Abstracts.

[9] Kurland, LT., 1988. Amyotrophic lateral sclerosis and Parkinson's disease complex on Guam linked to an environmental neurotoxin. Trends Neurosci. 11: 51-54.

[10] Martyn, C.N., 1994. Epidemiology. In Motor Neuron Disease, edited by Williams, A.C., Chapman and Hall Medical. New York, pp. 393-426.

[11] Mckeith, I.G., Galasko, D., Kosaka, K., Perry, E.K., Dickson, D.W., Hansen, L.A., Salmon, D.P., Lowe, J., Mirra, S.S., Byrne, E.J., Lennox, G., Quinn, N.P., Edwardson, J.A., Ince, P.G., Bergeron, C., Burns, A., Miller, B.L., Lovestone, S., Collerton, D., Jansen, E.N., Ballard, C., De Vos R.A., Wilcock, G.K., Jellinger, K.A., Perry, R.H., 1996. Consensus

guidelines for the clinical and pathologic diagnosis of dementia with Lewy bodies (DLB): report of the consortium on DLB international workshop. Neurology. 47(5): 1113-1124.
[12] Morris, H.R., Crook, R., Lees, A.J., Wood, N.W., McGeer, P., & Gwinn-Hardy, K., 1999. Genetic and clinical comparison between Parkinsonism dementia complex on Guam and Western neurodegenerative diseases. Soc. Neurosci. Abstr. 25: 1305.
[13] Reye, P., Sullivan, R., Scott, H., & Pow, D.V., 2002. Distribution of two splice variants of the glutamate transporter GLT-1 in rat brain and pituitary. Glia 38: 246-255.
[14] Sharma, N., Mclean, P.J., Kawamata, H., Irizarry, M.C., & Hyman, B.T., 2001 Alpha-synuclein has an altered conformation and shows a tight intermolecular interaction with ubiquitin in Lewy bodies. Acta Neuropathologica 102: 329-334.
[15] Tanner, C.M., Ottman, R., Goldman, S.M., Ellenberg, J., Chan, P., Mayeux, R., & Langston, J.W., 1999. Parkinson Disease in twins JAMA 281: 341-346.
[16] Williams, A.C., 1994. Motor Neuron Disease, Chapman and Hall Medical, 393-426.
[17] Wilson, J.M.B, Khabazian, I., Wong, M.C., Seyedalikhani, A., Bains, J.S., Pasqualotto, B.A., Williams, D.E., Andersen, R.J., Simpson, R.J., Smith, R., Craig, U.K., Kurland, L.T., & Shaw, C.A., 2002. Behavioral and neurological correlates of ALS-parkinsonism dementia complex in adult mice fed washed cycad flour. J. Neuromol. Med 1: 207-222.
[18] Wilson, J.M.B., Khabazian, I., Pow, D.V., & Shaw, C.A., 2002b. Loss of GLT-1 transporter subtypes and EAA receptor downregulation in CNS of mice fed washed cycad flour. Society for Neuroscience Abstracts.

# Chapter 9

# A Multi-Level Synthesis of Dyslexia

**Chris Phoenix**
cphoenix@crnano.org

Dyslexia has been studied from many angles. Researchers have obtained seemingly contradictory results and created widely varying theories and treatments. A complete understanding of dyslexia requires recognition of neurological and psychological components and their interaction, and could therefore benefit from a complex systems approach. This paper surveys and synthesizes results from many theoretical, experimental, and clinical approaches to dyslexia, including Galaburda, Davis, Geiger, and Merzenich. The magnocellular hypothesis combined with the Davis theory of "triggers" appear to explain nearly every experimental result, observation, and successful treatment of which the author is aware. Dyslexia can be understood as an accretion of simple symptoms in multiple sensory modalities, each symptom having the same neurological basis; each individual has a different combination of symptoms, and the symptoms are created and maintained through mental/psychological interaction with the individual's efforts to perform. There is strong observational evidence, confirmed by pilot studies carried out by the author, that the symptoms can change momentarily. Although such rapid change is not recognized by many dyslexia researchers, it has been demonstrated with PET scans in the case of stuttering; this finding is crucial to a full understanding of the interaction between neural function and mental state. The recognition of the diversity of symptoms, their common neurological basis, and their extreme plasticity in response to high-level mental state, may help to focus research and to develop increasingly effective and rapid treatments.

# 1 Introduction

Over the past several decades, much work has been done to investigate the causes, symptoms, and treatments of dyslexia. Several opposing theories have been pursued [Ramus 2001], and several successful treatments have been developed and evaluated [McAnally, Castles, & Stuart 2000]. While each approach attempts to explain all the features of dyslexia, there has not been an attempt to synthesize the various approaches. At first sight, this would be an impossible task, since the theories contradict each other at many points. This paper presents an approach that attempts to cover all the observations, although it differs from current theories at several important points.

The present approach requires an understanding of dyslexia as a multi-level phenomenon: specifically, a set of high-level responses to low-level disruptions of perception. It is notable that several dyslexia correction methods, grounded in contradictory theories, claim success rates well over 50%. Also striking is the variety of symptoms and findings of dyslexics. In today's educational environment, a failure to read adequately is crippling, and thus forms a natural target for intensive study. However, the symptoms and syndromes associated with dyslexia are much broader than simple failure to read. These diverse symptoms can all be understood as arising from a common low-level cause, with the caveat that the low-level mechanism can affect, and be affected by, high-level mental state. This leads to a complexity that can be confusing or misleading to researchers attempting to study a single aspect of dyslexia.

There is no universally accepted definition of dyslexia. Most studies attempt to define it in terms of reading deficit vs. general intelligence, with various methods of establishing reading ability and intelligence level. Some studies require that remedial education has been unsuccessful. This paper will not attempt to establish a single, universal definition of dyslexia; instead, it will assume that when researchers intended to study dyslexia, they generally succeeded in forming a suitable study population. Another factor that is important in some papers is that of subtypes of dyslexia, for example, auditory and visual subtypes. As will be seen later, this paper recognizes the existence of auditory and visual syndromes, but does not support the idea of subtypes that must be studied separately (although the syndromes may need to be treated individually in clinical practice). Therefore, studies of any dyslexic subtype will be considered.

# 2 The Low-level Basis Of Dyslexia

Many studies have found differences between dyslexics and non-dyslexics in low-level perception. For example, many researchers have found differences in visual perception such as the width and acuity of letter recognition field [Geiger & Lettvin 1987], or the sensitivity to contrast and motion under various conditions [Lovegrove, Garzia, & Nicholson 1990]. Tallal, Miller, and Fitch [1995] have found differences in auditory processing, such as a reduced sensitivity to rapid sound transitions. Any theory of dyslexia must recognize and account for the multisensory nature of the disorder.

It is tempting to suggest that the magnocellular system is responsible for these sensory problems. Livingstone, Rosen, Drislane, and Galaburda [1991] have found physical malformation of the cells in the visual magnocellular pathway in dyslexics. In addition, many studies have found visual deficits consistent with a deficit in magnocellular processing. Unfortunately, the first proponents of the "magnocellular hypothesis" made a supposition that appears to be incorrect about how a magnocellular deficit could affect the visual system [Skottun 1999], and other studies have obtained inconsistent results [Skottun 2001].

Grossberg and Cohen [1997] have found clear evidence of distinct fast and slow processing channels in the auditory system. Tallal, Miller, and Fitch [1995] and Shaywitz and Shaywitz [1999], have found that dyslexics have a deficient ability to detect rapid transitions in normal speech. As noted by Stein [1993, p. 83], "The magnocellular component of visual processing that is impaired in dyslexics, does have anatomical counterparts in the somaesthetic, auditory, and motor systems..."

Given the low-level nature of these symptoms, and (in some studies, at least) the unambiguous character of the difference between dyslexics and normals, it seems clear that dyslexia must involve a difference in the pre-cortical neural pathways of more than one sense. The best candidate seems to be the pathways responsible for maintaining information about timing and rapid change. In the visual system, this is the magnocellular pathway. This should not be read as support for the "magnocellular hypothesis", especially its original suggestion that the faulty mechanism was suppression of the parvocellular pathway by the magnocellular pathway. In hindsight, that suggestion was only one of many possibilities.

The claimed effectiveness of colored lenses (Ihrlen lenses) or colored overlays for books may provide additional evidence for a low-level visual basis in at least some cases. However, it should be noted that such claims are controversial [Evans & Drasdo 1991].

## 3 Evidence For Rapid Controllable Variability In Brain Function

Dyslexia has been viewed historically as resulting from some form of brain damage. More recently, many researchers have found that even the low-level symptoms of dyslexia may be improved with appropriate training [Geiger & Lettvin 1997]. This improvement typically takes place over a period of weeks or months. A key point of the present approach is the idea that the symptoms can vary on a far more rapid time scale.

Geiger [1999, personal communication] has described a phenomenon of "conditional dyslexia", in which a person's ability to read depends on external factors, and may thus change rapidly. Davis has described his own variable dyslexia: "When I was at 'my artistic best', I was also at 'my dyslexic worst'" [1997, p. 127]. At this point, it seems unarguable that dyslexia is not a permanent condition. The only question is how rapidly it can change.

That aspects of language processing can change with extreme rapidity is demonstrated by the phenomenon of stuttering. Stuttering appears to involve faulty pathways between various parts of the brain; the activation of these pathways can be detected with PET scans. Notably, these pathways are not permanently damaged. When a person is not stuttering, they work normally. However, they show a deficit if the person even thinks about stuttering [Sobel 2001]. It is well known that stuttering is conditional--it depends on a person's emotional state and past experience. The recent PET studies show that these conditions, including imaginary exercises, can affect even low-level signal pathways in the brain [Ingham, Fox, Ingham, & Zamarripa 2000]. The studies also underscore the point that the low-level mechanisms of stuttering can change in a matter of seconds. It should be noted that stuttering is the result of an interaction between memory, rapid cognitive events, and rapid low-level changes.

An alternate approach to controllability of dyslexic symptoms is provided by Nicolson and Fawcett [2001] who hypothesize that dyslexia is based in cerebellar deficit. They note that "Difficulties in skill automatization correspond directly to the traditional role of the cerebellum." Also, as noted at "What is dyslexia?" [Internet], "The cerebellum plays a significant role in the timing system of the brain; the magnocellular system," and, "Dyslexics can only do acts, that normally are automatized, if they get the opportunity to conciously compensate for their deficit in automatization." This implies that dyslexics can perform these acts under the proper mental conditions, but that a condition such as confusion that interferes with conscious mental processing would degrade their performance more than that of non-dyslexics.

Many dyslexia researchers have not looked for extremely rapid changes. As mentioned above, Geiger is an exception. Another notable exception is Davis, who has described several exercises for teaching a dyslexic person to deliberately control their perceptual distortion [1997, pp. 149-177]. All of these methods take only a few minutes to teach, and the results are claimed to be immediate: a person doing these exercises will briefly have undistorted perception.

The author has tested these claims and has found support for them. Subjects were taught two variations of the exercise, one of which is designed to decrease perceptual distortion, and the other designed to increase it. Subjects were then presented with stimuli in a range, and asked to classify the stimuli to one of the endpoints. The point of division appears to vary depending on which exercise the subject has most recently done. Intervals between successive trial sets were usually less than 20 seconds. At this point only pilot studies have been done, but the effect was found for both auditory stimuli (Klatt-synthesized speech sounds) and visual stimuli (Phi stimulus).

# 4 The Dyslexic Feedback Loop

Dyslexics manifest a multitude of symptoms. Different individuals may demonstrate primarily auditory [Schulte-Koerne, Deimel, Bartling, & Remschmidt 1999] or primarily visual symptoms, or a mix of the two. Dyslexia is often associated with other problems, such as ADD [Richards 1994], dyspraxia (clumsiness), and dyscalculia and dysgraphia [Temple 1992]. In addition, as noted above, the symptoms may vary from time to time in a single individual. It seems clear that low-level neural damage, although present, is not the whole story.

Again we can take a clue from stuttering. It has been noted that stutterers frequently stutter on words they expect to have trouble with, and it has been suggested that the experience of previous failures acts to exacerbate the stutter on troublesome words. Given that dyslexia also involves a low-level deficit that can be controlled by high-level mental state, it seems clear that a similar mechanism is probably involved; in fact, this is suggested by Davis in his theory of "trigger words" [1997 p. 21]. According to this theory, a dyslexic person suffers an initial failure when confronted with a confusing task such as letter or word identification. The memory of this failure may persist, causing confusion to occur whenever the individual is confronted with the same symbol. This confusion then activates the perceptual distortion, causing repeated failures and reinforcing the confusion.

By this account, a dyslexic person will have an almost random set of "triggers" that cause the perceptual problem to occur. However, in a task as complicated as reading, the chance of encountering a trigger is very high. This accounts for the consistency of reading failure observed in many dyslexics. Conversely, the fact that the perceptual problem in fact depends on mental state accounts for the cases of conditional dyslexia.

Other tasks, such as handwriting and coordination, are also complicated enough to cause confusion and may suffer from faulty perceptions. Thus a person with variable perceptions may suffer from dyspraxia, dyscalculia, dysgraphia, etc. If early confusion associated with a task is resolved quickly, it is quite possible for a person to avoid a dysfunction on that task. However, if the confusion is allowed to persist and grow, the person will develop a disability in that area as a result of accumulated triggers.

As found by Nolander [1999], the expectation of improvement is an important factor in reading improvement in dyslexics. Certainly anyone who expects to improve will be less stressed at the thought of failure. Also, a change of context may reduce the strength of recall of former failures. The expectation factor should be considered in evaluating results reported for any treatment method. However, that does not damage the present argument; in fact, Nolander's observation that, "The interaction of treatment group and presence of dyslexia indicated that dyslexics are more sensitive to expectation," indicates the presence of a feedback loop, while the fact that, "Few dyslexics improve in the general, 'nonresearch' population regardless of the treatment," indicates a real underlying problem.

The feedback loop is strengthened by the fact that perceptual distortion is not always a bad strategy. As noted by Merzenich, Schreiner, Jenkins, and Wang [1993, p. 15], "Could dyslexia and developmental dysphasia arise because some infants adopt more global hearing or looking strategies? .... Note that very early visual practice with a wider field of view does not necessarily represent a 'deficient' or 'negative' or 'dysfunctional' behavior. To the contrary, it would probably represent a practice strategy that ... presents advantages for the rapid processing of relatively complex, spatially distributed inputs .... Once a 'bad' looking (or listening) strategy is in place, by this hypothetical scenario, visual scene representation at every cognitive level as well as eye movement representations would be powerfully reinforced by the many tens or hundreds of thousands or millions of input repetitions." Davis [1997, pp. 72-74] describes a similar process, with imagination being used to supplement distorted vision: "Let's make little P.D. [Potential Dyslexic] three months old and put him in a crib. From his perspective, all little P.D. can see is ... a chest of drawers with someone's elbow .... If little P.D. happens to trigger the brain cells that alter his perception, he will no longer see what his eyes see, he will see something else .... little P.D. actually saw a face in his mind .... So here is little three-month-old P.D. recognizing things in his environment that he shouldn't be able to recognize for three more years. This ability he has for recognizing real objects in his environment will influence the rest of his early childhood development."

By the time dyslexics start learning to read, they will have been using a strategy of perceptual distortion quite successfully for many years. In particular, they will use it to resolve confusion when confronted with novel stimuli. This strategy is unsuitable for the task of reading, which involves many symbols that are meaningless without training and must be perceived accurately. However, as noted by both Davis and Merzenich, the strategy will be so ingrained that only a deliberate effort can countermand it. Absent that effort, the dyslexic will continue to be confused when trying to read, and the perceptual distortion strategy will only add to the problem.

# 5 Dyslexia Correction Methods

There are several dyslexia correction methods in current use that have a demonstrably high degree of success. It is worth noting that these methods are based on incompatible theories and generally address only one sensory modality. Any theory of dyslexia should explain how they can all be successful.

Geiger and Lettvin [1997] achieved notable improvement in 100% of 27 children using simple visual and visual-kinesthetic exercises. Having observed that non-dyslexic readers have a narrow and asymmetric field of letter recognition while dyslexics have a broader and more symmetric field, they have students move a small window over the text they are reading. In a few months they read significantly better, and their letter recognition field became narrow and asymmetric.

Paula Tallal and Michael Merzenich, among others, have developed a program that trains subjects to listen to distorted speech sounds. Subjects improve in their ability to distinguish the sounds. Their reading also improves ["Fast ForWord Outcomes", Internet]; [Tallal, Miller, Bedi, Byma, Wang, Nagarajan, Schreiner, Jenkins, & Merzenich 1997].

These results cannot be reconciled unless there is an underlying mechanism that affects both the auditory and visual senses, so that training either sense improves the overall function. It is also worth mentioning that Delacato [1966] and others have claimed success with kinesthetic training, although the effectiveness of these programs has been questioned [Cummins & Coll 1988]. Although each result is inconsistent with the theory of the other approaches, all results are consistent with a timing-pathway theory. If a transitory timing-pathway deficit exists, then practicing any task that requires stable perceptions (without engendering additional confusion) will train the dyslexic to control the timing path problems to some extent.

Rather than focusing on any specific sensory pathway, the Davis program focuses on finding and correcting sources of confusion, after teaching the subjects how to temporarily control their perceptions. Davis claims a 97% success rate for his original program ["Davis Dyslexia Correction", Internet]; a variation for younger children in a school setting has also been found to be effective [Pfeiffer, Davis, Kellogg, Hern, McLaughlin, & Curry 2001]. A theory of controllable timing pathways would predict that such an approach could be successful, and in fact could be applied to other disabilities as well. The author has provided Davis programs to many children and adults, and observed rapid improvement in reading (within one week) in most cases. Informal followup after approximately two years found that 17 out of 18 clients were reading significantly better than expected, and 15 out of 18 attributed this to the Davis program. The author also observed rapid changes in handwriting, math, and physical skills, in cases where clients experienced difficulty in these areas.

# 6 Discussion

This theory covers a lot of ground, so there is a lot of room for discussion and even speculation. I've arranged this section in several parts. Each of the parts stands on its own, and a failure of one speculation should not reflect on the others.

## 6.1 The Synthesis of "Real" and Imaginary Universes

During normal vision, the eye takes repeated "snapshots" of the environment, and moves rapidly between the snapshots. The rapid movements are called saccades. The fragmented images are integrated to form a complete picture of the visual field. It seems plausible that one effect of a visual magnocellular deficit would be a loss of information about how far the eye moved during a saccade: a reduced ability to track the area the eye has swept across. This could cause difficulty in the integration of the images. Such integration is vital to the proper functioning of the visual system, so the brain would have to compensate. One method of compensation would be for the visual system to gain competence in rotating images and identifying images from fragments--a function performed by the parvocellular system. Davis [1997] and West [1997] describe visual thinking as a common concomitant of dyslexia. This parvocellular competence could account for the frequent observation that dyslexics are "creative."

Some authors have speculated on a link between dyslexia and ADD. One effect of a timing-pathway deficit would be a reduced ability to estimate the passage of time, or to keep one's internal "clock" stable. A person with a timing deficit would thus be susceptible to one of two dysfunctional modes in response to the experience of boredom, confusion, or frustration. If boredom caused the internal clock to speed up, then the person would perceive time as dragging even more. This would intensify the boredom, which would cause a further speedup. Conversely, if confusion or frustration caused the internal clock to slow down, then an overwhelming experience such as too-rapid presentation of knowledge would cause the clock to slow, which would make the experience even more overwhelming. In either case, a vicious cycle would develop, reinforced by the expectations built by previous cycles.

Davis [1997, p. 72] has speculated that earlier and more severe onset of the same perceptual distortions that lead to dyslexia may be the cause of autism. Merzenich, Saunders, Jenkins, Miller, Peterson, & Tallal [1999] have noted that their Fast ForWord training results in the same performance gains in children with PDD as in children with specific language impairments. Delacato [1974] has also noted that autism appears to involve severe perceptual distortion, and has applied similar treatments to autistics and dyslexics. Davis [personal communication, 1995] has described unpublished experiments in which a variation of his orientation procedure has produced marked improvement in some autistic children. It seems likely that methods which train dyslexics to stabilize their perceptions may also help children with autism--especially given the preliminary success by Merzenich, Davis, and Delacato with their widely varying methods.

## 6.2 Investigating the Mechanisms Of Dyslexia

Dyslexia research is full of observations that are directly contradictory. Examples include [Livingstone, Rosen, Drislane, and Galaburda 1991] and [Johannes, Kussmaul, Muente, & Mangun 1996], and the results surveyed in [Skottun 2000]. Many good researchers have spent much time and effort only to find opposite results. In addition, such contradictions hamper the effort to develop consistent and useful theories, or worse yet, allow the development of theories that only account for fragments of the available evidence.

Any investigation of low-level brain function must take several new factors into account. First, individual dyslexics are likely to react differently to the same task. One dyslexic may be confused by a task and suffer perceptual distortions, while another finds it easy and approaches it with perceptions in "non-dyslexic" mode, and a third finds a way to use non-standard perceptions to make the task even easier. Second, some people who have no reading problem may nevertheless have a timing path deficit, and so should be included with the dyslexic population or excluded from the study. Third, when planning or attempting to replicate a study, the set and setting, and other factors affecting the emotional and cognitive state of the participants, may have a significant impact on the results obtained.

There are several ways to compensate for these difficulties. To detect a timing-path deficit in control subjects, subjects could be screened for multiple disabilities, not only dyslexia. It is likely that a timing path deficit will be apparent in at least one skill or activity. However, this method is difficult and ad-hoc. Another way to compensate would be to ensure that all dyslexics (and controls) are in the same mental state. The "orientation" exercise of Davis may be helpful here, although it may mask some dyslexic effects. Unfortunately, the effects of the exercise on mental state have not been well established; this is an area for further study. At this point, there is strong anecdotal evidence and some experimental evidence that the exercise has a marked effect on multiple perceptual mechanisms, including auditory, visual, and kinesthetic, in dyslexics. Effects on non-dyslexics are unknown.

The fact of "conditional dyslexia" needs to be definitively established and widely recognized. It seems likely that PET, fMRI, or possibly QEEG [Chabot, di Michele, Prichep, & John 2001] applied to dyslexics under appropriate conditions would be able to detect a difference between a "dyslexic" and "non-dyslexic" mode, as has already been done for stuttering with PET. Again, the Davis "orientation" exercise may be helpful here.

## 6.3 Related Conditions

There are a number of conditions (beginnning with the syllable "dys") that appear to be explainable by the present theory. A complete survey is beyond the scope of this paper; however, a partial list may be instructive and may suggest further research directions. The author has observed the Davis method providing rapid improvement (a few days) in cases of each of these conditions.

Dyscalculia is difficulty with arithmetic or other mathematics. It seems obvious that the same conditions leading to reading difficulty could also cause a problem when perceiving or interpreting the symbols used in math. There is even more opportunity for confusion because of the increased abstractness and reduced contextual clues available in math as opposed to text.

Dyspraxia, or clumsiness, may result from visual perceptual distortion. Balance is partially dependent on accurate vision, and a visual distortion may magnify or obscure the effects of subtle head movements. In addition to the obvious problems created by a timing deficit, a feedback loop may appear, in which the stress of expectation of failure (e.g. falling off a balance beam) may cause increased perceptual distortion, resulting in new failure and reinforcing the expectation.

Handwriting involves many fine motions and much feedback between eye and hand. According to Davis, the confusion feedback can occur with regard to certain motions or letters. Dysgraphia, or poor handwriting, may result.

## 6.4 Improved Dyslexia Correction

As noted previously, dyslexics frequently do poorly on a variety of tasks, and several programs have been designed that involve practicing a certain task; improvement in the task generally coincides with improvement in reading. Presumably the reading improvement depends on the subject first learning to do the task better and then practicing the new skill to "lock in" the improvement. A faster improvement on the task might translate into faster progress in reading, since the subject would be able to spend more time doing the task correctly. The Davis orientation exercise may provide such accelerated improvement in a variety of tasks.

Likewise, the confusion and frustration often felt by dyslexics could be ameliorated by directly addressing the confusion encountered by dyslexics during attempts to read. If dyslexia is based on a feedback loop between low-level perceptual distortion and high-level confusion, then explicit instruction in language symbols may be beneficial in any dyslexia program. Davis [1997, pp. 197-212] suggests a visual-kinesthetic learning method, echoing the observation [West 1997] that dyslexics frequently think visually.

Finally, it should be universally recognized that regardless of whether subtypes of dyslexia exist, a varitey of types of exercises work for the vast majority of dyslexics. Every clinical dyslexia program should be willing to test and integrate exercises from a variety of approaches. For example, Merzenich uses a fine-motor exercise in addition to the visual exercises, and Davis uses a large-motor exercise involving catching Koosh balls while balancing on one foot.

# 7 Summary

Dyslexia has been investigated from many angles. However, investigation has been hampered by contradictory theories and even contradictory results. The notable success of several treatment programs based in incompatible theories suggests an underlying mechanism that can be affected by a variety of interventions.

Much evidence has accumulated to suggest that the mechanism involves the timing pathways of the brain, probably at a cellular level. Many researchers have reported lasting results from interventions over a period of months, suggesting either that the problem is correctable or that dyslexics can learn to compensate with appropriate training (although they do not learn to compensate without training). Some researchers have gone farther and demonstrated improvement in a period of days, and perceptual changes requiring only a few seconds.

A theory in which the perceptions are unstable (flexible) and controllable by conscious or subconscious mental state on a time scale of seconds appears to account for all of the observations, including the contradictory results obtained by numerous experiments. Stuttering research appears to lead dyslexia research in this approach, having established that brain activity related to stuttering does in fact change rapidly depending on mental state.

The verification of rapid perceptual change in dyslexia, and the application to experiments and treatments, should be a high priority. A variety of new short-time-scale techniques for observing brain function, and the Orientation exercise of Davis which directly affects perceptual distortion in dyslexics, will be helpful. Dyslexia research, and treatments for dyslexia and many related conditions, will benefit greatly from the verification and control of this rapid change mechanism.

# Bibliography

[1] Brooks Chabot, R. J., di Michele, F., Prichep, L., & John, E. R. (2001). "The clinical role of computerized EEG in the evaluation and treatment of learning and attention disorders in children and adolescents", Journal of Neuropsychiatry & Clinical Neurosciences, 13(2), 171-186.

[2] Cummins, R. A., Coll, V. (1988). *The neurologically-impaired child: Doman-Delacato Techniques reappraised.* New York, NY, US : Croom Helm.

[3] Davis, R. (1997). *The Gift of Dyslexia* (2nd. ed.), Berkley Publishing Group, New York, NY.

[4] "Davis Dyslexia Correction" http://dyslexia.com/program.htm

[5] Delacato, C. H. (1966). *Neurological Organization and Reading*. Springfield, Ill. : Charles C Thomas.

[6] Delacato, C. H. (1974). *The Ultimate Stranger: The Autistic Child.* Doubleday.

[7] Evans, Bruce J. ; Drasdo, Neville (1991). "Tinted lenses and related therapies for learning disabilities: A review", Ophthalmic & Physiological Optics 11(3), 206-217.

[8] "Fast ForWord Outcomes" http://www.scientificlearning.com/scie/index.php3 ?main=graphs

[9] Geiger, G., & Lettvin, J. Y. (1987). "Peripheral vision in persons with dyslexia", The New England Journal of Medecine, 316 (20), 1238-1243.

[10] Geiger, G. & Lettvin, J. Y. (1997, June). "A View on Dyslexia", CBCL Paper #148/AI Memo #1608, Massachusetts Institute of Technology, Cambridge, MA. http://www.ai.mit.edu/projects/cbcl/people/geiger/memo_complete.pdf

[11] Grossberg, S. & Cohen, M., (1997). "Parallel Auditory Filtering by Sustained and Transient Channels Separates Coarticulated Vowels and Consonants", IEEE Transactions on Speech and Audio Processing 5(4), 301-317.

[12] Hannell, G., Gole, G. A., Dibden, S. N., Rooney, K. F., et al. (1991). "Reading improvement with tinted lenses: A report of two cases", Journal of Research in Reading 14(1), 56-71.

[13] Ingham, R. J., Fox, P. T., Ingham, J. C., Zamarripa, F. (2000). "Is overt-stuttered speech a prerequisite for the neural activations associated with chronic developmental stuttering?" Brain & Language 75(2), 163-194.

[14] Johannes, S., Kussmaul, C. L., Muente, T. F., & Mangun, G. R. (1996). "Developmental dyslexia: Passive visual stimulation provides no evidence for a magnocellular processing defect", Neuropsychologia 34(11), 1123-1127.

[15] Livingstone, M., Rosen, G., Drislane, F., & Galaburda, A. (1991). "Physiological and anatomical evidence for a magnocellular defect in developmental dyslexia", Proceedings of the National Academy of Science, 88, 7943-7947.

[16] Lovegrove, W. J., Garzia, R. P., & Nicholson, S. B. (1990). "Experimental evidence for a transient system deficit in specific reading disability", American Optometric Association Journal, 61, 137-146.

[17] McAnally, K. I., Castles, A., & Stuart, G. W. (2000). "Visual and Auditory Processing Impairments in Subtypes of Developmental Dyslexia: A Discussion", Journal of Developmental and Physical Disabilities, 12(2), 145-156.

[18] Merzenich, M. M., Saunders, G., Jenkins, W. M., Miller, S., Peterson, B., & Tallal, P. (1999). "Pervasive developmental disorders: Listening training and language abilities", In Broman, S. H. & Fletcher, J. M. (Eds.) The changing nervous system: Neurobehavioral consequences of early brain disorders. (pp. 365-385)

[19] Merzenich, M. M., Schreiner, C., Jenkins, W., and Wang, X. (1993) "Neural mechanisms underlying temporal integration, segmentation, and input sequence representation: Some implications for the origin of learning disabilities", In

Tallal, P., Galaburda, A. M., Llinas, R. R., & von Euler, C. (Eds.) Temporal Information Processing in the Nervous System: Special Reference to Dyslexia and Dysphasia. The New York Academy of Sciences (pp. 1-22)

[20] Nolander, C. R. "The effect of expectation and tinted overlays on reading ability in dyslexic adults. (scotopic sensitivity)", Dissertation Abstracts International: Section B: The Sciences & Engineering 59(11-B), (p. 6076).

[21] Pfeiffer, S., Davis, R., Kellogg, E., Hern, C., McLaughlin, T. F., & Curry, G. (2001). "The Effect of the Davis Learning Strategies on First Grade Word Recognition and Subsequent Special Education Referrals", Reading Improvement, 38(2)

[22] Ramus, F. (2001). "Dyslexia: Talk of Two Theories", Nature 412, 393-395.

[23] Richards, I. L. (1994). "ADHD, ADD and dyslexia", Therapeutic Care & Education 3(2), 145-158.

[24] Schulte-Koerne, G., Deimel, W., Bartling, J., & Remschmidt, H. (1999). "The role of phonological awareness, speech perception, and auditory temporal processing for dyslexia", European Child & Adolescent Psychiatry 8(3), 28-34.

[25] Shaywitz, S. & Shaywitz, B. (1999). "Cognitive and neurobiologic influences in reading and in dyslexia", Developmental Neuropsychology 16(3), 383-384.

[26] Skottun, B. C. & Parke, L. A. (1999). "The Possible Relationship Between Visual Deficits and Dyslexia: Examination of a Critical Assumption", Journal of Learning Disabilities, 32(1), 2-5.

[27] Skottun, B. C. (2000). "The magnocellular deficit theory of dyslexia: the evidence from contrast sensitivity", Vision Research 40, 111-127.

[28] Skottun, B. C. (2001). "On the use of metacontrast to assess magnocellular function in dyslexic readers", Perception & Psychophysics, 63(7), 1271-1274.

[29] Sobel, R. K. (2001, April 2). "Anatomy of a Stutter", U.S.News & World Report, 44-51.

[30] Stein, J. (1993). "Dyslexia--Impaired Temporal Information Processing?" In Tallal, P., Galaburda, A. M., Llinas, R. R., & von Euler, C. (Eds.) Temporal Information Processing in the Nervous System: Special Reference to Dyslexia and Dysphasia. The New York Academy of Sciences (pp. 83-86)

[31] Tallal, P., Miller, S. L., Bedi, G., Byma, G., Wang, X., Nagarajan, S. S., Schreiner, C., Jenkins, W. M., & Merzenich, M. M. (1997). "Language comprehension in language-learning impaired children improved with acoustically modified speech", In: Hertzig, M. E. & Farber, E. A (Eds.) Annual progress in child psychiatry and child development (pp. 193-200)

[32] Tallal, P., Miller, S., & Fitch, R. H (1995). "Neurobiological basis of speech: A case for the preeminence of temporal processing", Irish Journal of Psychology Special Issue: Dyslexia update. 16(3), 194-219.

[33] Temple, C. M. (1992). "Developmental dyscalculia", In Segalowitz, S. J. & Rapin, I. (Eds.), Handbook of neuropsychology (Vol. 7, pp. 211-222).

[34] West, T. G. (1997). *In the Mind's Eye : Visual Thinkers, Gifted People With Dyslexia and Other Learning Difficulties, Computer Images and the Ironies of Creativity* Prometheus Books.

[35] "What is dyslexia?" http://www.dys.dk/eng/arso.html

# Chapter 10

# Emergent computation in CA: A matter of visual efficiency?

**Mathieu S. Capcarrere**[1]

Logic Systems Laboratory
School of Computer Science and Communication
Swiss Federal Institute of Technology, Lausanne
1015 Lausanne, Switzerland
mathieu@capcarrere.org

Cellular Automata as a computational tool have been the subject of interest from the computing community for many years now. However, WHAT is meant by computation with cellular automata remains elusive. In this paper, I will argue that emergent computation with CAs is a matter of visual efficiency. Basing our argument on past, recent and also previously unpublished results around the density task, I will propose a tentative definition of *emergent behavior*, in this limited scope, and thsu envisage differently the whole question of what may be sought in computing research in CAs. The practical consequences of this approach will alter the HOW question answer, and most notably how to evolve computing CAs.

[1]Now at: Computing Laboratory, University of Kent, Canterbury CT2 8DF, UK.

## 1.1 Introduction

Cellular Automata as a computational tool have been the subject of interest from the computing community for many years now. More precisely, the development of the Artificial Life field led many to wonder on how to do computation with such tools. Artificial Evolution which gave good results on specific tasks, like density or synchronisation was often given as an answer. However, it appeared that the limitations of such an approach were severe and really the question of WHAT meant computation with cellular automata became pregnant.

The answer to this question is far from obvious. Mitchell, Crutchfield, Hanson *et al* proposed an analysis of "particles" as a partial answer[7, 11]. Wuensche more recently developed the Z parameter as a paraphernalia to treating this question[22]. Before this question appeared in its full-blown form in the A-life/Computer science community, there were already interrogations going this way with Wolfram's class III and, related, Langton's computing at the edge of chaos.

In this tentative paper, I will argue that computation of CAs is a matter of visual efficiency. Basing our argument on past, recent and also previously unpublished results around the density classification task, I propose a definition of what is computation by means of CAs. This will be the occasion to (re)define emergent behaviour, in this limited scope, but also to envisage differently the whole question of what may be sought in computing research in CAs. The practical consequences of this approach will alter the HOW question answer, and most notably how to evolve computing CAs. Our point in this discursive paper is only to bring new paths of research rather than discarding old ones.

First, in section 1.2 I will shortly review the different ways computation in CAs has been tackled and define more precisely the kind of use CAs we are concerned with. Then in section 1.3, I will expose what is the density classification task and present the important results that are behind the reflections developed in this paper. Section 1.4 is, somehow, the core of this paper, where my thesis is detailed and argued for. This section will also be the occasion to open paths for future research in computing CAs.

## 1.2 Computation in cellular automata

There are four ways to view the computational aspects of Cellular Automata. *The first* and most natural one is to consider them as abstract computing systems such as Turing machines or finite state automata. These kind of studies consider what languages CA may accept, time and space complexity, undecidable problems, etc. There has been a wealth of work in this field [8, 13], most of which is based on or completes the mathematical studies of CAs. *The second way* is to develop structures inside the CAs (a specific initial configuration) which are able to perform universal computation. A prime example of this is the *Game of Life*, which was proved to be a universal computer using structures like gliders and glider guns to implement logic gates [3]. This was also applied to one-dimensional CAs where structures were found to be a universal Turing machine [15]. A *third way* is to view CAs as "computing black boxes". The initial configuration is then the input data and output is given in the form of some spatial configuration after a certain number of time steps. This includes so-called soliton or particle or collision based computing. In these systems, computing occurs on collision of particles carrying information [21]. In these works, usually, computation really occurs in some specific cells, i.e., most cells are only particle transmitters and some do the computation, this even if factually all cells are identical. Hence, parallelism is not really exploited as such. However, there are CAs belonging to this "black box" type which are arguably different. In these, like the density task solver that I will present in section 1.3 and which is demonstrated in Figure 1.1, computation is intrinsically parallel. Their solving property relies on each and every cell. All are involved in the computation and there is no quiescent state as such. The result is then to be considered globally. The frontier between these two sub-categories is not always clear, and in fact, the core of this distinction depends on the ill-defined notion of emergence. The latter type being so, the former not. We will come back to this discussion in section 1.4. Finally, there is a *fourth* way which is to consider CA computational mechanics. This kind of study concentrates on regularities, particles and exceptions arising in the spatial configuration of the CAs considered through time. This is really a study of the dynamical global behaviour of the system. This research does not concentrate on what is computed but rather how it is computed. Though it often took as object of study CAs of the emergent type of the third category, it was also applied to CAs with no problem solving aim, dissecting their behaviour with no further consideration. There has been much work in this domain, accomplished mainly by Hanson, Crutchfield and Mitchell [7] and Hordijk [12].

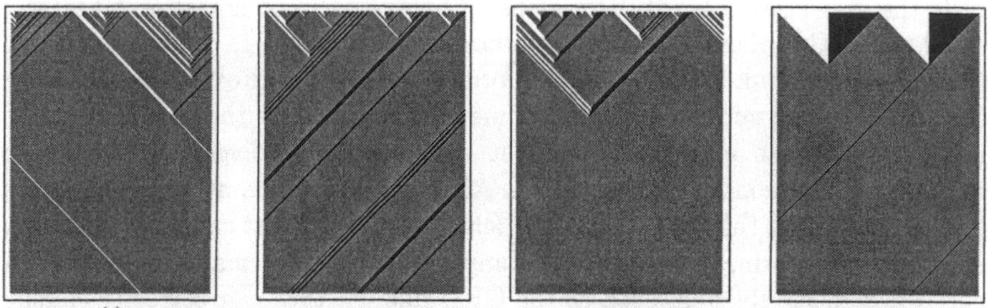

(a) (b) (c) (d)

**Figure 1.1**: This figure presents an example of emergent computation in a one-dimensional, binary CA demonstrated on the density classification task. The rule in all four figures is 184 in Wolfram's notation. Time flows downward. When there are more ones than zeros, only left-going black lines remain, (b) and (d), while when there are more zeros than ones, only right-going white lines remain, (a). No lines are left when the density is exactly 0.5, (c).

In this paper, I propose to concentrate on the third kind of study. More particularly on the second type, the research of problem solving, emergent CAs. The interrogation on what is an emergent, global behaviour will be tackled more precisely in section 1.4. Let's first define more precisely through a few characteristics the kind of computation we are concerned with. – First, all cells are input (and output). This is an important property, in that it guarantees that there is no "storage" area and no separated "computing" area and that all cells are equally involved and necessary. The corollary of this is that there is no quiescent state as such. – Second, there is no special cell,i.e, no cell is actually doing the computation on its own. The previous characteristics prevent this to happen in most cases, but it is essential to explicitly express this as otherwise emergent computation, according to the definition to come in section 1.4 would not occur. – Third, the behaviour is *perfect*. This characteristic is all the more important as it has been often neglected, but by definition if computation occurs effectively, then it must be perfect. Otherwise we are facing some kind of pseudo-computation,some kind of cheating.

## 1.3 The density task example

In this section I shortly present the density task, demonstrated in figure1.1, as it is a prime example, simultaneously, of the kind of computation that we are looking for in CAs and of the importance of the look we lay on the CA .

The one-dimensional density task is to decide whether or not the initial configuration of a binary CA contains more than 50% 1s. Packard was the first to introduce the most popular version of this problem [17] where the output is the CA relaxing to a fixed-point pattern of all 1s if the initial density of 1s exceeds 0.5, and all 0s otherwise (Figure 1.1). As noted by Mitchell *et al.* [16], the density task comprises a non-trivial computation for a small radius CA ($r \ll N$, where $N$ is the grid size). Density is a global property of a configuration whereas a small-radius CA relies solely on local interactions. Since the 1s can be distributed throughout the grid, propagation of information must occur over large distances (i.e., $O(N)$). The minimum amount of memory required for the task is $O(\log N)$ using a serial-scan algorithm, thus the computation involved corresponds to recognition of a non-regular language. After many fruitless attempts to find a perfect solution to this task, Land and Belew [14] proved that cannot be perfectly solved by a uniform, two-state CA of any radius However many imperfect CA performed relatively well, such as the GKL rule [9], that can correctly classify approximately 82% out of a random sample of initial configurations, for a grid of size $N = 149$ [1]. This lead some researchers to focus on the use of artificial evolution techniques to try to find *non-uniform* two-state CAs to solve this tasks ([20]). But these attempt remained unsuccessful. Actually, I proved recently that for this version of the density classification task, no two-state, non-uniform CA of any radius solves it perfectly[2].

Thus the density classification tasks seemed utterly impossible for two-state

[2]This unpublished results may be found in [4], chapter 4.

CAs (uniform and non-uniform). Nevertheless this was just an appearance and this impossibility only applies to the above statement of the problem, where the CA's final pattern (i.e, output) is specified as a fixed-point configuration. As we proved in [6], if we change the output specification, there exists a two-state, $r = 1$ uniform CA, rule 184 in Wolfram's notation, that can perfectly solve the density problem. Hence, this task, according to its definition, changes status dramatically in terms of "CA" computability. It moves from unsolvable for two-state CAs, to "basic" as it found a solution in the simplest class: the elementary CAs.

These results led us to wonder on what was required to solve that task in its essential form. Essential in the sense of stripping down its definition to the minimal form beyond which the task loses its meaning. From these minimal assumptions, we derived that a perfect CA density classifier must conserve in time the density of the initial configuration, and that its rule table must exhibit a density of 0.5 [5]. Thus, non-density-conserving CAs (such as the GKL rule) are by nature imperfect, and, indeed, any specification of the problem which involves density change precludes the ability to perform perfectly density classification.

The necessary condition of density conservation brings out the question of what is computation by means of cellular automata. As seen the computability of the task was more dependent on its definition rather than its inherent difficulty. Moreover, if we conserve density, then computation here is only re-ordering the "1s" among the "0s", which usually means simply *no computation*. In effect, usually computation is taken as a reduction of informational complexity. For instance, the original question was it not to reduce any configuration, $2^N$ bits, to an answer yes or no, 1 bit ? But here there seems to be no loss of information through time. However, this is not true of spatial information as rule 184 is not invertible. What we are looking for, hence, is a simplification of the input in terms of apprehensibility by an onlooker. So beyond the aid that this result on necessary conditions might give in the search for locally interacting systems that compute the global density property, it is its consequences on the question of computation that makes it important. Let's now extends on this question.

## 1.4 Emergent computation and visual efficiency

To answer the question of what is emergent computation with CAs, we shall first explicit the idea of emergence. For sake of brevity and of clarity, we will not engage in a discussion around this theme, but rather expose our own understanding of the term. Beneath the idea of emergence, there is always the key concept that *new properties* that emerge were not present at the level of the individual element. In fact, this is the only tangible aspect of emergence. But, by definition, a property has an effect. A property without effect has no existence. For instance one may say that the emergent property of social insect societies is robustness to attack and to environment hardship. The effect is then their ecological success. However, even if these properties and effects are objective, their consideration and definition is subjective. For instance, the necessity of describing a bunch of $H_2O$ molecules as being liquid is merely a question of viewpoints. It appears as a necessity, given the dramatic effects of the liquid property, but one should remember that the liquid state of the matter is still ill-defined as of today. This inevitable subjectivity lead many to introduce the idea of an observer [2, 19]. The question thus becomes what or who is the observer? In other words, what is the correct viewpoint?

These considerations on emergence call for a reflection on the question of what constitutes emergent computation in CAs. I am not pretending, here, to give a definitive answer nor a universal one, but rather some tracks for thought. As evoked in the preceding section, what constitute the result of rule 184 is two-fold. First, the appropriate look makes it a perfect density-classifier and second this look is possible because of the *visual efficiency* of the final (or rather the temporal) configuration of the CA. That is, if one watches Figure 1.1, one can instantly say if the original configuration was holding more 1s than 0s or the contrary. Of course, this translates also into the fact that a simple three-state automaton can then classify the density, but the most convincing argument that CA 184 does the job is that it is visually efficient. It is hard to define formally what is this efficiency, but we can say, without doubt, at least in this case that it relies on patterns that become visibly obvious as they stand out of a regular background. This view is also what changed between the impossibility of the task in its original form to its evident solution. So the question of computation is finally the hazardous meeting of an actual computation by CAs with the "good" look from an outside observer. "[We] have to act like painters and to pull away from [the scene], but not too far. How far then? Guess!"[3].

The argument in favour of an observer, besides the result on the density classification task are multiple.– First and foremost, most CAs designed in the past have most often been (inappropriately ?) aimed at human, knowingly or unknowingly. For instance, Langton's loop, which does not lie in this category of computation, still relies on the impression of life it leaves in the eye of the spectator. Wolfram's classes are subjectively defined according to the judgement of the human beholder. And even the original definition of the density classification task was taking into account that fact. Effectively, even though it was defined according to a model close to the one of figure 1.2.a, that is of a classical computing machine taking an input and finishing in a yes or no state, the specification of the output itself, all 0s or all 1s, was clearly aimed at a human. – Second, beyond these historical considerations, and more importantly, in emergent systems, the observer is necessary to establish the emergent properties and its effect, but usually not for the property to happen. However in the case of CAs, computation only happens in the eye of the beholder. The CA in itself cannot get from any feedback, such as the environment for social insect and thus the effect of the property can only take place in the observer. This is due to the simple fact that no element, especially in simple, basic binary CA could "grasp" the global, emergent result...by definition. Maybe this is the main

[3] ¡¡Il faut que je fasse comme les peintres et que je m'en éloigne, mais non pas trop. De combien donc ? Devinez!¿¿ Blaise Pascal, pensées 479 in [18].

point of this article, CAs can't be self-inspecting and thus it is meaningless to consider its computation *in abstracto*, *in absentia* of an observer. In other word, **emergent computation in CA is epistemic.** So what we are looking for is really a system made out of a couple CA/beholder. – Finally, third, taking these considerations allows the result not to be a final configuration but rather the global temporal dynamic of the CA, such as illustrated in figure 1.2.b. This last point is interesting in two ways. On the one hand, it discards the need to look after $n$ time steps, which solves one of the usual paradox. Solving the density task without a global counter, but requiring it to get the final result is rather paradoxical. On the other hand, now considering cycle and the global dynamic provides a much richer pool of potentially interesting CAs.

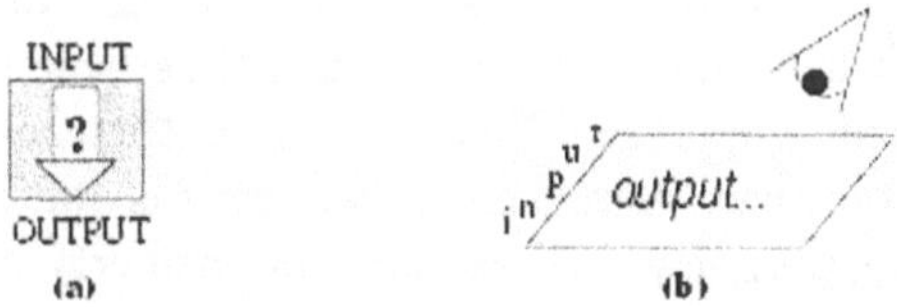


**Figure 1.2**: We need to change our look on CA computation to grab its full potentialities.

The necessity to consider the couple/CA beholder established[4], this says nothing on the nature of the observer. As said earlier, the natural observer is the human being. However the possibility to create an artificial observer is by no means not discarded. Actually, one could imagine a mechanical search (either exhaustive of by means or evolutionary computation) where the success criteria would be given by an artificial observer. Wuensche [22] proposed a measure of input entropy, based on the frequency of look up of the different neighbourhoods. This measure was devised in order to find "interesting" CAs, interesting in the sense of Wolfram's class III. Wuensche followed the ideas originated by Gutowitz and Langton: is there a quantitative way to define class III CAs and can we find them by means of artificial evolution [10]? All these works were somehow seeking an artificial observer, as Class III is only defined on criteria based on observation. We pursue the same goal but for a different purpose. As said before, interesting CAs for us were CAs which produced regularity, or more exactly irregularity (patterns) on regularity, and these are surely to be found in class II rather than class III. Nevertheless, Wuensche's frequency measure may be a good start for future work. He introduced the idea of filtering the output of the CAs by omitting in the *visual representation* the neighbourhoods that were used the most often. For instance, this would lead in CA 184 to discard patterns 101 and 010. This is definitely an interesting path to follow to discard regularity, and thereby provides a first step in the direction of an automatic observer.

[4]To the least proposed.

## 1.5 Concluding remarks

The nature of computation in cellular automata is a complex one. As we saw, there are four ways to consider this question, but only the global, emergent type, the second kind of the third type therein, was of real interest to us. In effect, only this type of computation (as defined in the last paragraph of section 1.2 presents the fascinating qualities that one seeks in cellular systems, namely, a global behaviour that goes beyond the capabilities of each of the elements and dismisses artifacts of sequential systems developed on a cellular substrate. But the nature, the essence of the computation of this kind is far from obvious. As seen in the past literature, the idea of the necessity of an observer to establish emergent behaviour is often present, but what we introduce here is the major distinction between ontological and epistemic emergence. In the former case, the system gets feedback from its environment and thus benefits directly from the effects of the emergent properties. The observer is then just there to state the property. In the latter case, however, the observer is the beneficiary from the effect of the emergent property, and as such the property cannot exist without it. We argued that CA lie in that latter category. Then the observer being mainly humans, emergent CA computation is thus merely a matter of visual efficiency.As gleaned from the density, this visually efficient computation occurs in creating order (regularity). To be more accurate, it creates a large "ocean" of regularity which allows the onlooker to see the distinguishing patterns of irregularity. Hence, one may conclude from this, and I would, that problem-solving truly emergent CAs are to be found in class II rather than Wolfram's class III CAs. I believe that this conclusion goes well beyond the density task. So a problem-solving CA is the fortunate meeting of a good look with a good local rule. However, this computation is not elusive, i.e, CA computes really. We are unable to see in the input arrangement and the local rule what will happen after a few time steps. CA computation thus hinges on the weakness of our mind in a certain way. This is not degrading at all and in fact is common to all sorts of computation. If one would see the same thing in $\sqrt{27225}/5$, and in 33, then there would be no need for calculators. This view of problem-solving CAs could have great influence on how one may find such CAs. For instance, by evolution, whereas today how the CA should behave globally is decided a priori, in the future an artificial observer could actually judge the fitness of the CA, the result being then the pair observer/CA.

# Bibliography


[1] David Andre, Forrest H. Bennet III, and John R. Koza. Discovery by genetic programming of a cellular automata rule that is better than any known rule for the majority classification problem. In John R. Koza, David E. Goldberg, David B. Fogel, and R. L. Riolo, editors, *Genetic Programming 1996: Proceedings of the first conference*, pages 3–11, Cambridge, MA, 1996. The MIT Press.

[2] Nils A. Baas and Claus Emmeche. On emergence and explanation. *Intellectica*, 25:67–83, 1997.

[3] Elwin R. Berkelamp, John H. Conway, and R. K. Guy. *Winning ways for your mathematical plays*. Academic Press, 1982.

[4] Mathieu S. Capcarrere. *Cellular Automata and Other Cellular Systems: Design & Evolution*. Phd Thesis No 2541, Swiss Federal Inst. of Tech., Lausanne, 2002.

[5] Mathieu S. Capcarrere and Moshe Sipper. Necessary conditions for density classification by cellular automata. *Phys. Rev. E*, 64(3):036113/1–4, December 2001.

[6] Mathieu S. Capcarrere, Moshe Sipper, and Marco Tomassini. Two-state, r=1 cellular automaton that classifies density. *Phys. Rev. Lett.*, 77(24):4969–4971, December 1996.

[7] James P. Crutchfield and Melanie Mitchell. The evolution of emergent computation. *Proceedings of the National Academy of Science*, 23(92):10742, 1995.

[8] Karel Culik II, Lyman P. Hurd, and Sheng Yu. Computation theoretic aspects of cellular automata. *Physica D*, 45:357–378, 1990.

[9] Peter Gacs, G. L. Kurdyumov, and L. A. Levin. One-dimensional uniform arrays that wash out finite islands. *Problemy Peredachi Informatsii*, 14:92–98, 1978.

[10] Howard A. Gutowitz and Christopher Langton. Mean field theory of the edge of chaos. In F. Morán, A. Moreno, J. J. Merelo, and P. Chacon, editors, *Advances in Artificial Life, Proceedings of the 3rd European Conference on Artificial Life*, volume 929 of *LNAI*, pages 52–64. Springer-Verlag, 1995.

[11] James E. Hanson and James P. Crutchfield. Computational mechanics of cellular automata. *Physica D*, 103:169–189, 1997.

[12] Wim Hordijk. *Dynamics, Emergent Computation and Evolution in Cellular Automata*. Computer Science Dept, University of New Mexico, Albuquerque, NM (USA), Dec. 1999.

[13] Oscar Ibarra. Computational complexity of cellular automata: an overview. In M. Delorme and J. Mazoyer, editors, *Cellular Automata: A parallel model*. Kluwer Acadamecic Publishers, 1999.

[14] Mark Land and Richard K. Belew. No perfect two-state cellular automata for density classification exists. *Physical Review Letters*, 74(25):5148–5150, June 1995.

[15] K. Lindgren and M. G. Nordahl. Universal computation in simple one-dimensional cellular automata. *Complex Systems*, 4:299–318, 1990.

[16] Melanie Mitchell, James P. Crutchfield, and Peter T. Hraber. Evolving cellular automata to perform computations: Mechanisms and impediments. *Physica D*, 75:361–391, 1994.

[17] Norman H. Packard. Adaptation toward the edge of chaos. In J. A. S. Kelso, A. J. Mandell, and M. F. Shlesinger, editors, *Dynamic Patterns in Complex Systems*, pages 293–301. World Scientific, Singapore, 1988.

[18] Blaise Pascal. *Penses*. Folio/Gallimard, 1977. Edition de Michel le Guern.

[19] Edmund Ronald, Moshe Sipper, and Mathieu S. Capcarrre. A test of emergence, design, observation, surprise! *Artificial Life*, 5(3):225–239, 1999.

[20] Moshe Sipper and Ethan Ruppin. Co-evolving architectures for cellular machines. *Physica D*, 99:428–441, 1997.

[21] Kenneth Steiglitz, R. K. Squier, and M. H. Jakubow. Programmable parallel arithmetic in cellular automata using a particle model. *Complex Systems*, 8:311–323, 1994.

[22] Andrew Wuensche. Classifying cellular automata automatically: Finding gliders, filtering, and relating space-time patterns, attractor basins, and the z paramater. *Complexity*, 4(3):47–66, 1999.

# Chapter 11

# Emergent Probability

## A Directed Scale-Free Network Approach to Lonergan's Generic Model of Development

**Michael Bretz**
Department of Physics
University of Michigan
mbretz@umich.edu

## 1.1 Introduction

An intriguing heuristic model of development, decline, and change conceived by Bernard J.F. Lonergan (BL) in the late 1940's was laid out in a manner now recognizable as representing an early model of complexity. This report is a first effort toward eventually translating that qualitative vision, designated *Emergent Probability** (EP), into a viable network computer study.

In his study of human understanding, Lonergan [1992] saw the task of constructing a cohesive body of explanatory knowledge as a convoluted building process of *recurrent schemes* (RS) that act as foundational elements to further growth. Although BL's kernal RS was composed of the cognitional dynamics surrounding *Insight* [Bretz, 2002], other examples of recurrent growth schemes abound in nature: resource cycles, motor skills, biological routines, autocatalytic processes, etc. The corresponding growing generic *World Process* can alternatively be thought of as chemical, environmental, evolutionary, social, organizational, economical, psychological, or ethical [Melchin, 2001], and its generality might be of particular interest to complex systems researchers.

*Schemes of Recurrence* are conjoined dynamic activities where, in simplest form, each element generates the next action, which in turn generates the next, until the last dynamic regenerates the first one again, locking the whole scheme into long term stable equilibrium. BL modeled generic growth as the successive appearance of *conditioned Recurrent Schemes* (RS), each of which comes into function with high probabilistically once all required prior schemes have become functional. RS's can be treated as dynamic black cells of activity which themselves may contain internal structures and dynamic schemes of arbitrary complexity. Emergent Probability is a generic heuristic model. Applications to specific physical problems requires detailed knowledge of the recurrent schemes' makeup and of their interrelationships (an elementary example is provided in the Appendix).

The universe of possibilities available to any open-ended development process is vast, and the interrelations among the elements can become arbitrarily convoluted as the process proceeds. All growth in that universe, no matter how complex, depends upon a suitable underlying environmental "situation", and an ecology, or niche, in order to thrive. Any full simulation of EP must allow for adaptation to a changing situation and ecology, so that recovery can occur when events disrupt underlying schemes, or when separate growths vie among themselves for dominance within a stressed environment

Here, I present first results obtained from an exploratory toy model of EP development. This MATLAB simulation uses a scale-free, directed network (nodes as RS's, inward links from the conditional RS's) to represent the universe of relational possibilities. Growth occurs when nodes of the network are sequentially activated to functionality. The appearance of RS clusters (*Things*), their dependence on the underlying ecology and situation, and their interplay are sought.

## 1.2 Modeling the Potential World

A sparse adjacency matrix was grown that defines all of the possible nodes of the toy model simulation, their link dependencies on selected previously grown nodes and their role in constraining younger nodes. Each row/column number names a specific, unique node and values of 1 in the sparse matrix elements represent individual links between adjacent nodes of the network. The network was grown from 3 core nodes by combining aspects of the scale-free growing network methods of [Dorogovtsey, 2000] and [ Krapivsky, 2001]. For each succeeding iteration, with probability q = .83, a new node without links was created having an attractiveness A = 1. With the complementary probability, p = .17, m = 3 new links were added between statistically chosen existing nodes. Selection of the m target nodes and of originating nodes used weightings directly proportional to $(k_{in} + A)$ and $(k_{out} + A)$, where the k's refer to the number of inward and outward links associated with individual nodes, respectively.

The resulting square matrix containing a total of 700,000 undirected nodes was altered to ensure that all of the links pointed toward younger nodes. This was accomplished by superimposing the transpose of the upper triangular portion of the matrix onto its lower triangular portion before isolated nodes and nodes having zero or one in-link were winnowed from the matrix. After compacting, the final triangular adjacency matrix representing the universe of possibilities for our toy model contained 99,059 nodes and 277,493 directed links, with zeros along the diagonal. Although not huge, the adjacency matrix, named WORLD, is still large enough that elements of the RS growth activity can be explored before being unduly quenched by finite size limitations

### 1.3 Characterizing the WORLD Matrix

The node link distributions, P(k), representing the total number of nodes having specific $k_{in}$ and $k_{out}$ are presented in a log-log graph shown in Fig. 1. Both distributions follow power law behavior from $k \cong 4$ with exponents $\gamma \sim 1.64$ and $\gamma \sim 8$ for the out-link and in-link distributions, respectively. Significantly, the oldest two nodes possesses an appreciably larger number of outlinks than is consistent with the out-degree scale-free trend. These are interpreted as defining the 'situation' under which the recurrent scheme nodes will grow upon WORLD. Traversing all directed link paths emanating from node 1 reveals a degree of separation, d, of 5 steps to all existing nodes in WORLD. Following the link paths from node 6 gives d = 7 steps. Probing other link paths indicates a maximum of $d \cong 18$ directed steps for the WORLD matrix.

Figure. 1. Degree Distibutions, P(k)

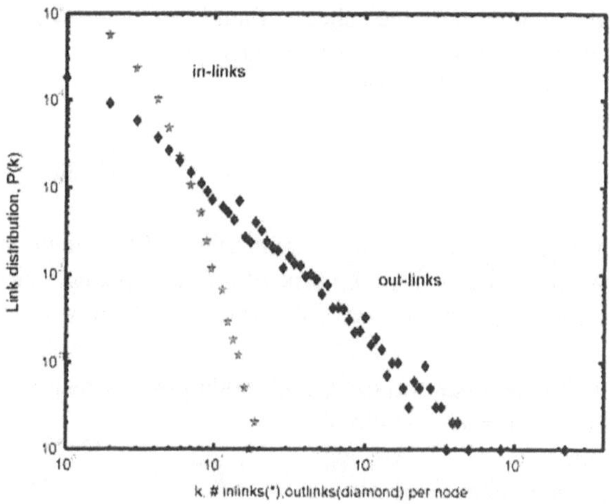


Another parameter that characterizes networks is the correlation coefficient, CC = (# links between parent nodes)/(# possible links), which measures the local connectivity between constraining in-link (or out-link) nodes. A coefficient near 1 would indicate a highly connected network. The WORLD matrix, however, possesses very low correlation coefficients for parent nodes and also for child nodes, with values $CC_{in} = .0028$, $CC_{out} = 7.0 \times 10^{-6}$, respectively.

## 2.1 Development as World Process

Our *World Process* in *Emergent Probability* is the probabilistic activation to functionality of nodes upon the WORLD matrix as the situation, ecology (to be defined), and node conditioning permit. Recurrent Schemes become *virtually unconditioned* toward activation when all of their *conditions* have been satisfied by the actual functioning of their originating in-link nodes. Starting from the three *formally unconditioned* WORLD nodes, #'s 1, 2 and 6 that possess no in-links, each iteration of EP visits and interrogates all non-functioning nodes.

Activation of RS's to functionality is probabilistic. All RS's have some small probability, $p_i$, of irregularly receiving actions from their non-functioning constraints. The probability, P, for all k actions to occur simultaneously and bring that particular RS to functionality is

$$P = \prod_i (p_i). \tag{1}$$

Should that RS instead become virtually unconditioned, the probability for functionality of its scheme suddenly jumps to a much larger value, for if any one of the several conditional node actions starts the scheme, then the scheme will emerge to functional stability. P can then be better written as one minus the product of probabilities for conditional nodes not initiating action (see Appendix),

$$P = 1 - \prod_i (1-p_i) \sim \sum_i (p_i). \tag{2}$$

On other words, once the parent schemes are functioning, the probability for a scheme's appearance leaps from the product of the separate action probabilities of its parents to their sum, making the chances of its appearance hugely more likely!

For calculational purposes, each node possesses an individual action probability, $p_i$, which was assigned randomly within a Rayleigh distribution envelope having mean value = 0.1 and a long low tail to higher values (just choosing the probabilities as random numbers between 0 and some small number < 1 seems quite unphysical). Unlike Lonergan's EP, the assumptions made for this study are that i) the P's from eq. (1) are all vanishingly small, ii) only a single RS of each type can be functional, but iii) all individual RS's may share their dynamics with many dependent RS's at the same time

Figure 2. WORLD Logic Network (partial plot)

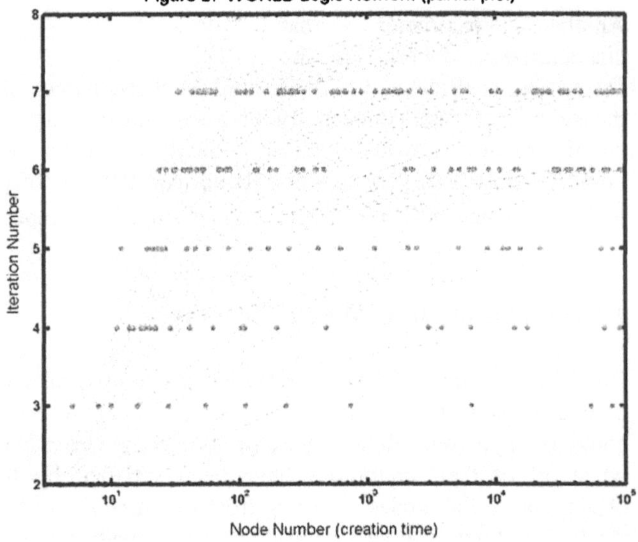

### 2.2 Characterizing the Growing Node Network

It is instructive to initially grow RS's upon the WORLD matrix with all appearance probabilities set to 1. This corresponds to a logic matrix where all activations to function that might occur at every iteration, will actually occur. Figure 2 is a map of such a logical run. The respective functional node numbers lie along the log(node #) axis and the iteration # is along the y-axis. Thus, every potential node of WORLD can be classified by its logical iteration number. Also shown in the figure are all connecting links to nodes activated during the first 8 iterations. (Notice that all links point toward higher numbered (younger) nodes and higher iteration #s.) Since the links already dominate the scene after several iterations, showing the entire plot with associated links would not be useful. (The full plot occupies a triangular area extending to iteration 47 and node numbers $\sim 10^5$ at which point all nodes are finally functional.) The effective EP network diameter, then, is much larger than the WORLD matrix's maximum degree of separation ($d \cong 18$) that was determined by just jumping along directed nodes .

Figure 3. Error Tolerance of WORLD Network

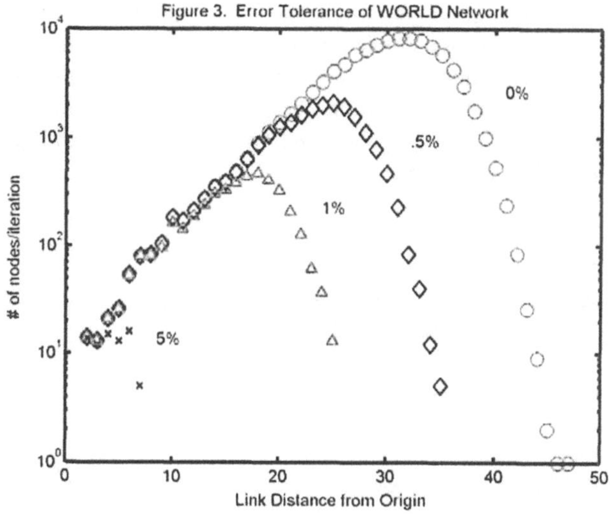


A second instructive procedure is to randomly specify a small number of nodes as permanently non-functional. These nodes strongly poison and fragment the whole growth process, as is expected for scale-free networks [Albert, 2000]. Activation growth rates under these logic conditions are shown in fig. 3 for four impurity concentrations (0%, 0.5%, 1% and 5%). For the pure case all nodes are eventually accessed by iteration 47, but even slight impurities appreciably cut the accessibility to nodes.

Now we shall activate nodes probabilistically (rather than logically), starting with the formally unconditioned nodes 1, 2 and 6. With each iteration more and more nodes spring to functionality (as recorded by 1's along the sparse WORLD diagonal), as they are driven by an ever increasing state space of possibilities. The log of this simple nodal function growth rate is shown as the black line in Fig. 4. The rate rises as a steep exponential for the first 55 or so iterations (to 2004 total functioning nodes), before settling in to a lesser, but still substantial exponential growth rate. But then, after another 55 iterations and some 44,000 functioning nodes later there appears a rounding off (and eventual descent, not shown). The ever-decreasing state space density of the WORLD matrix starts to come into play from this point forward in the simulation. Results up to iteration 55 are kept as the starting point for subsequent runs, since initial growing pains seem to have subsided by then. One can think of the early iterations as a baseline representing the world situation and ecology upon which further activity occurs.

## 2.3 Toward Complexity

Although Lonergan envisioned the interplay of vast numbers of recurrent schemes operating across almost limitless regions of space and through enormous ranges in time, we are constrained here to a simple toy model without spatial dimensions and operating for only a few hundred time iterations. Here, we shall only perform a few simulations to observe an interplay between two equivalent growing clusters as a probe of Emergent Probability dynamics.

Figure 4. Emergent Probability Runs

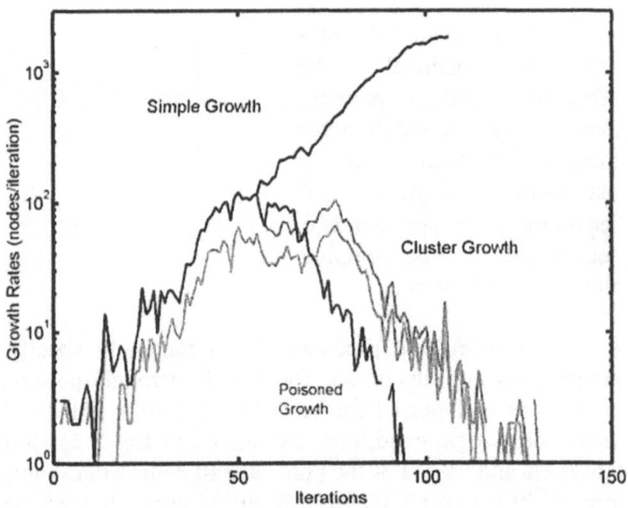


All ecology nodes are randomly allotted node strengths = +/-1 along the WORLD diagonal. The sign identity of each further node activation will be governed by a simple majority rule. That is, the activation of nodes are accompanied by sign designations determined by the sum of constraining nodes. Should that sum be zero for nodes having an even number of parents, then no sign decision is made and those sign-unresolved nodes are inactivated remain thereafter. The simulation (blue line in Fig. 4) demonstrates that the discarded nodes (923 in all) act as errors that poison the growing EP network, whose low error tolerance quenches the growth rate!

Simple competition between the two +/- signed clusters (*Things*) can only play out if the simulation contains dynamic mechanisms to thwart such quenching. A host of cleaver scenarios can be designed for the purpose, but presenting only one of them is sufficient to demonstrate a route toward increasing complexity for the system. The node identity problem is partially resolved by envisioning an ongoing process of merging two similar nodes into a single new signed node. When trying to activate a sign-unresolved node, a search is made among that node's parents to see if any of them have a pair (or more) of common out-links to a second, different node. If found, then with 20% probability, the in-links and out-links of the two nodes are merged, respectively, and copying the sign of the second node (if functional), is recorded in the WORLD matrix overwriting the unresolved node. The second node, if not yet functional, is barred from further participation.

Results are shown as the red line in Fig. 4. The growth rate recovered from an initial drop, but after another 10 iterations it collapsed, as if the node merging process itself limited the state space for growth. Also, the collapse is less steep than for the poisoned run (blue line). The two signed clusters grew at non-identical rates, the magenta line denoting the growth rate for the + cluster in Fig. 4. Until about iteration 75 its size remained somewhat more than half that of the total active nodes, but thereafter it grows in strength relative to the negative cluster (not shown), ending up as dominant.

Duel cluster growth can be further enhanced by assigning random strengths between +/-1 to each ecology node (those nodes functioning by iteration 55). Cluster designation takes place as before, with sign-unresolved nodes being those with $|\Sigma(\text{parent strengths})| <$ weakness, where this threshold is arbitrarily chosen as weakness = 0.05. Instead of inactivating below-threshold nodes, they are activated with randomly chosen strengths = (+/-)weakness (black dots). The resulting growth rate is shown as the red curve in Fig. 5. The rate now follows the simple exponential growth rate curve of Fig. 4 up to iteration 75 where there are 600 nodes/iteration, before dropping back exponentially. A total of 18840 nodes were activated to functionality by iteration 140. The growth rate for the + cluster is also displayed (magenta line in Fig. 5). Integrated growth totals rise exponentially for both clusters before leveling out to a cluster node ratio of 1.27, reflecting the bias from the 3 formally unconditioned situational nodes. The average strength/node for each iteration, a good measure of node fitness, becomes stable at 0.2+/-0.05, but after iteration 100 it gets very noisy. A repeat run allotting all unassigned nodes to the negative cluster eliminated the node ratio bias.

Figure 5. Cluster Growth Rates

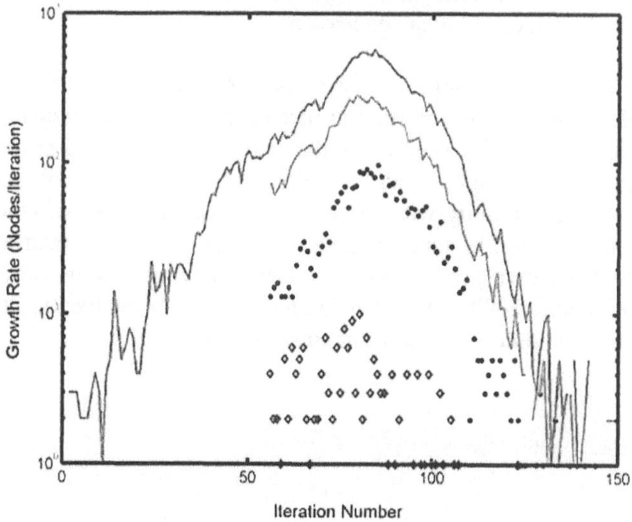


The dynamics of cluster growth becomes clearer when the appearance rate of merged nodes is added to Fig.5 (diamonds). These new nodes apparently redirect future growth by strongly affecting the state space for development. In the present case when merging opportunities dry up, the cluster growth rate quenches, since it can no longer continue along the original simple trajectory of Fig 4.

Node merging (or alternate scenarios), then, will slowly evolve the highly sparse adjacency WORLD matrix, fundamentally altering the world logic network of Figure 2. The matrix will no longer remain fully triangular, but will record in-links to older nodes, creating closed directed adjacency matrix loops. A simple MATLAB eigen routine that seeks such pathways produced about as many degenerate eigenvalues as the number of merged nodes in our cluster simulation. The sudden appearance of these eigenvalues can be interpreted as the emergence of a higher order set of recurrent schemes that substantially increases world process complexity. The remerging of merged nodes, although not implemented in this network study, will act to *accelerate* complexity growth, enriching and deepening the ongoing Emergent Probability dynamics. A similar change is known to mark the emergence of autocatalytic sets in a recent complexity model of evolution [Jain, 1998].

## 3.1 Conclusion and Outlook

The heuristic model designated Emergent Probability introduced a half century ago by Bernard Lonergan is a powerful model of development, decline and change (aspects of which have evidently been reinvented as key elements in various present day approaches to complexity [Padgett, 1996] [Edelman,1987]). Although we have explored some of its basic underlying dynamics and glimpsed but one simple route toward complexity, the full EP model represents a genetic framework for the understanding of process. Moreover, the model greatly expands in reach when human, social and dialectical interactions are subsequently addressed. Emergent Probability methods, then, yield a vantage point for furthering the unification of disparate interdisciplinary fields and for effectively modeling the dynamics of human endeavors.

Work is continuing to enlarge the network simulation size, scope and efficiency. Key elements - growth disruption mechanisms, resource generation and competition, multiple niches, etc.- still need computer implementation. To properly simulate *Things* and their complex interactions also requires variable populations and the continual expansion of possibilities along with the ever-*higher integrations* of emergent complexity patterns. In future network models nodes of the WORLD matrix will be assigned inherent multiplicative growth fitness values, allowing late appearing nodes to bloom more fully [Bianconi, 2000]. Additionally, techniques for the tracking of EP's projected *upwardly but indeterminately directed dynamism* will be sought.

# References

The author wishes to thank D.W. Oyler and K.R. Melchin for conversations regarding Lonergan's Emergent Probability model.

* key terms used by Lonergan are *italicized* in this paper

MATLAB is a product of The MathWorks, Inc., Natick, MA 01760

Albert, R., Jeong, H., and Barabasi, A-L., 2000, *Error and attack tolerance of complex networks*, Nature, **406**, p378.

Bianconi, G., and A.-L. Barabasi, 2000, *Competition and multiscaling in evolving networks* lanl.arXiv.org: cond-mat/0011029.

Bretz, M., 2002, Physics Today(*letters*), **55**, #2, pg 12.

Dorogovtsev, S.N., Mendes, J.F.F., and Samukhin, A.N., 2000, *Structure of Growing Networks with Preferential Linking*, Phys. Rev. Lett., **85**, p4633.

Edelman, G.M., 1987, *Neural Darwinism*, Basic Books, New York.

Jain, S. and Krishna, S. 1998, *Autocatalytic Sets and the Growth of Complexity in an Evolutionary Model*, Phys. Rev. Lett. **81**, p5684.

Krapivsky, P.L., Rodgers, G.J., and Redner,S., 2001, *Degree Distributions of Growing Networks*, Phys. Rev. Lett., **23**, p5401.

Lonergan, Bernard J.F., 1992, *Insight, A Study of Human Understanding*, Collected Works(**3**), edit. F.E.Crowe and R.M. Doran, Toronto Press, 1992 (Longman, Greens & Co., London, 1957).

Melchin, K.R., 2001, *History, Ethics and Emergent Probability*, 2nd Ed., The Lonergan Web Site (ISBN: 0-9684922-0-7).

Padgett, J., 1996, *The Emergence of Simple Ecologies of Skill: A Hypercycle Approach to Economic Organization,* Santa Fe Institute, Working paper 96-08-053

# Appendix: Learning to ride a bike

As a simple example of how a nest of recurrent schemes may constrain, but sustain skill development, we schematically model the process of learning to ride a bicycle by a well-coordinated, inexperienced student (see diagram). Once learned, bike riding sustains a host of further activities.

The essential underlying *situation* is that: i) the student has good health, strength and motivation, ii) there is a suitable terrain and sufficiently good weather to ride, iii) an operable bicycle is available that is properly sized and adjusted, and iv) effective instruction is provided. The recurrent *functions* to be mastered are peddling, balancing, turning, braking and gear shifting. They allow their respective *actions* to recur over and over during bike rides.

First the instructor, Bernie, verbally lectures student, Mike, on how to perform each action. There is some probability, Bernie reasons, that he will start riding immediately, since each rider action is commonly known to first occur with finite probability, $p_i$. Let's assign $p_i$'s as peddling ~.75, balance ~ .1, turning ~ .3, braking ~.4, and shifting ~.4. Mike is given a push, tries to do all actions at once, and after a few wobbles, falls with a crash to the ground! That's because the combined probability of all required actions being performed correctly on the first try is miniscule ($P = \Pi\, p_i < .004$).

So, now Bernie reasons that in order to lift constraints to riding each of the five *actions* must be made to actually *function* through training the individual skills. He therefore has Mike thoroughly practice each separate activity by holding the bike seat and pushing him around. After a half hour of training, Mike again attempts to perform all actions at once without assistance after the initial push. This time he starts to wobble along on his own, quickly gaining stability and confidence.

Why has he succeeded on this second try? Because with all requisite skills now functional, Mike must successfully perform only ONE action, and all the rest will naturally fall into place. For instance, supposing Mike turned his handlebars a wee bit, this action could produce an increase in balance. The extra balance would allow him a chance to peddle some, which would increase his speed, which would complement his newly acquired balance skills, allowing other skills to also happen automatically. After a few moments he would be able to steer around obstacles and brake appropriately for safety as needed, completing the task. This cooperative function, viewed as a new emergent scheme of recurrence, is given the name "Biking".

Bernie is impressed and realizes that his calculations for the probability of Mike's success should be calculated quite differently now that the underlying skills have been mastered. Although Mike started out by trying to perform all of his new skills at once (having respective first occurrence probabilities as given above), one took hold and all other actions followed naturally. There is only a small probability of rider failure, $(1 - P)$, since any other action could have also initiated the task. Now, P equals one minus the product of all the separate failure probabilities, $(1 - p_i)$, or simplifying, $P \sim \Sigma p_i = .94$.

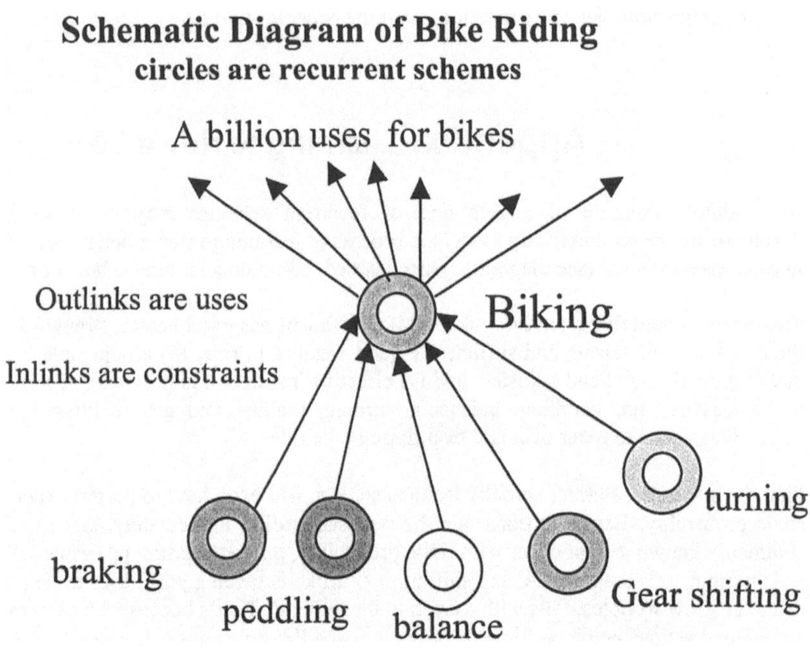

# Chapter 12
## Virtual Stability: A Principle of Complex Systems

**Burton Voorhees**
Center for Science
Athabasca University
1 University Drive
Athabasca, AB
CANADA T9S 3A3
burt@athabascau.ca

We introduce the concept of virtual stability, defined as a system's ability to gain in flexibility and maneuverability by using self-monitoring in order to maintain itself in a state that would normally be unstable. After presenting a general description of virtual stability we look at a number of examples and outline a simple cellular automata model that is being used to explore virtual stability.

## 1. Introduction

A system is said to be in a virtually stable state if it is maintaining itself in an unstable state through self-monitoring and small (virtual) corrective actions. The advantage purchased by the energy expended in these corrective actions is an increase in flexibility or maneuverability. Virtually stable systems are ubiquitous in nature and the concept is closely related to W. Ross Ashby's law of requisite variety [1].

We suggest that virtual stability provides an answer to the question of whether or not there is a direction to evolution. To the extent that flexibility of action within an environment gives an advantage, there will be a selective pressure favoring the evolution of species with the capacity for self-monitoring and adaptation. As a corollary, given a rich enough environment with sufficient sources of energy, the emergence of conscious life is almost certain.

This conclusion does not contradict writers like Stephen J. Gould [2] who have argued for the contingency of human evolution. It may well be that our presence on earth is contingent on a number of historical accidents. All that is implied on the basis of an argument from virtual stability is that, with high probability, some form of intelligent life would appear at some point in the evolution of life on earth.

Perhaps the most obvious area where examples of virtually stable systems can be found, however, is in the realm of human society. The fact that human beings consciously monitor their behavior, and can keep different course of action in mind without making an immediate commitment to any of them, is a prime example of the capacity to maintain virtually stable states.

# 2. Examples

## 2.1 Standing

A simple example of virtual stability is a person standing. The upright position is maintained by feedback circuits in the vestibular system, connected to our kinesthetic sense of body position and to muscle groups in the legs and feet that make the small adjustments necessary to maintain the standing position. In a recent conference presentation, Bach-y-Rita [3] describes a sensory substitution experiment in which a woman with neurological damage affecting her vestibular system was unable to stand until she was provided with a special hat containing accelerometers and motion sensors that would provide a tactile signal whenever her head began to move away from the vertical. With this assistance, she was able to consciously control her body posture while standing and walking, something that is automatic for most of us.

The flexibility we gain from the fact that standing is a virtually stable state is the ability to move quickly in any direction. There are even degrees of flexibility. This shows up, for example, in boxing, in the distinction between the stalkers, who plant their feet firmly on the floor, and the dancers who, in the words of Mohammed Ali, "float like a butterfly, sting like a bee". Of course, for most of us, standing does not <u>seem</u> to be an unstable state. It has become automated to the point that we only notice the instability when something such as alcohol interfere with the control systems involved.

## 2.2 Aircraft Design

In the early 1980's, the United States Air force was testing an experimental fighter called the X-29. The unusual feature of this aircraft was that its wings were swept forward rather than back. This configuration is aerodynamically unstable so the plane required a triply redundant computer monitoring system that checked the plane's motion 40 times per second and made the control adjustments necessary to keep it on course. If this system were to fail for even one-quarter of a second, the X-29 would have tumbled out of control. The advantage gained was maneuverability. While an ordinary fighter with swept-back wings requires energy to change course, the X-29 would simply "fall" in the direction indicated. Although this particular plane was never produced, aircraft designers are well aware of the trade-off between stability and maneuverability. Fighter planes today are, by design, very close to being unstable, while passenger planes are designed for stability.

## 2.3 Conversational Positioning

Feminist theorists and social psychologists that follow the social constructionist point-of-view, consider many apparently ordinary conversations as power struggles based on the idea of "positioning". The idea is that each party to the conversation is consciously or unconsciously attempting to position themselves favorably with respect to the others in the moral universe defined by the elements of discourse that are in play. As described by Burr [4], "in any interchange between people, there is a constant monitoring of the 'definition of the situation' that each participant is struggling to bring off". Each player in this game can choose to accept a "position" that is offered them, or to resist it. The statement "you've really been taken advantage of," for example, offers the person it is addressed to the role of victim and places the person making the statement in the superior position of sympathetic friend or advisor. A response of "Yes, it's really been tough" accepts this position while a reply like "No, I've got everything under control," resists it. Within such a situation, it is important to be aware of the continuing interplay in order to avoid placing others into unintended, undesirable positions, and also being placed in such positions by others. In such rhetorical battles of wit the advantage goes to the person who is best able to maintain him/herself in the unstable state of having no position, or at least a very flexible position, until the opportunity arises to seize the high ground. Keeping one's options open is an important aspect of many forms of social behavior.

## 2.4 Mate Selection

In sociobiology, male and female sexual strategies are explained in terms of the attempt to maximize the transmission of genetic material to the next generation. Thus, males are seen as being genetically programmed to seek multiple sex partners, while a female seeks to keep a man sexually dependent so that he will stay and provide for her children. Be this as it may, we can consider mate selection strategies in a monogamous culture. One strategy is to marry the first available person. This leads to stability in one's sex life, a stability that is reinforced by the bonds of romantic love. Such stability is very helpful in allowing a person, male or female, to "get on with the rest of their life," but it may also end up locking a person into an ultimately destructive relationship. Another strategy is to experiment, to "play the field," and maintain a number of relationships until a choice of permanent mate can be made on a basis of compatibility and affection. Following this strategy requires self-monitoring and emotional control in order to avoid obsessive attachments. It is, in this sense, unstable, but it also provides a better chance over the long run of finding a more suitable mate.

## 2.5 Global Finance

The behavior of money managers in the emerging global economy provides an economic example of the advantage conferred by virtual stability. Thomas L. Friedman describes the behavior of these individuals in his book, The Lexus and the Olive Tree [5], which provides many examples that can be understood in terms of virtual stability. The most obvious is the fact that the rapid availability of information about world financial markets means that money managers must have the flexibility to rapidly move capitol from one market to another, a state that would be highly unstable without the continual monitoring of the performance of their investments. The other side of this is that money can move in and out of a national economy so quickly that any minor instability in that economy can be greatly magnified if the national government is not also engaged in high frequency self-monitoring.

## 3. Models of Virtual Stability

Simple one-dimensional cellular automata can be used to construct model systems that display virtually stable behavior. One such model is currently being developed. It is based on a binary rule defined on a six-site lattice with periodic boundary conditions. The rule table is

| 00 | 01 | 10 | 11 |
|---|---|---|---|
| 0 | 1 | 1 | 0 |

(This is Rule 102 in the usual labeling convention.) This rule has four basins of attraction, which will be labeled A, B, C, and D.

To model a virtually stable system based on a CA rule three discrete time scales are employed. Iterations of the CA rule itself take place on the intermediate scale. The fast scale is used to introduce fluctuations, and the slow scale corresponds to the control function that specifies which basin the system is to be in at any given (slow scale) time. The basin labeled D is called death.

For the simplest case, the control function will be defined by a sequence that specifies either basin A or basin B at each (slow scale) time step. The initial conditions will specify the control sequence, an initial state in the required basin (say A for definiteness), and set the values of three integer functions, c, m, and k to zero. The value of m will count the number of iterations of the CA rule, and when $m = M$ a slow scale time step is taken, possibly introducing a new target basin. The value of k counts the number of slow scale time steps and when $k = K$ the system stops and prints the accumulated value of c, which is a cost function.

The CA rule operates on the initial state, producing a new state that is still in the target basin for $k = 0$, and an integer function n is set to 0. The value of n counts the number of fast scale time steps. The system state is now subjected to a perturbation based a probability matrix. The entries in this matrix are what determine the stability or instability of the system.

For this first case, the probability of an A → C or B → C transition is set to 0. There is a small but positive probability, however, for the A → D and B → D transitions (i.e., there is always a small chance the system will die). If one of these transitions occurs, the program stops and prints out m, k and dead.

Whether the system itself is stable or unstable depends on the A → B and B → A transition probabilities. If the A → B probabilities are small while A → A and B → B are large the system is relatively stable, while if the A → A and B → B probabilities are small while the A ←→ B probabilities are large, then the system is unstable.

The perturbed state is first tested to see if it is in basin D. If not, it is tested to see if it has remained in the specified target basin (e.g. A). If so, n is incremented by one and the state is perturbed again and tested again. If a transition to the other basin (e.g., B) has occurred, then it is necessary to return the state to the target basin. There is a cost associated to this. The return is carried out through use of the same probability matrix as before. A loop is established in which the state is perturbed and tested until a state in the target is achieved. The cost of this return is equal to the number of cycles through the loop. At this point, n is increased by one and the system reenters the fast time scale process of perturbation and testing until n = N and another CA step is taken. At the end, assuming that the system has not died, the total accumulated cost is printed.

For stable systems the probability of a perturbation out of the target basin is small, but once it occurs, the cost of a return is likely to be high. For unstable systems, on the other hand, there is a high probability of perturbation out of the target basin, but the cost of return will be low, modeling virtually stable states.

By adjusting the transition probabilities a variety of different situations can be studied as a means of gaining some insight into the conditions under which a capacity for virtual stability provides an advantage.

## 4. Some General Considerations

There are a number of aspects of complex systems that relate to the capacity to maintain, and the desirability of, virtually unstable states.

One obvious parameter involved will be the frequency of self-monitoring. In maintaining a virtually stable state there is a trade-off between a small but ongoing energy expenditure and the advantage gained from the ability to change states quickly without excessive energy expenditure. Thus, the self-monitoring frequency must be high enough that corrective actions remain small and require minimal energy expenditure. This requires that this frequency be syntonized with the spectrum of external fluctuations that produce destabilizing effects. Too high a self-monitoring frequency and energy and attention become monopolized, too low and the system looses its virtual stability.

Another factor involved is illustrated by the example of the X-29. This aircraft was so maneuverable that it was dangerous to the life of the pilot. In other words, it is possible for a system to be too flexible. Roughly stated Ashby's law of requisite variety says that the variety of control possibilities must match the variety of the external disturbances if the outcome is to be uniquely controlled. Thus, the degree of flexibility required is no greater than what is sufficient to deal with the spectrum of ordinary environmental fluctuations, with a cut-off for very low probability events. The location of this cut-off itself becomes a question for theoretical investigation.

The basic lesson of virtual stability is that life is not about stability, it is about managing instability so as to produce the illusion of stability.

## Acknowledgements

Work on this paper has been supported by NSERC Operating Grant OGP 0024817

# Biblography

[1] W. Ross Ashby (1958) Requisite variety and its implications for the control of complex systems. Cybernetica 1, 83 – 99.

[2] Stephen J. Gould (1989) Wonderful Life: The Burgess Shale and the Nature of History. NY: Norton.

[3] Paul Bach-y-Rita (2002) Tactile-sensory substitution in blind subjects. Presented at Tucson 2002: Toward a Science of Consciousness, April 8 – 12, 2002.

[4] Vivien Burr (1995) An Introduction to Social Constructionism. London: Routledge. p. 146.

[5] Thomas L. Friedman (2000) The Lexus and the Olive Tree. NY: Anchor.

# Chapter 13

# Fungal Colony Patterning as an Example of Biological Self-Organization

**Elena Bystrova, Evgenia Bogomolova, Ludmila Panina, Anton Bulianitsa*, Vladimir Kurochkin***
A.A. Ukhtomsky Institute of Physiology
Saint-Petersburg State University
*Institute for Analytical Instrumentation RAS
St.-Petersburg, Russia
elena@EB5029.spb.edu

## 1. Introduction

The form, the size and the color of a fungal colony are the features which can characterize genera and species of fungi. Besides the individual properties of fungal cultures (such as the existence of circadian rhythms, the ability to produce metabolites - growth inhibitors - antibiotics, organic acids, exoenzymes, etc.) the environmental parameters (nutrient concentration, temperature, light availability, etc.) also have a significant influence on the colony form [4,12,20,21].

On the other hand the process of a certain colony type formation can be considered as a phenomenon of biological self-organization, since the fungal system undergoes a number of spontaneous changes which include increase and/or complication of the system elements, modification of the system functioning conditions, etc. In the terms of complex systems approach we treat the fungal system as having a discrete number of macroscopic stable states on the colony level. The variety of these states is determined by a morphological potential of a certain fungal species. Transitions between states are managed by external control parameters (environmental factors). The process of pattern formation in fungi can be therefore regarded as an adaptation of the colony during its development to the changing environment due to the collective interactions of the system elements with each other and with environment as well as due to the existence of feedback between fungal system state and control parameters.

In this paper we describe the main types of stationary dissipative structures, which can arise in colonies of mycelial fungi. We investigate conditions required in order for patterns to appear and explain how changes in fungal environment influence the morphology of the colony, or, in other words, how the system interacts with its environment. On the basis of experimental data obtained we have developed a mathematical model for description of observed non-linear phenomena in colonies of mycelial fungi. By means of the model and experimental results we estimate the value of metabolite diffusion coefficient, which was shown to be one of the main parameters defining the colony shape.

## 2. Materials and Methods

Experiments were carried out on mycelial fungi *Ulocladium chartarum*, *Ulocladium botrytis*, *Ulocladium consortiale*, *Alternaria alternata*, *Penicillium chrysogenum* (Deuteromycotina, Hyphomycetes: Hyphales). The given species were cultivated at 25°C and 8°C on standard and modified nutrient media in Petri dishes (d = 9 cm). We varied glucose concentration (0 – 3%) and volume of nutrient medium in a Petri dish (5 – 20 ml). Fungal growth patterns and their sizes were registered by a digital camera Casio-QV100. For confirmation of the mycelial property to produce metabolites (growth inhibitors) a pH-sensitive indicator was added to the nutrient medium, which caused the changes of medium color according to its pH. The spatial dynamics of metabolite distribution was estimated by means of vertical photometry on the device "Chicken" (Institute of Analytical Instrumentation, Russia). The optical density of medium is proportional to the metabolite concentration ($\lambda$ = 610 nm) which allowed us to evaluate average radial distribution of metabolites around the colony and hence the values of metabolite diffusion coefficient.

## 3. Results and Discussion

The colony of mycelial fungi is a uniform multicellular structure developing radially by growth and branching of mycelium. The colony is able to consume substrate (carbon source, i.e. glucose) and to produce diffusible metabolites which can suppress fungal development. The final stage of fungal morphogenesis is the formation of spores.

Different growth patterns can arise in colonies of mycelial fungi while they are cultivated on solid agar media of various thickness and nutrient content. These macroscopic spatiotemporal patterns were classified into four main types: concentric rings (zones), "sparse lawn" and "dense lawn" (continuous mycelial growth), ramified (fractal-like) structures (Fig.1). The experimental data obtained enable us to construct a two-dimensional morphological diagram in which a characteristic colony form corresponds to each pair of given parameters (Fig.2) (in our case, we varied two parameters - nutrient concentration and volume of nutrient medium, i.e. medium thickness).

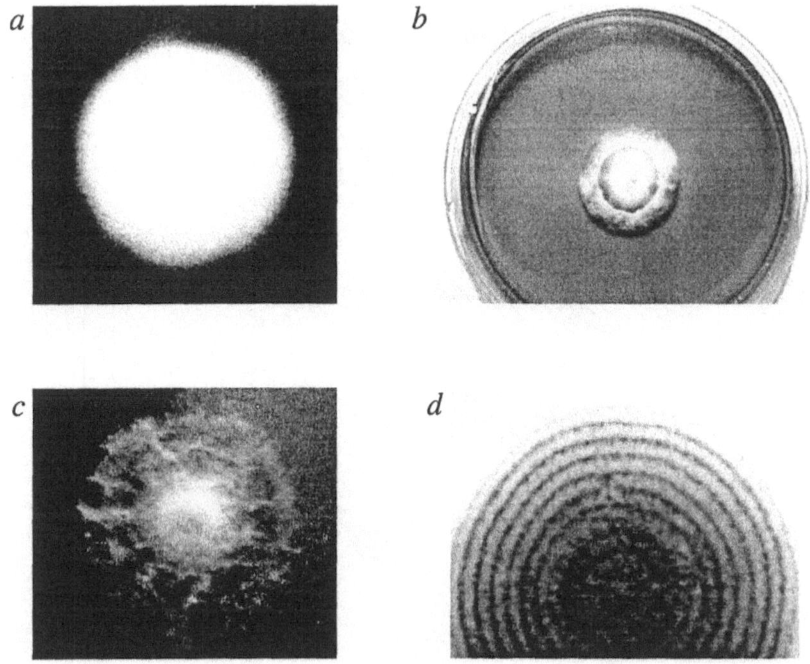


**Figure 1.** Different growth patterns which can develop in colonies of mycelial fungi: *a* – "dense lawn", *b* – "sparse lawn", *c* – fractal-like structures, *d* – concentric rings.

When fungi are grown on optimal nutrient media (glucose 1-3%, agar 2%, volume of medium in a Petri dish 5-20 ml), hyphal branching is maximal and the colony represents a continuous surface of well-developed mycelium ("dense lawn"). Variations of mycelium or spore density are not visible on the colony surface. On thick poor nutrient media (glucose < 0.1%, agar 2%, 15-20 ml of medium in a Petri dish) hyphal branching is minimal and the colony grows in the form of a weakly-developed lawn ("sparse lawn"). Cultivation of fungi on thin poor media (glucose 0-0.05%, agar 2%, 5-10 ml of medium in a Petri dish) can result in the emergence of fractal-like structures. There is a limited range of substrate concentrations (glucose 0.1-0.5%, agar 2%, medium volume 5-10 ml) in which zone formation occurs. Microscopic examination of concentric rings formed in colonies of *Hyphomycetes* has shown them to be regions of high mycelium or spore density intermitted by ones of lower density. We have found out that zone formation in fungal colonies can be effected by external synchronizing stimuli, for example, by alteration of temperature and light incubation conditions. In particular, the decrease of cultivation temperature causes the formation of distinctly-shaped concentric rings, and visible light can stimulate the emergence of synchronous wave structures.

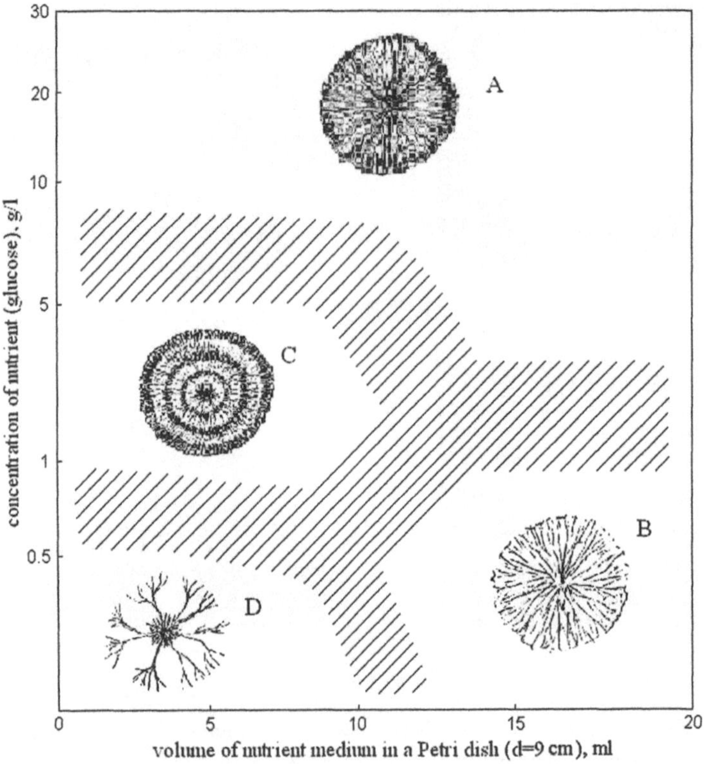


**Figure 2.** Morphological diagram of fungal colony patterns: A – "dense lawn", B – "sparse lawn", C – concentric rings, D – fractal-like structures.

We have proposed a general mechanism of fungal pattern formation, which is based on two simultaneous processes: consumption of substrate ($s$) (activator) by mycelium and suppression of mycelial growth by diffusible metabolites ($m$) (growth inhibitors). At the beginning of fungal colony development metabolite concentration is zero. As the colony grows metabolite concentration increases and substrate concentration decreases due to its consumption by mycelium. It is assumed also that there is a threshold $(s/m)_t$ at which growth of mycelium stops. In case of "lawn" formation (both, "sparse lawn" and "dense lawn" types) a certain ratio of substrate and metabolite concentrations $s/m > (s/m)_t$ is maintained so these patterns are relatively stable. In the range of concentrations suitable for zone formation spatial alternation of $s/m > (s/m)_t$, $s/m < (s/m)_t$ is established since diffusion coefficient of metabolite exceeds diffusion coefficient of substrate and the gradient of metabolite concentration is large enough. In other words during radial growth of the colony ratios of substrate and metabolite concentrations are broken alternately, which result in periodic changes of fungi growth modes. The essential factors, which are supposed to be responsible for the emergence of fractal-like fungal patterns, are diffusion of nutrients (in this case diffusion coefficient of substrate exceeds diffusion coefficient of metabolite) and unequal accessibility of nutrients to separate hyphae. Thus the conditions are created when several actively growing branches are always in a maximal nutrient gradient and inhibit competing branches development.

***A mathematical model*** has been developed for description of observed non-linear phenomena [8], which is a system of reaction-diffusion-type equations [1,24] and takes into account the models of spatiotemporal order generation, for example [9,13,18,23]. The system of differential equations describes spatiotemporal distribution of mycelium ($\xi$), spore ($\chi$), substrate ($s$) and metabolite ($m$) concentrations:

$$\frac{\partial m}{\partial \tau} = \alpha\xi^{2} U_1(s,\xi) + D_m \Delta_\rho m. \tag{1}$$

$$\frac{\partial s}{\partial \tau} = -\gamma\xi K(s) + D_s \Delta_\rho s, \qquad K(s) = \frac{s}{s+1}. \tag{2}$$

$$\frac{\partial \xi}{\partial \tau} = \lambda\xi(1-\varepsilon\xi)K(s)U_2(m), \qquad U_2(m) = 1[-m(\tau-\tau^0)+\mu^0]. \tag{3}$$

$$\chi(\tau) = \sigma\xi(\tau-\tau^*)U_3(s,m). \tag{4}$$

The model contains non-dimensional parameters which characterize such quantities as radial growth rate of mycelium ($v$), specific growth rate of mycelium concentration ($\lambda$), substrate consumption rate ($\gamma$), scaling ratio of metabolite production rate ($\alpha$), diffusion coefficients of metabolite ($D_m$) and substrate ($D_s$), initial mycelium concentration ($\xi^0$), maximal mycelium concentration (resource concentration) ($\varepsilon^{-1}$), initial substrate concentration ($s^0$), threshold of metabolite concentration* ($\mu^0$) at which growth of mycelium stops, time of mycelium response delay on presence of metabolites ($\tau^0$), etc. Trying different values of the above-mentioned parameters it is possible to model the main ways of fungal colony development and to determine parameters which have the strongest influence on pattern formation processes.

** The values of $\mu^0$ can differ depending on cultivation conditions and individual properties of fungal species.*

In ***computational experiments*** we have demonstrated that proposed model is able to describe properly the emergence of such macroscopic patterns as "concentric rings", "dense lawn" and "sparse lawn". The analysis of computer simulation results has shown $\alpha$, $\mu^0$, $D_m$ and $\tau^0$ to be among the most important factors influencing the distribution of mycelium concentration. Examples of fungi growth simulation are presented on Fig. 3 and Fig. 4.

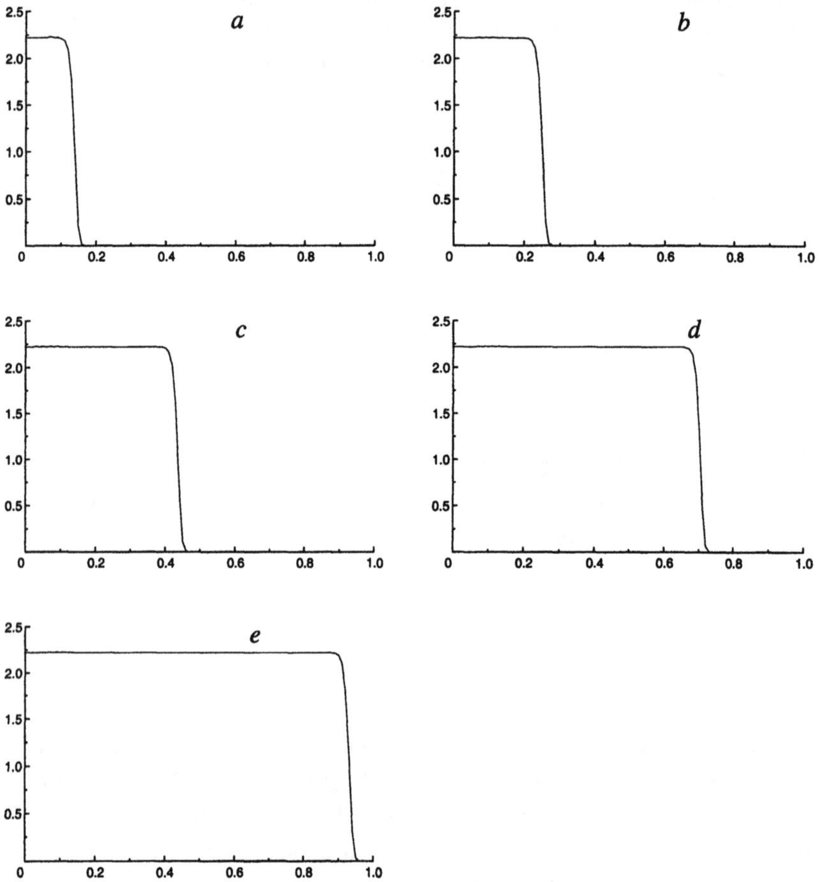


**Figure 3.** Dynamics of fungal colony growth on optimal nutrient media bringing to the development of "lawn" ($x$-axis is relative radius, $y$-axis is relative mycelium density). Parameters of the model: $\nu = 1.2$; $\rho^0 = 0.05$; $\xi^0 = 0.1$; $\lambda = 5$; $\varepsilon = 0.45$; $s^0 = 10$; $D_s = 1.5 \cdot 10^{-5}$; $\gamma = 1$; $D_m = 0.003$; $\alpha = 0.6$; $\mu^0 = 0.5$; $\tau^0 = 0.50$; $\eta = 0.80$. Time: $a$ - $\tau = 4.2$; $b$ - $\tau = 8.3$; $c$ - $\tau = 16.7$; $d$ - $\tau = 25.0$; $e$ - $\tau = 33.3$.

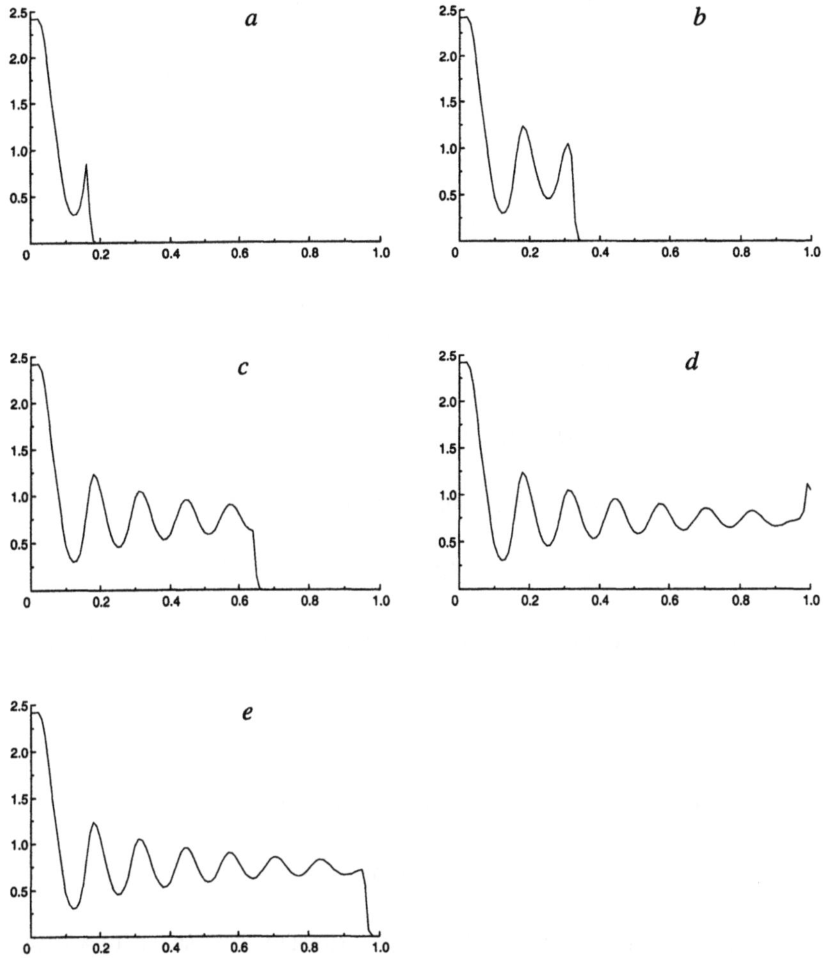


**Figure 4.** Dynamics of fungal colony development resulting in concentric rings formation (*x*-axis is relative radius, *y*-axis is relative mycelium density). Parameters of the model: $\nu = 1.2$; $\rho^0 = 0.03$; $\xi^0 = 0.1$; $\lambda = 7$; $\varepsilon = 0.40$; $s^0 = 10$; $D_s = 5.0 \cdot 10^{-5}$; $\gamma = 1$; $D_m = 0.001$; $\alpha = 0.5$; $\mu^0 = 0.1$; $t^0 = 0.50$; $\eta = 0.90$. Time: *a* - $\tau = 4.2$; *b* - $\tau = 8.3$; *c* - $\tau = 16.7$; *d* - $\tau = 25.0$; *e* - $\tau = 33.3$.

As it was mentioned above metabolite diffusion coefficient is one of the main parameters defining the colony morphology. The values of the given coefficient have been estimated by means of the model and experimental results.

We have proposed the so-called one-dimensional model of metabolite production and distribution, which assumes that a dominating direction of metabolite diffusion is its radial distribution in substrate due to a relatively thin layer of medium.

The value of metabolite concentration can be represented as follows [22]:

$$y(r,t)=\sum_{i=0}^{\infty} C_i J_0(\mu_i r / r_{\max})\exp(-\mu_i^2 / r_{\max}^2 D_m t)\,. \qquad (5)$$

Due to the boundary conditions, which correspond to a case of impermeable external walls:

$$J_1(\mu_i)=0. \qquad (6)$$

Besides,

$$C_i \sim J_0^{-2}(\mu_i).$$

Here $J_0$, $J_1$ - cylindrical Bessel functions of the orders 0 and 1, $r$ – a radial coordinate, $r_{\max}$ - a radial coordinate, which corresponds to the external wall of a Petri dish, $t$ – time. The first roots of the equation (6) are [22]:

$$\mu_0 = 0;\ \mu_1 = 3.83;\ \mu_2 = 7.02.$$

Since all measurements were carried out at the same time, then

$$y(r_i, t) = y(r_i) = y_i,\ (i = 1, 2, 3).$$

Having chosen the first three items (the largest ones) from equation (5) it is possible to estimate the expression

$$\exp(-(\mu_2^2-\mu_1^2)/r_{\max}^2 D_m t)$$

and respectively the value of $D_m$ on the basis of ratio

$$(y_2 - y_1) / (y_3 - y_1).$$

The absolute values of $D_m$ (Table 1) approximately correspond to a case of low molecular compounds diffusion in a liquid medium. The results of calculations presented in Table 1 demonstrate also a relatively weak influence of cultivation temperature on the value of metabolite diffusion coefficient. This fact allows us to give a possible explanation for the experimentally observed phenomenon of zone patterning stimulation due to the decrease of cultivation temperature. The point is that radial growth rate of mycelium considerably depends on temperature conditions. In particular, the decrease of temperature to 6-8 °C causes 2-4 times mycelium growth rate reduction. Thus, when the colony is grown at low temperatures $v$ decreases while the value of $D_m$ changes insignificantly, therefore metabolite concentration at the colony growth front can easily reach its threshold $\mu^0$ which results in concentric rings formation or makes them more distinct.

Apparently, in case of "lawn" formation relative values of $D_m$ are rather low and metabolite concentration doesn't reach the threshold; consequently the colony develops evenly.

**Table 1.** The values of metabolite diffusion coefficient and radial growth rate of mycelium at different cultivation temperatures (for *U. chartarum*).

| Cultivation temperature $T$, $^0$C | Time of cultivation $t$, days | Radial growth rate of mycelium $v$, mm per day | Metabolite diffusion coefficient $D_m \cdot 10^{-6}$, cm$^2$/s |
|---|---|---|---|
| 8 | 14 | $0.75 \pm 0.15$ | $0.40 \pm 0.10$ |
| 25 | 8 | $2.05 \pm 0.35$ | $0.38 \pm 0.11$ |

## 4. Conclusion

It should be noted that there is a number of alternative approaches and models for description of fungal colonies patterning [3,10,14-17,19]. For example, concentric rings formation in *Neurospora crassa* is explained as a result of circadian clock functioning [5,10,19]. Other models also take into account the effect of toxic metabolites accumulation, which play an important role as inhibitors for cell division [14,15]. Anyway, fungal colonies patterning deserves consideration as an interesting example of biological self-organization since fungi are extremely labile and possess a potential for a variety of growth forms development. In turn, the understanding of general features and mechanisms of spatiotemporal self-organization in fungal colonies, as well as in colonies of bacteria [2,6,7,11] and other microorganisms, which are able to generate order on a macroscopic level displaying collective behavior, may help us to reveal the main laws of morphogenesis of multicellular tissues, organs and organisms.

## References

1. Belintsev, B.N. "Physical basis of biological morphogenesis". Moscow: Nauka, 1991.
2. Ben-Jacob, E., Cohen, I., Shochet, O., Tenenbaum, A. "Cooperative formation of chiral patterns during growth of bacterial colonies". *Phys.Rev.Lett.* **75** (1995): 2899-2902.
3. Bogokin, S.V. "A mathematical model of morphological structure of fungi". *Biophysika* **41** (1996): 1298-1300.
4. Boltianskaya, E.V., Agre, N.S., Sokolov, A.A., Kalakutskiy, L.V. "Concentric rings formation in colonies of *Thermoactinomyces vulgaris* as a result of temperature and humidity changes". *Mikrobiologia* **41** (1972): 675-679.
5. Brody, S. "Circadian rhythms in *Neurospora crassa*: the role of mitochondria". *Chronobiol.Int.* **9** (1992): 222-230.
6. Budrene, E., Berg, H. "Complex patterns formed by motile cells of *Escherichia coli*". *Nature* **349** (1991): 630-633.

7. Budrene, E., Berg, H. "Dynamics of formation of symmetrical patterns by chemotactic bacteria". *Nature* **376** (1995): 49-53.
8. Bulianitsa, A.L., Bogomolova, E.V., Bystrova, E.Yu., et al. "The model of spatiotemporal periodical patterns formation in the colonies of mycelial fungi ". *Journ. Of General Biology* **61** (2000): 400-411.
9. Cross, M.C., Hohenberg, P.C. "Pattern formation outside of equilibrium". *Rev.Mod.Phys.* **65** (1993): 851-1112.
10. Deutsch, A., Dress, A., Rensing, L. "Formation of morphological patterns in the ascomycete *Neurospora crassa*". *Mech.Dev.* **44** (1993): 17-31.
11. Fujikawa, H., Cohen, I., Shoket, O., et al. "Complex bacterial patterns". *Nature* **373** (1995): 566-567.
12. Jerebzoff, S. "The fungi, an advanced treatise". N.Y.; London: Acad. Press, 1966.
13. Koch, A.J., Meinhardt, H. "Biological pattern formation: from basic mechanisms to complex structures". *Rev.Mod.Phys.* **66** (1994): 1481-1507.
14. Lopes, J.M., Jensen, H.J. "Generic model of morphological changes in growing colonies of fungi". *Phys.Rev.E* **65** (2002): 021903-021907.
15. Lopes, J.M., Jensen, H.J. "Nonequilibrium roughening transition in a simple model of fungal growth in 1+1 dimensions". *Phys.Rev.Lett.* **81** (1998): 1734-1737.
16. Matsuura, S. "Colony patterning of *Aspergillus oryzae* on agar media". *Mycoscience* **39** (1998): 379-390.
17. Matsuura, S., Miyazima, S. "Formation of ramified colony of fungus *Aspergillus oryzae* on agar media". *Fractals* **1** (1993): 336-345.
18. Polegaev, A.A., Ptysin, M.O. "Mechanism of spatiotemporal order generation in bacterial colonies". *Biophysika* **35** (1990): 302-306.
19. Ruoff, P., Vinsjevik, M., Mohsenzadeh, S., Rensing, L. "The Goodwin model: simulating the effect of cycloheximide and heat shock on the sporulation rhythm of Neurospora crassa". *J.Theor.Biol.* **196** (1999): 483-494.
20. Saveliev, A.P. "The influence of the environmental changes on zone formation in the colonies". *Izvestia RAS, Ser. "Biology"* **4** (1996): 460-466.
21. The Mycota. Growth, differentiation and sexuality. Wessels, J.G.H., Meinhardt, F. eds. Berlin; Heidelberg: Springer-Verlag, 1994.
22. Tihonov, A.N., Samarsky, A.A. "The equations of mathematical physics". Moscow: Nauka, 1977.
23. Tsyganov, M.A., Kresteva, I.B., Medvinsky, A.B., Ivanitsky, G.R. "New mode of bacterial waves interactions". *Doclady Ac.Sci.USSR* **333** (1993): 532-536.
24. Turing, A. "The chemical basis of morphogenesis". *Phil.Trans.R.Soc.London, Ser.B* **237** (1952): 37-72.

# Chapter 14

# Production-rule complexity of recursive structures

**Konstantin L Kouptsov**
New York University
klk206@panix.com

Complex recursive structures, such as fractals, are often described by sets of production rules, also known as L-systems. According to this description, a structure is characterized by a sequence of elements, or letters, from a finite alphabet; a production rule prescribes replacement of each letter by a sequence of other letters.

Practical applications of rewriting systems however are quite restricted since the real structures are not exactly recursive: they involve casual deviations due to external factors. There exist fractals corresponding to the strings, invariant under the rewriting rules. Some strings which are not invariant may also characterize a fractal structure. Sometimes the sequence of symbolic descriptions of a structure is known while the rewriting rules yet to be discovered. Following the Kolmogorov's idea to separate regular and random factors while considering the system's complexity it may be possible to filter out the underlying recursive structure by solving an inverse L-systems problem: given a sequence of symbolic strings determine the simplest (if any) rule set performing the transformation.

The current paper is a work-in-progress report, presented at the Fourth International Conference on Complex Systems (Nashua, 2002), in which I give a preliminary account for issues involved in solving the inverse problem, and discuss an algorithm for finding the shortest description of the symbolic strings (analogous to the Chaitin's algorithmic complexity).

# 1 Introduction

Biological structures are well known for their hierarchical organization which is dictated by the efficiency of functioning of the organism as a whole. Their description sometimes admits a recursive approach: the structure of a particular component at one level of hierarchy is a composition of lover-level parts; the structure of a single component largely resembles the structure of a whole.

For phenomenological description of plants structure, a biologist Aristid Lindenmayer proposed in 1968 a mathematical formalism [7] called *L-systems*. The L-systems language well suites for definition of fractal structures such as Koch and Hilbert curves, Sierpinski gasket, fern leaf. The language provides a simplified version of the object structure by abstracting from the random variations. In spite of that, it catches the very essence of the geometry: the computer model of the object based on the L-description looks quite like a live photograph (recursive generation of landscapes and plants is fast enough to be used in computer animations, movies, and games).

The role of L-description of a recursive structure can be understood in the context of the Kolmogorov-Chaitin complexity [2, 5]. An individual finite object (which may be encoded by a finite binary string) contains the amount of information equal to the length $l(p)$ of the shortest (binary) program $p$ that computes the object. An object encoded by a trivial binary string such as $00000\ldots0$ contains minimal amount of information, whereas the completely random string (obtained by flipping the coin) represents the minimal program by itself, so the complexity of a random string is proportional to its length. For the intermediate objects the description contains both regular (meaningful) information and the random (accidental) one. At a Tallinn conference in 1973 A. N. Kolmogorov has proposed to consider the separation of these two parts in the case of finite set representation. This idea was later generalized [4, 11]. The approach is the following: given the data object $D$, identify the most probable (optimal) finite set $A$ of which the object is a typical example. The shortest description of the set $A$ is a "program" $p$, containing no redundant information. The choice of $D$ within $A$ is the "data" $d$, the random part. The minimization of description length is performed over all sets $A \supset D$. The complete description of the object is given by the pair $\langle p, d \rangle$ which can be represented as a single binary string.

Despite the seemingly machine-oriented definition, the algorithmic complexity is, on the contrary, a universal and absolute characteristics of an object, up to an additive constant, and is machine independent in the asymptotic sense. Expression of the shortest program computing the object using different languages may involve an overhead which can be attributed to the redundancy of the language and contain information for translation of one language into another. Practical computer languages are highly redundant, as they contain long words and combinations of symbols that themselves are not chosen randomly, but follow some predefined pattern. In order to compare "lengths" of the programs written in these languages one has to introduce quite sophisticated probability measures, putting less weight on pattern words. On the other hand, these measures are completely irrelevant to the problem and complicate things.

Kolmogorov and independently Chaitin [2] proposed to consider binary programs in which every bit is 0 or 1 with equal probability. Definition of algorithmic (Kolmogorov-Chaitin) complexity may, for example, be done by means of specialized (prefix) Turing machines. Let $T_0, T_1, \ldots$ be a set of Turing machines that compute $x$. The *conditional complexity* of $x$ given $y$ is

$$K(x|y) = \min_{i,p}\{l(p) : T_i(\langle p, y\rangle) = x\}. \tag{1.1}$$

The unconditional *Kolmogorov complexity* of $x$ is then defined as

$$K(x) = K(x|\epsilon), \tag{1.2}$$

where $\epsilon$ denoted the empty string. The variables $i$ and $p$ are usually combined into $\lambda = \langle i, p\rangle$, and the above definition is given in terms of the universal Turing machine $U(\langle\lambda, y\rangle) = T_i(\langle p, y\rangle)$, where the minimum is calculated over all binary strings $\lambda$. Every (self-delimiting) binary string is a valid program that, if halts, computes something. In [3], Chaitin demonstrates quite practical LISP simulation of a universal Turing machine and discusses the properties of algorithmic complexity.

More intuitive approach to complexity suggests that a complexity of two completely irrelevant objects must be the sum of complexities of each of them. In order for the algorithmic complexity to correspond to this approach, the binary program must be *self-delimiting*, such that two programs appended into a single string can be easily separated.

## 2 Binary versus extended alphabet

The reasoning behind the choice of binary description is quite clear. We need a description (or a programming language, or a computational model) that is:

(a) *Non-redundant, or syntactically incompressible.* Every symbol must bear maximum amount of information. There must be no correlation between symbols in the program, as any correlation would allow reduction of the length of the program. However, if a shorter program yields exactly the same object that a larger program, the shorter one may be considered as *semantically-compressed* version of the larger one.

(b) *Syntax-free.* Any combination of symbols of an alphabet must be a valid program. Then a simple probability measure can be introduced. Existence of a syntax also means that we assume some structure of an object beforehand, which gives incorrect complexity measure.

There is a contradiction between self-delimiting and syntax-free properties of a program. Two of the possible delimiting schemes either use a special closing symbol (LISP [2]), or a length-denoting structure in the beginning of the program [3]. Both introduce a bias in symbol probabilities.

(c) *Problem-oriented.* Binary numbers seem natural choice for information encoding, especially considering the current state of computer technology. However, they are not applicable to description of tri-fold or other odd-fold symmetry objects, since there is no finite binary representation of $1/(2n+1)$ numbers. Thus, the binary description of a Koch curve (Fig. 1.2), for example, is quite awkward. However, the curve is very elegantly encoded using 3-letter alphabet.

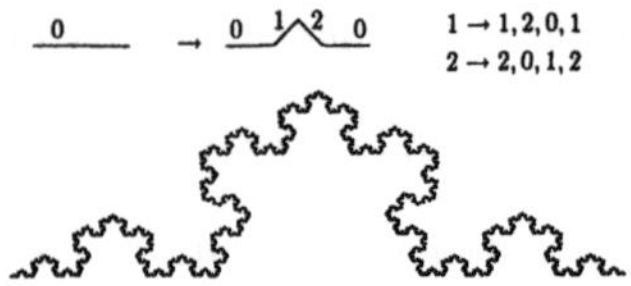


**Figure 1.1**: Ternary encoding of the Koch curve

Thus the extended alphabet is sometimes preferable to $\{0, 1\}$ alphabet.

Recursive structures, such as fractals, are conveniently described by a set of production rules, the L-systems, operating on a finite alphabet with $N$ letters. It was shown (see overview in [10], Section 3.2) that the *parallel derivation grammar* may have the same generating power as the formal grammar of the Chomsky type. According to the Church's Thesis, the recursively enumerable languages describe the largest set obtainable by any effective computational model. Remarkable illustration of this point is the computation performed by a cellular automaton (an example of an L-system) [12].

Thus, the L-system complexity measure, satisfying all the requirements above, is equivalent to the Chaitin's one and convenient for our purposes.

# 3 Parallel derivation systems

The parallel derivation systems (string rewriting systems, L-systems) have received due attention over the past years. Among many publications in this subject I would like to mention the comprehensive works by Book and Otto [1] and P. Vitányi [10].

An L-system is a string rewriting system. Let *alphabet* $A = \{a, b, \ldots\}$ be a finite set of elements, or *letters*. The *free monoid* $A*$ generated by $A$ is a set of all finite strings over $A$, i.e.

$$A* = \{\lambda, a, b, aa, ab, bb, ba, aaa, \ldots\},$$

where $\lambda$ is the empty string. Elements of $A*$ are called *words*. Let $A+ = A * \setminus \{\lambda\}$. A *rewriting system* is a set of *rules* $(u, v) \in A+ \times A*$, written as $u \longrightarrow v$.

We start from the initial string $S_0 \in A+$. At every consecutive state $n+1$ the string $S_{n+1}$ is obtained by simultaneous replacement of each letter $u \in S_n$ by a string $v$ which may be empty. The rewriting may depend on $l$ left and $r$ right neighbors of the letter — hence $(l,r)L$ notation. The resulting string $S_{n+1}$ is a concatenation of all $v$'s. The process is repeated thereby generating an infinite number of strings.

It is also possible to divide letters of the alphabet into *terminals* and *non-terminals* requiring the final strings to consist only of terminals. This action makes great attribution to the derivation power of an L-system.

# 4 Invariant strings and limit process

The Cantor set description can be generated by an L-system over a binary alphabet using starting string "1" and rules: $0 \to 000$, $1 \to 101$. Further, a geometrical meaning to the 0,1 symbols – a line of length $1/3^n$ colored white or black may be assigned. The infinite string 101000101000000000101... appears to be invariant under these generating rules.

Other examples of invariant strings are

a) the rabbit sequence 1011010110110... with the rules $0 \to 1$, $1 \to 10$,

b) Thue-Morse sequence 011010011001... with $0 \to 01$, $1 \to 10$,

c) Thue sequence 11011011111011011111... with $1 \to 110$, $0 \to 111$.

The invariance property of an infinite string $S_\infty$ means its self-similarity in the fractal-geometry sense: one can regenerate the whole string from a specially chosen piece by applying the rules (instead of just amplification). The subsequent finite strings $S_n$ in this context are the *averaged* description of the object. Application of the rules to every letter of $S_n$ may then be seen as a *refinement* of the structure of the object. Thus, for the Cantor set, $1 \to 101$ reveals the central gap, and $101 \to 101000101$ further gaps in the left and right segments.

We have seen that $S_n$ converges to $S_\infty$ in the refinement process. There has been observed [8, 9] another example of convergence in which the symmetry of the approximating (finite) strings in no way resembles the symmetry of the limiting structure (Fig. 1.4 (a)).

In this example, the generating process starts from a very irregularly shaped curve which suggests nothing about the final curve. After few steps of "refinement" the curve starts acquiring the right shape, however the convergence is only in some average sense. The final curve is a quite symmetric fractal-looking line having its own generating rules (and a fractal dimension of $D = 1.246477\ldots$) (Fig. 1.4 (b)).

By combination of the subtle analysis of the underlying model [6] and a brute force, it is possible to find the generating rules for approximating curves. There are 32 "letters" (numbers) in the *primary* alphabet. Each number is associated with a vector representing one of possible 14 directed line elements composing

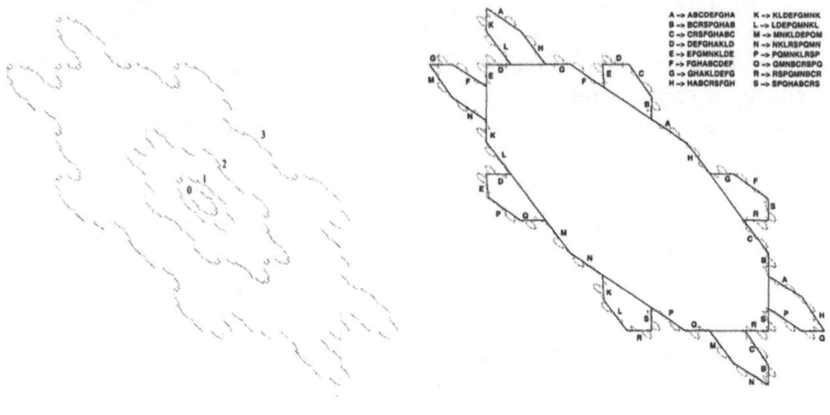


**Figure 1.2**: a) Successive approximations $S_n$ to a fractal curve $S_\infty$. b) $S_\infty$ with its own generating rules.

the curves,

$$\begin{array}{lllll} \{\begin{pmatrix}0\\0\end{pmatrix}_{3,7,10,15,19,23,27}, & \begin{pmatrix}1\\0\end{pmatrix}_{14,26}, & \begin{pmatrix}0\\1\end{pmatrix}_{18,25,31}, & \begin{pmatrix}1\\1\end{pmatrix}_{16,17,21}, & \\ \begin{pmatrix}2\\1\end{pmatrix}_{13,28}, & \begin{pmatrix}1\\2\end{pmatrix}_{30}, & \begin{pmatrix}2\\2\end{pmatrix}_{29} & \begin{pmatrix}1\\-1\end{pmatrix}_{1}, & \begin{pmatrix}0\\-1\end{pmatrix}_{2,12}, \\ \begin{pmatrix}-1\\0\end{pmatrix}_{11,20,24}, & \begin{pmatrix}-1\\-1\end{pmatrix}_{6,8,9}, & \begin{pmatrix}-1\\-2\end{pmatrix}_{0,4}, & \begin{pmatrix}-2\\-1\end{pmatrix}_{22}, & \begin{pmatrix}-2\\-2\end{pmatrix}_{5}\}, \end{array}$$

where the subscripts denote associations. The vectors can be grouped into sequences (*words*) thus forming the *secondary* alphabet.

$$\begin{array}{llll} A=(27,16,21) & B=(26,12,20,28) & C=(29,21,9) & D=(19,22) \\ a=(5,2,15) & b=(7,14,1,8) & c=(22,5,11) & d=(9,3) \\ E=(30) & F=(24) & U=(26,12,20,14,1,18) & o=() \\ e=(25) & f=(6) & u=(22,5,10,21,9) & \end{array} \tag{1.3}$$

Capital and small letter notation represent the symmetry of the curves with respect to the long diagonal line, as one can see by plotting sequences of vectors for each letter. It appears that these sequences are the forming blocks of the curves. Note that there is a *null* symbol in the alphabet.

Initial string for the 0-th level becomes

$$\begin{aligned} &14,1,18,5,2,15,9,3,6,22,5,10,21,9,24,19,22,27,16,21,26,12,20 \\ &\quad = adfuFDAU, \end{aligned}$$

and the rewriting rules become

| | | | | | |
|---|---|---|---|---|---|
| $A \to dfca$ | $B \to dfue$ | $C \to Ua$ | $D \to dba$ | $E \to oAUa$ | $F \to o$ |
| $a \to ACFD$ | $b \to EUFD$ | $c \to Au$ | $d \to ABD$ | $e \to Auao$ | $f \to o$ |
| $U \to Eue$ | $u \to U$ | $o \to dfUFD$ | | | |

(1.4)

Emergence of the short-diagonal symmetry of the curves is quite remarkable, since it is not at all reflected in the rewriting rules.

One can think that the shape of the limit curve $S_\infty$ is determined only by the rewriting rules (1.4). However, using the same rewriting rules and a different $S_0$, we obtained different fractal curves with other fractal dimensions. Thus the encoded information about the shape of $S_\infty$, and its properties in general, is shared by the initial string and the rewriting rules. The author finds it challenging to make a connection between the properties of $S_0$ or (1.4) and the final object.

The generating rules given above are not unique. They are given in the above form for illustration purposes, and have not been optimized or "spruced up" in any way. Other, neater and more concise rule systems are possible. How can one possibly find another set of rules, or enumerate all of them, or find the shortest one? (Of course, approach with "isomorphisms" would do part of the job). Generally, how to systematize the brute force approach?

# 5 Guess-Match brute-force algorithm

Given a set of strings $\{S_n\}$, we would like to answer a few questions. Is there a finite *dictionary*, i.e. a set of letter sequences acting as building blocks for all the strings (in a way that english clauses are made up from words)? Are these strings related by substitution rules? If yes, what is the substitution system having a particular property (e.g. the shortest)?

The answer to the first question is a different matter and is not addressed in the current paper. Here we assume that the dictionary and the rewriting rules exist.

From a sequence $\{s_n\}_{n=1}^{\infty}$ of strings we form a set of pairs $\mathcal{S} = \{\dots, (S_n, S_{n+1}), (S_{n+1}, S_{n+2}), \dots\}$. Generally, the algorithm should work for any set of pairs, and this property is used in the recursion.

Let $\mathcal{V} = \{u_n \to v_n\}$ be a set of rules. In the proper rule $r = u \to v \in \mathcal{V}$ the $u$ part is non-empty. The set $\{u_n\}$ must be *prefix-free*, so there is no confusion whether $0 \to \dots$ or $00 \to \dots$ is to be used in $\dots 00 \dots$.

Suppose the rule $r = u \to v$ is identified. Then there are pairs in $\mathcal{V}$ such that

$$(S_n, S_{n+1}) = (S'_n u S''_n, S'_{n+1} v S''_{n+1}) \tag{1.5}$$

The rule $r$ is added to $\mathcal{V}$ and all pairs of the form 1.5 are elimitated from $\mathcal{S}$ by replacing $(S_n, S_{n+1})$ by $(S'_n, S'_{n+1})$ and $(S''_n, S''_{n+1})$ in $\mathcal{V}$. The process of elimination is continued until either the set $\mathcal{S}$ is empty, or the only possible rule to be chosen is improper. In the first case the problem is solved and the rule set $\mathcal{V}$ is found. In the last case the algorithm simply backtracks.

The algorithm consists of two parts that call each other recursively. One part tries to determine a next rule to try, either enumerating all possible rules or using heuristics. The chosen rule is then applied to a pair of strings, and splits them into two pairs. The rule is applied at all possible places, and is eliminated if any of the resulting pairs is improper. The process is repeated until no more rules can be found. The total number of pairs may vary with every found rule, but the length of pairs in $S$ is reduced with each step. The total number of rules to be tried is also bounded. Thus the algorithm will finally terminate whether yielding result or not.

All above said works when initial strings do not contain a random component. In the contrary case, the search algorithm needs to be expanded to allow for nonexact matches of substrings. This is done by studying the probabilty distributions of substrings of approximately the same length, or by introducing the notion of *editing distance.* The probabilistic formulation of the problem goes beyond the goals of the current article and will be published elsewhere.

# 6 Conclusion

When measuring Chaitin's algorithmic complexity of a recursive structure, one has to choose the language properly reflecting the symmetry of the object, for example by using the extended alphabet rather than the binary one. L-system approach seems very convenient for description of the fractal structures due to its inherent recursive property and adequate expressive power, so the idea of defining algorithmic complexity in terms of length of rewriting rules is being conveyed by this article. When dealing with rewriting systems we may encounter interesting phenomena like convergence of the refinement process.

This paper demonstrates the first steps in studying applications of parallel derivation systems to measure the complexity of recursive objects.

# Bibliography


[1] BOOK, R., and F. OTTO, *String Rewriting Systems*, Springer-Verlag Berlin (1993).

[2] CHAITIN, Gregory J., *Algorithmic Information Theory*, Cambridge University Press (1987).

[3] CHAITIN, Gregory J., *Exploring Randomness*, Springer-Verlag (2001).

[4] GÁCS, Péter, John T. TROMP, and Paul M. B. VITÁNYI, "Algorithmic statistic", *IEEE Transactions on Information Theory* (2001).

[5] KOLMOGOROV, A. N., "Three approaches to the quantitative definition of information", *Problems Inform. Transmission* **1**, 1 (1965), 1–7.

[6] KOUPTSOV, K. L., J. H. LOWENSTEIN, and F. VIVALDI, "Quadratic rational rotations of the torus and dual lattice maps", *MPEJ*, 02-148, `http://mpej.unige.ch/cgi-bin/mps?key=02-148`, Genève mirror.

[7] LINDENMAYER, A., "Mathematical models for cellular interactions in development", *J. Theor. Biol.* **18** (1968), 280–315.

[8] LOWENSTEIN, J. H., and F. VIVALDI, "Anomalous transport in a model of hamiltonian round-off errors", *Nonlinearity* **11** (1998), 1321–1350.

[9] LOWENSTEIN, J. H., and F. VIVALDI, "Embedding dynamics for round-off errors near a periodic orbit", *Chaos* **10** (2000), 747–755.

[10] VITÁNYI, Paul M. B., *Lindenmayer Systems: Structure, Language, and Growth Functions*, Mathematisch Centrum Amsterdam (1980).

[11] VITÁNYI, Paul M. B., "Meaningful information", arXiv: *cs/0111053* (2001).

[12] WOLFRAM, S., *A new kind of science*, Wolfram Media, Inc. (2002).

# Chapter 15

# Iterons: the emergent coherent structures of IAMs

**Paweł Siwak**
Poznań University of Technology
60-965 Poznań, Poland
siwak@sk-kari.put.poznan.pl

## 1.1 Introduction

Iterons of automata [15, 18] are periodic coherent propagating structures (substrings of symbols) that emerge in cellular nets of automata. They are like fractal objects; they owe their existence to iterated automata maps (IAMs) performed over strings. This suggests that the iterating of (automata) maps is a fundamental mechanism that creates localized persistent structures in complex systems.

Coherent objects appear in the literature under various names; there are waves, building blocks, particles, signals, discrete solitons, defects, gliders, localized moving structures, light bullets, propagating fronts, and many other entities including those with the characteristic suffix –on, like fluxons, cavitons, excitons, explosons, pulsons, virons, magnons, phonons, oscillons, peakons, compactons, etc., etc.

In this paper we present a unified automaton approach to the processing mechanisms capable of supporting such coherent entities in evolving strings.

The iterons comprise of particles and filtrons. The particles, or signals, are well known [2, 3, 6, 10, 12] in cellular automata (CAs) where iterated *parallel* processing of strings occurs. They spread and carry local results, synchronize various events, combine information, encode and transform data, and carry out many other actions necessary to perform a computation, to complete a global pattern formation process in extended dynamical systems, or simply to assure stability of a complex system.

The filtrons form another new [14-18] class of coherent objects supported by IAMs. They emerge in iterated *serial* string processing which is a sort of recursive digital filtering (IIR filtering) [13]. In many aspects the filtrons are like solitons known from nonlinear physics; e.g. they pass through one another, demonstrate elastic collisions, undergo fusion, fission and annihilation, and form breathers as well as other complex entities. The first observation of filtron type binary objects has been done by Park, Steiglitz and Thurston [13]. They introduced the model called parity rule filter CA, and showed that it is capable of supporting coherent periodic substrings with soliton-like behavior. Now, a number of particular models exist that support filtrons. These are iterating automata nets [4], filter CAs [1, 8, 11], soliton CAs [7, 11, 21], higher order CAs [2], sequentially updated CAs [4], integrable CAs [5], iterated arrays [6], IIR digital filters or filter automata [14-18], discrete versions of classical soliton equations (KdV, KP, L-V) [5, 20, 22], and fast rules [1, 11]. Some new models like box-ball systems [20, 22] and crystal systems [9] were introduced quite recently.

All these models and their coherent structures can be described by automata and their iterons [14-18]. In our approach the automata are a sort of medium (or complex system) and the passing strings resemble disturbances that propagate throughout this medium (or system). We identify the iterons by particular sequences of automaton operations. One could call these sequences an active mode of automaton medium.

For the filtrons we consider one-way 1-d homogeneous net with an automaton $M$. The symbols of evolving strings are distinguished from the states of automata. In such nets the strings flow throughout automata and evolve, the IAM is performed over a string in a natural way. The filtrons are special $M$-segments that involve certain sequences of operations of automaton $M$. These sequences are related to a class of paths on the automaton state diagram.

For the particles we consider de Bruijn graph $G$ of their CA. This graph represents the constraints on possible sequences of elementary rules (ERs) used to update any segment of symbols. Again, the particles (as $G$-segments) are related to special sequences of ERs, which form paths on de Bruijn graph.

The characteristic sequences of operations associated with iterons lead to analytical tools in the analysis (and synthesis) of coherent structures in complex systems. So-called ring computation [14, 16, 17] has been already proposed to this end.

We present some examples of various phenomena of interacting filtrons, like multifiltron collisions, fusion, fission, and spontaneous decay or quasi-filtrons, and also some automata capable of supporting these events.

## 1.2 Automata and filtrons

Automata maps can be used to perform the processing of strings either in serial or in parallel manner. In both cases we characterize automata maps by some elementary operations; state-implied functions in serial IAMs and ERs in parallel IAMs.

### 1.2.1 Automaton and its state-implied functions

A Mealy type automaton $M$ with outputs and an initial state is defined to be a system $M = (S, \Sigma, \Omega, \delta, \beta, s_0)$, where $S$, $\Sigma$ and $\Omega$ are nonempty, finite sets of—respectively—states, inputs and outputs, $\delta: S \times \Sigma \rightarrow S$ is called the next state (or transition) function of $M$, and $\beta : S \times \Sigma \rightarrow \Omega$ is called the output function of $M$. Symbol $s_0 \in S$ denotes the initial state of $M$.

The automaton converts sequences of symbols (finite or infinite words). For each symbol $\sigma_i$ read from an input string it responds with an associated output symbol $\omega_i$ which is a consecutive element of the resulting string. The input string is read sequentially from left to right, one symbol at each instant $\tau$ of time, in such a way that $\delta(s(\tau), \sigma(\tau)) = s(\tau+1)$ and $\beta(s(\tau), \sigma(\tau)) = \omega(\tau)$ for all $\tau = 1, 2, \dots$ .

Next state and output functions of automata are presented in tables or in a graph form that is called the state diagram of automaton. For any $s \in S$ and $\sigma \in \Sigma$ that imply $t = \delta(s, \sigma)$ and $\omega = \beta(s, \sigma)$ in Mealy model, there is a directed edge on the graph going from node $s$ to node $t$, and labeled by $\sigma/\omega$. In Moore model the output function is defined by $\lambda(s(\tau)) = \omega$ thus the outputs $\lambda(s)$ are attached to the nodes.

To allow the iterations we apply a unified set of symbols $A = \Sigma = \Omega = \{0, 1, \dots, m\}$. Converted strings, when listed one under another, form an ST (space-time) diagram. It is often convenient to shift each output string by $q$ positions to the left with respect to its input string. We say in such a case that the shift $q$ has been applied to a diagram.

We also describe the automaton's operation by (state-implied) functions $f_s : A \to A$. They depend on states and are such that $f_s(a_i) = \beta(s, a_i)$ for all $s \in S$ and $a_i \in A$. The succession of outputs of the automaton are then: *next* $[f_s(a_i)] = f_{\delta(s, a_i)}(a_{i+1})$.

It is clear that the labeled path on state diagram of the automaton implied by any input string can be viewed as a sequence of operations $f_s$. Any input string determines the sequence of automaton operations; in a sense, the string and automaton interact.

### 1.2.2 Filtrons

We treat the automata lines as a medium or system and assume that its mode (idle or excited) decides on the existence or absence of coherent structures. Thus we use automata in the role of substring recognizers. The idea is as follows. Suppose that the automaton $M$ reads a string $\dots a_1 \dots a_L \dots$ . Each time when $M$ leaves a fixed (starting) state $s$ under a symbol $a_1$ this transition is treated as the beginning of a substring, and activation of $M$. Also, each time when $M$ enters some fixed state $t$ (lets call it final state) under a symbol $a_L$ we say that the end of the substring $a_1 \dots a_L$ is recognized, and $M$ is extinguished. The substring $a_1 \dots a_L$ is said to be the $M$-segment.

Consider now some special $M$-segments. We assume strings $\dots 0a_1 \dots a_L 0 \dots$ where symbol 0 represents a background; $\delta(s_0, 0) = s_0$. For given $M$ we choose initial state $s_0$ to be the starting state as well as the final state. In general case one can use another more complicated selection; e.g. the subsets of automaton states can be chosen as starting states and/or as final states, or even these sets can evolve in time.

Our basic coherent structure, the filtron is defined as follows [14, 15]. By a $p$-periodic filtron $a^t\_$ of an automaton $M$ we understand a string $a_1^t a_2^t \dots a_{L_t}^t$ of symbols from $A$ with $a_1^t \neq 0$, such that during the iterated processing of configuration $a^t$ = $\dots 0a^t\_0 \dots$ by the automaton $M$ the following conditions are satisfied for all $t = 0, 1, \dots$:

- the string $a^t\_$ occurs in $p$ different forms (filtron's orbital states), with $0 < L_t < \infty$,
- the string $a^t\_$ is an $M$-segment.

When a number of extinctions of given $M$ still occurs before the last element of the string segment $a^t\_$ is read by $M$, we say that $a^t\_$ is a multi-$M$-segment string. Multi-$M$-segment strings lead to complex filtrons.

### 1.2.3 The models that support filtrons

The first model shown to be capable of supporting coherent periodic substrings with soliton-like behavior was parity rule filter CA [13]. The PST (Park-Steiglitz-Thurston) model consists of a special ST-window (called here FCA window) and a parity update function. The string processing, $a^t \to a^{t+1}$, proceeds as follows.

Assume a configuration at time $t$, $a^t = \ldots\, a_i^t \,\ldots\, = \ldots 0\, a_1^t \ldots a_{L_t}^t\, 0 \ldots$, of elements from $A = \{0, 1\}$ such that: $0 \le t < \infty$, $-\infty < i < \infty$, $1 \le L_t < \infty$ and $a_1^t \neq 0$. The model ($f_{PST}$, $r$), with $r \ge 1$, computes the next configuration $a^{t+1}$ at all positions $i$ ($-\infty < i < \infty$) in such a way that: $a_i^{t+1} = f_{PST}\,(a_i^t,\, a_{i+1}^t, \ldots, a_{i+r}^t,\, a_{i-r}^{t+1},\, a_{i-r+1}^{t+1}, \ldots, a_{i-1}^{t+1}) = 1$ if $S_{i,t}$ is even but not zero, and otherwise $a_i^{t+1} = 0$; $S_{i,t}$ is the sum of all arguments (window elements). Zero boundary conditions are assumed, which means that the segment $a^t\_$ is always preceded in configuration $a^t = \ldots 0a^t\_0\ldots$ at the left side by enough zeros.

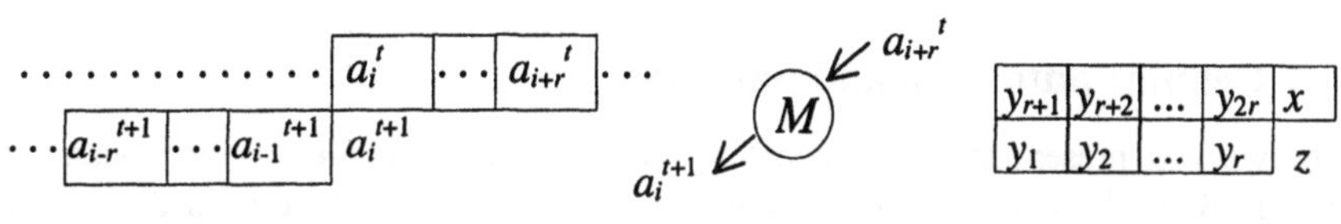

**Figure 1.** FCA window, and its automaton view: input ($x$), state ($y_i$) and output ($z$) variables.

The map $f_{PST} : A^{2r+1} \to A$ is a Boolean function. With the variables shown on the right in figure 1, the function $f_{PST}$ is $z = f\,(y_{r+1}, y_{r+2}, \ldots, y_{2r}, x, y_1, y_2, \ldots, y_r)$ and is given by: $z = y_1 \oplus y_2 \oplus \ldots \oplus y_{2r} \oplus x \oplus b$ where $b = \bar{y}_1 \cdot \bar{y}_2 \cdot \ldots \cdot \bar{y}_{2r} \cdot \bar{x}$ and $\oplus$ is *XOR* operation. The FCA window slides to the right, and $f_{PST}$ is a nonlinear function.

The PST model is an automaton [14, 16]. Processing of strings is performed in the cycles of operations (N, A, ..., A) or NA$^r$ where N is the negate operation and A is the accept operation. The processing starts with the first nonzero symbol $a_i = 1$ entering the window, and stops when the substring $*0^r$ (* is an arbitrary symbol) coincides with the cycle. Such set $\{*0^r\}$ is called a reset condition. The example of IAM for $r = 3$ is shown below. The applied shift is $q = 0$. The collision is nondestructive.

```
 0   ..1001----...111-----1111----.....................
 1   .....11------.1101----1111----....................
 2   ......1001----.1011----1111----...................
 3   .........11------11100001111-----.................
 4   ..........1001----110100101101----................
 5   .............110000101101001011------.............
 6   ..............10010100101101001001----............
 7   .................11100001111-----11------.........
 8   ..................110100101101----1001----........
 9   ...................101101001011----..11-------....
10   .....................1111----111------.1001----...
11   ......................1111----1101----...11----..
```

We show the filtrons on ST diagrams using a special convention. The symbol $0 \in A$ in a string represents a quiescent signal or a background, but sometimes it belongs to an active part of a string ($M$-segment). Thus, we use three different characters to present it on the ST diagrams. A dot "." denotes zeros read by the automaton, which is inactive. A dash "-" represents tail zeros of an $M$-segment, that is all consecutive zeros preceding immediately the extinction of the automaton. Remaining zeros are shown as the digit 0. Moreover, all those symbols that activate the automaton are printed in bold. This convention helps one to recognize whether any two filtrons are distant, adjacent or overlap and form a complex object.

In the last few years, there has been an increasing interest in looking for models that support filtrons. We mentioned them in introduction. For all of them the equivalent automata have been found [14-18]. Some examples of automata that represent the models which were based on FCA window are:

$z = y_1 \oplus \ldots \oplus y_{2r} \oplus \bar{x} \oplus \bar{y}_1 \ldots \bar{y}_r\, \bar{y}_{r+2} \ldots \bar{y}_{2r}\, \bar{x}$; ($r \geq 0$), for Ablowitz model [1],

$z = y_1 \oplus \ldots \oplus y_r \oplus y_{r+2} \oplus \ldots \oplus y_{2r+1} \oplus \bar{x} \oplus \bar{y}_1 \ldots \bar{y}_{2r+1}\, \bar{x}$; ($r > 0$), for Jiang model [11],

$z = (y_{r+1} = 0) \wedge (y_{r+1} + \ldots + y_\infty > y_1 + \ldots + y_r)$; ($r > 0$), for TS model [21],

$z = y_3 \oplus y_1\, y_4 \oplus y_2\, x$; ($r = 2$), for BSR-1 model given by formula 1 in [5], and

$z = y_4 \oplus y_1\, y_2 \oplus y_2\, y_3 \oplus y_5\, y_6 \oplus y_6\, x$; ($r = 3$), for BSR-23 model (formula 23 in [5]).

Some other automata for more recent models are given further.

## 1. 3 Cellular automata and particles

Now we will present the iterons that emerge in iterated parallel string processing performed in cellular automata. These are widely known [2, 6, 10] and called particles or signals. They are associated with some segments of strings. Usually, these are periodic objects which can be localized on some chosen area of ST diagram. In the simplest cases, when the background is steady, it is not difficult to extract and identify these objects. In the computing theory such particles are considered frequently. They are treated as functional components in the hierarchical description of a computation. The geometrical analysis of possible results of the interactions of particles dominates in this approach [6]. Another technique aimed at dealing with particles on periodic background was based on a filtering of ST diagrams by a sequential transducer [10]. Also, statistical analysis of ST diagram segments has been proposed [3]. However, there are many more complicated periodic entities. Some particles are like defects on a spatially periodic background or boundaries separating two phases of it. The defects can join into complexes. The background of what one would call particle can be periodically impure, and the particle itself may even contain some insertions. Such complicated boundary areas can be distant, adjacent or may overlap. Even more difficult is the case when neighborhood window is not connected (there are "holes" in a processing window). Moreover, in many cases the complexes of various entangled particles occur. This is why the description of particles and the complete analysis of their collisions results is not a trivial task [2, 3, 6, 10, 12].

### 1.3.1 Cellular automaton and its elementary rules

Cellular automata are defined by $CA = (A, f)$ where $A$ is a set of symbols called the states of cells, $f: A^n \rightarrow A$ is a map called the local function or rule of CA, and $n = 2r + 1$ is the size of neighborhood (or processing window) with $r$ left and $r$ right neighbors. Typically, especially when $|A| = 2$ and $r$ is small, the rule is determined by the number $\sum_{j=0}^{j=2^n-1} f(w_j) \cdot 2^j$ ; $w_j$ denotes the neighborhood state (contents of the window).

The 1-d CA model converts the strings of symbols. We denote them by $a^{\tau} = ..., a_i^{\tau}, a_{i+1}^{\tau}, a_{i+2}^{\tau}, ...$, and call the current configuration of a CA. The next configuration $a^{\tau+1}$ is a result of updating simultaneously all the symbols from $a^{\tau}$; for all $-\infty < i < +\infty$ we have $a_i^{\tau+1} = f(a_{i-r}^{\tau}, a_{i-r+1}^{\tau}, ..., a_i^{\tau}, ..., a_{i+r}^{\tau})$. The resulting global CA map $a^{\tau} \to a^{\tau+1}$ is denoted by $\gamma(a^{\tau}) = a^{\tau+1}$.

The function $f$ can be specified as the set of all $(n+1)$-tuples $(a_1, a_2, \ldots, a_{n+1}) \in A^{n+1}$, with $f(a_1, a_2, \ldots, a_n) = a_{n+1}$; these represent simply the single values of function $f$. Any such $(n+1)$-tuple is called here the elementary rule (ER) of the CA model. The sequences of ERs will be used to recognize the particles of CAs.

Note that we have a hierarchy of processing in CAs; there are ERs, sets of ERs (level of particles), local function $f$, global function $\gamma$, and iterations of function $\gamma$.

### 1.3.2 De Bruijn graph

De Bruijn graphs are used to build the automata that scan sequentially a string to detect some specific substrings of symbols. For a CA with $n$-wide window ($n = 2r+1$), the Moore automaton $G_n = (A^n, A, A, \delta, \lambda, s_0)$ that mimics the sliding window is defined by the next state function $\delta((a_1, ..., a_n), a_{n+1}) = (a_2, ..., a_{n+1})$, and the output function $\lambda(a_1, ..., a_n) = f(a_1, ..., a_n)$ identical with the rule $f$ of CA.

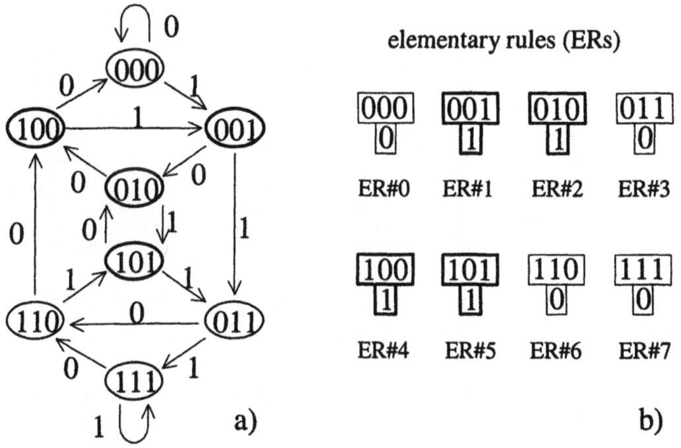


**Figure 2.** CA of rule 54; (a) de Bruijn graph as possible sequences of ERs, (b) the set of ERs.

We will use $G_n$ to detect the possible strings of ERs associated with particles. By a $p$-periodic particle $a^{\tau}\_$ of an automaton $CA$ we understand a string $a_1^{\tau} a_2^{\tau} ... a_{L_\tau}^{\tau}$ of symbols from $A$ such that during the iterated $CA$ processing the configuration $a^{\tau} = ...ua^{\tau}\_v...$ occurs in $p$ different forms. The strings $...u$ and $v...$ represent regular (spatially periodic) areas. The starting states of $G_n$ related with area $...u$, and its final states with area $v...$ (the roles of these sets can interchange) define the $G$-segments similarly to $M$-segments. However, all outputs of automaton $G_n$ can be determined simultaneously for the entire configuration. $G_n$ expresses the constraints on possible sequences of ERs and does not imply sequential detection of substrings of symbols or ERs involved in CA processing. Thus, for $G$-segments we use undirected graphs $G_n$.

Consider the CA with the rule 54 [3, 12]. Its local function $a_2' = f(a_1, a_2, a_3) = f(w_j)$ is given by $a_2' = 1 \Leftrightarrow w_j \in \{001, 010, 100, 101\}$. Below, we show three ST diagrams of CA processing. The second ST diagram is a recoded (in parallel, not sequentially) version of the first, and shows all ERs which are involved in processing. The position of each ER# is at its resulting symbol. We assumed four spatially periodic segments in the strings: (0111), (0001), (0) and (1). These are identified by the following sequences of ERs: $z = (5,3,7,6)$, $x = (4,0,1,2)$, o = (0) and | = (7), respectively. Thus we have four regular ER areas $\{z, x,$ o, $|\}$. The boundaries between these areas are shown in the third diagram. They are recognized as the paths between starting and final sets of states of the de Bruijn graph; these sets correspond to regular areas $\{z, x,$ o, $|\}$. These paths form $G$-segments on the undirected de Bruijn graph.

```
  1234567890123456789012345678  1234567890123456789012345678  1234567890123456789012345678

0 11.111...1...1..1...1...1...  2401253765376400137653765376  xxxx25zzzzzzz4oo1zzzzzzzzzzz
1 ..1...1.111.111111.111.111.1  7653764012401241240124012401  zzzzz64xxxxxxx41xxxxxxxxxxxx
2 .111.111...1......1...1...1.  0124012537653777765376537653  xxxxxx25zzzzz7||7zzzzzzzzzzz
3 1...1...1.111....111.111.111  5376537640124000012401240124  zzzzzzz64xxx40oo01xxxxxxxxxx
4 11.111.111...1..1...1...1...  2401240125376400137653765376  xxxxxxxx25zzz4oo1zzzzzzzzzzz
5 ..1...1...1.111111.111.111.1  7653765376401241240124012401  zzzzzzzzz64xxx41xxxxxxxxxxxx
6 .111.111.111......1...1...1.  0124012401253777765376537653  xxxxxxxxxx25z7||7zzzzzzzzzzz
7 1...1...1...1....111.111.111  5376537653764000012401240124  zzzzzzzzzzz640oo01xxxxxxxxxx
8 11.111.111.111..1...1...1...  2401240124012400137653765376  xxxxxxxxxxxxx4oo1zzzzzzzzzzz
9 ..1...1...1...1111.111.111.1  7653765376537641240124012401  zzzzzzzzzzzzz641xxxxxxxxxxxx
0 .111.111.111.1....1...1...1.  0124012401240137765376537653  xxxxxxxxxxxxx1377zzzzzzzzzzz
1 1...1...1...111..111.111.111  5376537653765240012401240124  zzzzzzzzzzzz524001xxxxxxxxxx
2 11.111.111.1...11...1...1...  2401240124013764137653765376  xxxxxxxxxxx13z6413zzzzzzzzzz
3 ..1...1...111.1..1.111.111.1  7653765376524013640124012401  zzzzzzzzzz52xx1364xxxxxxxxxx
4 .111.111.1...111111...1...1.  0124012401376524125376537653  xxxxxxxxx13zz524125zzzzzzzzz
5 1...1...111.1......1.111.111  5376537652401377776401240124  zzzzzzzz52xx137||764xxxxxxxx
6 11.111.1...111....111...1...  2401240137652400001253765376  xxxxxxx13zz5240oo0125zzzzzzz
7 ..1...111.1...1..1...1.111.1  7653765240137640013764012401  zzzzzz52xx13zz4oo1zz64xxxxxx
8 .111.1...111.111111.111...1.  0124013765240124124012537653  xxxxx13zz52xxxx41xxxx25zzzzz
9 1...111.1...1......1...1.111  5376524013765377776537640124  zzzz52xx13zzzz7||7zzzz64xxxx
0 11.1...111.111....111.111...  2401376524012400001240125376  xxx13zz52xxxx40oo01xxxx25zzz
1 ..111.1...1...1..1...1...1.1  7652401376537640013765376401  zz52xx13zzzzzz4oo1zzzzzz64xx
```

**Figure 3.** Three ST diagrams of rule 54 CA processing; string evolution, ERs involved in CA processing, and specific sequences of ERs (particles) that separate areas $x$, $z$, o and |.

Let us show some $G$-segments using the associated sequences of ERs. For a single particle we have: $a^{\tau}$ = ..xx**25**zzz.., $a^{\tau+1}$ = ..zzz**64**xx.., and for a complex particle: $a^{\tau}$ = ..xxx**41**xxx.., $a^{\tau+1}$ = ..zz**7**||**7**zz.., $a^{\tau+2}$ = ..x**40**oo**01**x.., $a^{\tau+3}$ = ..zz**4**oo**1**zz.. But other objects may be transient like ..zz**41**xx.. and ..zz**7**|||.. or irregular in some other way; e.g. they may annihilate or generate new entities.

In general, one has to determine how to filter out the particles, or which sequences of ERs are to be treated as regular areas to play the role of starting and final state sets (background) for $G$-segments. From the example it is seen that the central entity ..xxx**41**xxx.., which is known as $g_e$ particle of the CA 54 [3, 12], should be rather considered as a complex object, since it is a kind of breather – arises from interaction between two simple particles represented by ERs ..xxx**40**ooo.. and ..ooo**01**xxx.. .

## 1.4 Some automata and iteron phenomena

In this section we present some filtrons phenomena and IAMs of some special automata. Let us start with multifiltron collisions.

In figure 4 we show solitonic collisions of filtrons that vibrate in a way. These are supported by automata that perform the cycles of operations. Similar cycles were applied in the PST model. Automaton $M_{11}$ has the cycle (NNNNAAA) (N is the negate operation and A is the accept operation) and reset condition {***0000}, while automaton $M_{12}$ has the cycle (NNNAAAA) and the same reset condition. The cyclic processing starts with the first encountered nonzero element $a_i = 1$.

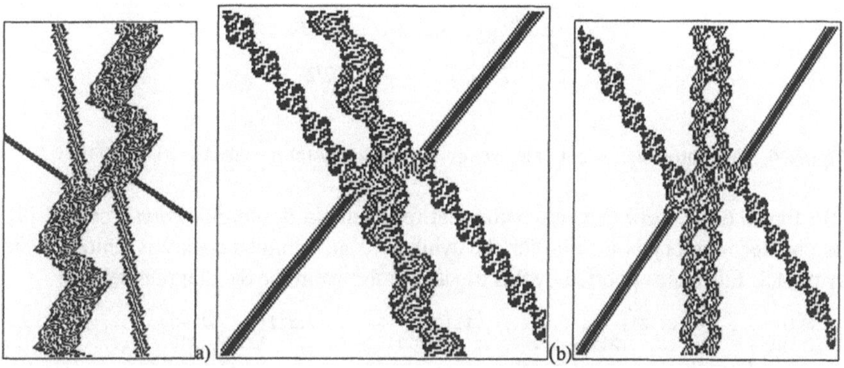


**Figure 4.** Colliding filtrons of automaton $M_{11}$ (a), and of $M_{12}$ – two ST diagrams (b); $q = 1$.

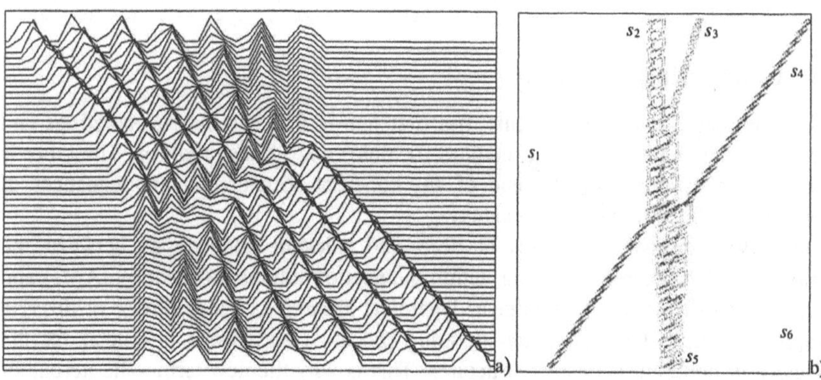


**Figure 5.** (a) Multi-object collision of filtrons of $M(16, 22)$; $q = 1$ [18]. (b) Quasi-filtron, $q = 2$.

In figure 5 (a) we show the multivalued filtrons over alphabet $A$ with $|A| = 17$. These collision is supported by automaton $M(16, 22)$. $M(m, n)$ is given by: $s' = \delta(s, \sigma) = s + \min(n - s, \sigma) - \min(s, m - \sigma)$, $\omega = \beta(s, \sigma) = \sigma + \min(s, m - \sigma) - \min(n - s, \sigma)$. These automata are equivalent to a box-ball system with carrier [16, 20].

Figure 5 (b) shows fusion, fission and a quasi-coherent object. Such quasi-filtrons remain coherent for a long number of iterations, and at some moment they decay. Here, two filtrons $s_2$ and $s_3$ get into fusion into an unstable object. After 160 iterations it decays onto filtrons $s_5$ and $s_6$. In the meantime this quasi-filtron takes a part in solitonic collision with two other filtrons ($s_1$ and $s_4$) approaching from its both sides.

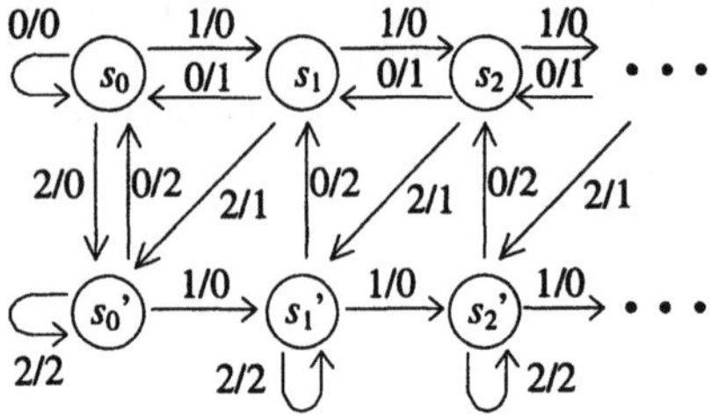


**Figure 6.** Automaton equivalent of a bozonic-fermionic crystal model; $A = \{0, 1, 2\}$[18].

In figure 6 we show the automaton that represents a *bozonic-fermionic* crystal [9]. Its counter memory is infinite for the symbol 1, and simultaneously is finite for the symbol 2. Filtrons supported by this model are shown in the ST diagram below.

```
0 .211---....1-.....211---....2-......111021---2-.......111---..21--2-................
1 ....211---..1-.......211---..2-........11021102--........111---.2102-...............
2 .......211---1-.........211---2-.........11--2121---........111---212--.............
3 ..........21101---.........21102--.........11--2-211---........11100221--...........
4 .............2-111---.........2121---........1102-..211---........1112021---........
5 ..............2-..111---........2-211---.......112--...211---........1210211---.....
6 ...............2-....111---......2-..211---......121--....211---......1-21--211---..
7 ................2-......111---....2-....211---....1-21--.....211---....1-.21--.211-.
```

## 1.5 Concluding remarks

In our approach, two issues occur. The first one is that automaton operations and converted strings interact like a medium (or field) and its disturbances. The other is that we identify the spatial extend of any coherent structure by the sequences of operations of the underlying automaton.

Automaton approach indicates at deep and relevant connections between the computational processes occurring within the nets of automata (given by IAMs) and the equations of motion of nonlinear dynamical systems or the behavior of discrete complex systems. The automaton iterating process over strings is crucial for the existence of coherent persistent structures. This is why the new term—the iterons of automata—has been proposed. The iterons seem to be as fundamental as are fractals.

The behavior of iterons is very rich and strongly depends on the underlying automaton. We have shown here only few examples. Others, like bouncing filtrons, trapped colliders, cool filtrons, repelling filtrons, annihilation, are shown in [14-18].

Potential applications of the theory of iterons cover, among others, simulating nonlinear physics phenomena, future solitonic computations [19], complex systems behavior analysis and synthesis, photonic transmission, and solitary waves prediction.

## References


[1] M. J. Ablowitz, J. M. Keiser, L. A. Takhtajan. A class of stable multistate time-reversible cellular automata with rich particle content. *Physical Review* A, **44**, No. 10, (15 Nov. 1991), pp. 6909-6912.

[2] A. Adamatzky. *Computing in Nonlinear Media and Automata Collectives.* Institute of Physics Publishing, Bristol, 2001.

[3] N. Boccara, J. Nasser, M. Roger. Particle-like structures and their interactions in spatio-temporal patterns generated by one-dimensional deterministic cellular-automaton rules. *Physical Review* A, **44**, No. 2 (15 July), 1991, pp. 866-875.

[4] C. L. Barrett, C. M. Reidys. Elements of a theory of computer simulation I: sequential CA over random graphs. *Applied Mathematics and Computation*, **98**, 1999, pp. 241-259.

[5] M. Bruschi, P. M. Santini, O. Ragnisco. Integrable cellular automata. *Physics Letters* A, **169**, 1992, pp. 151-160.

[6] M. Delorme, J. Mazoyer (eds.). *Cellular automata.* A Parallel Model. Kluwer, 1999.

[7] A. S. Fokas, E. P. Papadopoulou, Y. G. Saridakis. Coherent structures in cellular automata. *Physics Letters A*, **147**, No. 7, 23 July 1990, pp. 369-379.

[8] A. H. Goldberg. Parity filter automata. *Complex Systems*, **2**, 1988, pp. 91-141.

[9] Hikami K., Inoue R. Supersymmetric extension of the integrable box-ball system. *Journal of Physics A: Mathematical and General*, **33**, No. 29 (9 June 2000), pp. 4081-4094.

[10] W. Hordijk, C. R. Shalizi, J. P. Crutchfield. Upper bound on the products of particle interactions in cellular automata. *Physica* D, **154**, 2001, pp. 240--258.

[11] Z. Jiang. An energy-conserved solitonic cellular automaton. *Journal of Physics A: Mathematical and General*, **25**, No.11, 1992, pp. 3369-3381.

[12] B. Martin. A group interpretation of particles generated by 1-d cellular automaton, 54 Wolfram's rule. LIP Report No. RR 1999-34, Ecole Normale Superieure de Lyon, pp. 1-24.

[13] J. K. Park, K. Steiglitz, W. P. Thurston. Soliton-like behavior in automata. *Physica* D **19**, 1986, pp. 423-432.

[14] P. Siwak. Filtrons and their associated ring computations. *International Journal of General Systems*, **27**, Nos. 1-3, 1998, pp. 181-229.

[15] P. Siwak. Iterons, fractals and computations of automata. In: D. M. Dubois (ed.), *CASYS'98* AIP Conference Proceedings **465**. Woodbury, New York, 1999, pp. 367-394.

[16] P. Siwak. Soliton-like dynamics of filtrons of cyclic automata, *Inverse Problems*, **17**, No. 4 (August 2001), pp. 897-918.

[17] P. Siwak. Anticipating the filtrons of automata by complex discrete systems analysis. In: D. M. Dubois (ed.), *CASYS'2001* AIP Conference Proceedings **627**. Melville, New York, 2002, pp. 206-217.

[18] P. Siwak. Iterons of automata. In: *Collision-based computing*. A. Adamatzky (ed.), Springer-Verlag, London. 2002, pp. 299-354.

[19] K. Steiglitz. Time-gated Manakov spatial solitons are computationally universal. *Physical Review* A, **63**, 2000.

[20] D. Takahashi, J. Matsukidaira. Box and ball system with carrier and ultradiscrete modified KdV equation. *Journal of Physics A: Mathematical and General*, **30**, No.21, 1997, pp. L733-L739.

[21] D. Takahashi, J. Satsuma. A soliton cellular automaton. *Journal of the Physical Society of Japan*, **59**, No. 10, October 1990, pp. 3514-3519.

[22] T. Tokihiro, A. Nagai, J. Satsuma. Proof of solitonical nature of box and ball systems by means of inverse ultra-discretization. *Inverse Problems*, **15**, 1999, pp. 1639-1662.

# Chapter 16

# Criticality of the Brain and Criticality of Art

**Igor Yevin**
Mechanical Engineering Institute, Russian Academy of Sciences
yevin@list.ru

## 1. Criticality of the Brain

The ability of our brain to respond to small extrinsic or intrinsic perturbations points out that the brain as a complex system is operating close to instability, or criticality, because any system at the critical state has a very high sensitivity to tiny perturbations [Haken 1996]. Per Bak gives another reason why the brain should be critical: the input signal must be able to access everything that is stored in the brain. The brain cannot be in subcritical state. In this case input signal would be access to only a limited part of information. But the brain cannot be supercritical either: in this case any input would cause an explosive process in the brain, and connect the input with everything that is stored in the brain [Bak 1996]. Hence, the waking brain must operate strongly at the critical state, where a neural network reveals Weber-Fechner logarithmic law and Steves power law [Kinouchi 2006]. The critical point maximizes information transmission within a neural network.

## 2. Criticality of Art.

### 2.1. Instability of the Compositional Balance in Painting

Pictorial balance is the most important principle of compositional design in painting, sculpture, and architecture. R.Arnheim defined pictorial balance as the appropriate distribution of optical weight in a picture [Arnheim 1964]. Optical weight refers to the perception that some objects appear heavier than others. Dark colors, unusual shapes, larger objects, and unstable elements look heavier than light colors, regular shapes, small objects, and stable elements of a picture. Compositional balance is achieved when pictorial elements are grouped in such a way that their perceptual forces compensate one another.

Example of compositional balance between unstable state of the small girl staying on the ball and stable state of the athlete sitting on the cube on Pablo Picasso's painting *"The Young Girl on the Ball"* is shown on Figure 1.

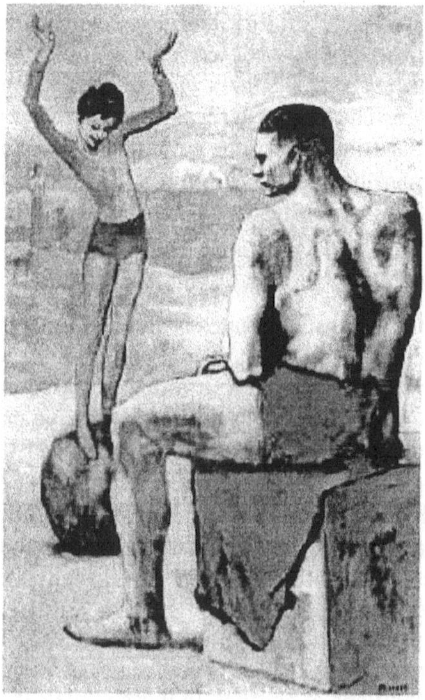

**Figure 1**. Compositional balance P.Picasso's painting *"The Young Girl on the Ball"*. Pushkin's Museum of Fine Art, Moscow, Russia

Compositional balance principle is a holistic principle resulting from the joint effects of the physical components of a stimulus field and therefore can not be afforded by any one element alone.

D.Chek Ling Ngo and L.Seng Teo proposed an objective measure for quantitative evaluation screen layouts balance for computer graphic but this method does not fit for investigation of painting pictorial balance [Chek Ling Ngo D 2001].

More promising for painting is empirical approach based on eyes-movements studies. As one gathered information from a picture, the eyes move in rapid jumps or succades followed by pauses of fixation. The average duration of fixation is typically between 200 and 300 milliseconds. The map of the location of fixations, called a scanpath, provides a graphic record of how information is selected and processed perceptually as a viewer looks at a picture.

One of the first eye-movement investigations of pictorial balance in painting was the work by Nodine in 1982 [Nodine 1982]. Nodine created a less-balanced version of each examined painting by cutting a copy of original apart and rearranging the pictorial elements. Comparison of fixation times of original versus altered version of paintings revealed that a change in the balance structure of the compositions produces a change in the pictorial weights (by changes in viewing time) of the key elements of the paintings.

As well known, balancing state may be stable or unstable. No doubt, R.Arnhiem considered compositional balance in painting as a stable state. Indeed, he wrote in his book "Art and Visual Perception": "It must be remembered that visually, just as physically, balance is the state of distribution in which everything has come into standstill.....In physics the principle of entropy, also known as the second law of thermodynamics, asserts that in any isolated system each state represents an irreversible decrease of active energy. The Universe tends toward a state of equilibrium, in which existing asymmetries of distribution will be eliminated" [Arnheim1964].

Of course, compositional balance in painting is always unstable, because adding or removal of any element in a canvas upsets this balance

## 3.2. Instability in Literature

Ability to excise control over unstable states is a basis of many circus arts (for instance, rope walking) and, of course, of ballet art. In other types of choreographic art - folk, ball and modern dances the unstable elements are also widely introduced, though very frequently they do not have such a perfectly expressive nature as in ballet. A great deal of elements of figure skating, gymnastics, acrobatics also saturated by unstable in their nature states.

Before we proceed to the discussion of instability in literature it is necessary to dwell on the meaning of this conception as regards the literary genre because generally it does not coincide with the mechanical meaning of instability that we deal with in the art circus and choreography.

The most adequate mathematical apparatus for the description of situation in belles-lettres may be the theory of games which is nowadays widely used in various models of behavior basically in conditions of conflict. In these models the parties making decision are called “players” and the action they choose are called “strategies”. When two players take part in a game, any pair of strategies is called “situation”.

In the theory of games the definition of instability named “the Nash”, after American mathematician John Nash. According to the Nash, some situation becomes apparently unstable, because it is about to break down, either because factors leading to that possibility exist or because one of the players obtains the best result by means of a unilateral choice of his strategy.

For example, a hare and a wolf living together is an unstable situation. Moreover, the essence of instability in the theory of games and in the mechanical models, for instance, of balletic art is the same: Unstable states are momentary by their very nature and rapidly break down. But the causes of their short duration are, certainly, different.

Now let’s look at instability in literature. Let us start with fiction, the literature of an entertaining genre. When in a detective story a crime is committed, the criminal according to the Nash appears to be in unstable position: because at this moment his life at large is incompatible with penal laws. The duration of the instability depends on adroitness and skill of both the criminal and organs of justice.

In general, instability in literature of an entertaining genre can be described as follows: Some value (life, honor, wealth, or power, for example) is in danger, but owing to personal traits of the heroes of the books and sometimes to favorable coincident this value is preserved.

This does not mean that instability can be met only in fiction; it is widely presented in classical literature as well. In Shakespeare's "Hamlet" the main character learns the name of his father's killer at the beginning of the first act, but he takes vengeance only at the end of the tragedy, thus, unstable state, King Claudius' life is preserved throughout the play.

In "Hamlet" there are instabilities in interpersonal relations that can be precisely defined with the help of the notion frustration from the theory of spin glasses [Toulouse 1979]. Let us consider the central characters of the play: King Claudius, Queen Gertrude, and Prince Hamlet.

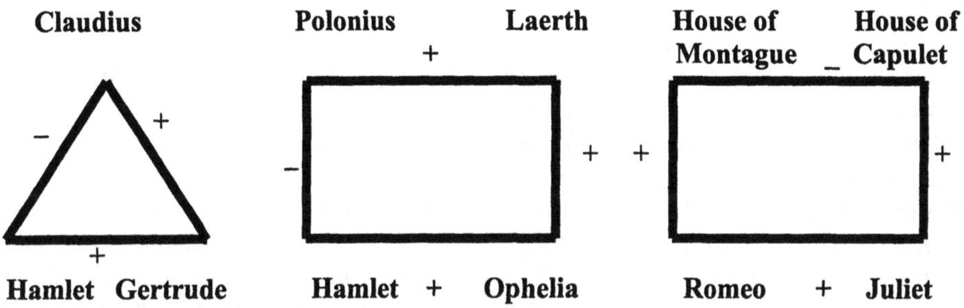


**Figure 2.** Frustrating bonds in plays by W. Shakespeare.

The triangle formed by the interrelations between these personages is obviously a frustration configuration (Figure 2a) and therefore this structure is metastable. Another frustrating contour is Hamlet's relations with Polonius's family (Figure 2b) at the beginning of the play. Shakespeare often brought such types of instabilities into his plays. Figure 2c is a graph of interrelations in "Romeo and Juliet" which is also frustrating configuration.

If we carefully analyze all the examples of unstable states in fine art considered above we may conclude that these instabilities perform the functions of the order parameter and subordinate the whole composition of the artwork.

The order parameter can be compared with a well-known conception of "dominant "in the studies of art. The most typical definition of the concept of dominant is [Vygotsky, 1971]:

*"Any story, poem, or picture is a complex unity made up of absolutely different elements that are differently organized and are in different hierarchic subordination and connection; in that complex unity exists a prevailing moment which determines the structure of the whole story, the meaning and the purpose of each part of the* story".

L.S. Vygotsky has brilliantly demonstrated the validity of the synergetical slaving principle through the example of I.A. Bunin's story "Light Breathing". Chronological progression of the events depicts a distressing, tragic story of a provincial life. Actually, the writer arranges all the events in such a way that the burdens of life are gone. All non-chronological skillful leaps from one event to another used to develop the plot are aimed at softening and eliminating the spontaneous impression of the events taking place and forming quite opposite impression.

While reading the story "Light Breathing" L.Vygotsky and his co-workers registered breathing. It turned out that even when reading about the murder and death of a hero the reader breathed lightly and freely, as if he did not perceive the horrible events but some liberation from the horror.

No doubt, this psychological state of lightness and freedom is unstable in this story.

## 3. Long Range Correlations in Artworks

Recent investigations show that the most interesting peculiarity of our visual world is the scale invariance of natural images. Mandelbrot have shown In the $1970^{th}$ that many of natural patterns: landscapes, mountains, coastlines, clouds, trees and so on are fractal [Mandelbrot 1982].

Natural landscapes are prime examples of complex systems with fractal properties. What are the general principles governing the formation of landscapes? Mathematical models in geomorfology using the theory of self-organized criticality explain fractal features of natural landscapes. According to Per Bak all the variability of Earth landscapes (including mountscapes) can be regarded as an self-organized critical phenomenon [Bak 1996]. .

R.Voss and J.Wyatt estimated fractal dimensions of early Chinese landscape painting analyzed of digital image intensities J(x,y) [Voss 1993]. Space representation changed drastically during Sung dynasty (960-1279 A.D.), when landscape painting reached the highest development in Chine. It turned out that landscapes painted at the beginning of this period have fractal dimension in the range 1.3-1.4, while the later paintings have fractal dimension in the range 1.07-1.13.

The paintings of US artist Jackson Pollock possess fractal properties. Richard Taylor and his colleagues analyzed Pollock painting produced between 1943 to 1952 and shown that fractal dimension of his works increased from 1.1 to 1.7 [Taylor 2003].

There are some preliminary data indicating that color palette in painting one can trace some characteristics that are subordinated to power low of distribution. In a painting were investigated some portions of areas painted in one color. From the reproduction of canvases were cut out the places visually perceived as one-colored and these places were sorted by color. Then each group of one color was weighted upon analytical scales. So far as the paper upon which were printed the reproductions was uniform in thickness, the weight of each heap appeared to be proportionate to a total area of the entire number of pieces in one heap.

It turned out, that in the most different pictures the colors occupying approximately 95 per cent of pictures area and about two-thirds of its color range can be satisfactory described by power law formula [Orlov 1980].

The Harvard linguistics George Kingsley Zipf was the first, who revealed power law in rank distributions of words in English language texts, including literature works.

$$P = \frac{a}{r}$$

where a - constant and r – rank.

A.Schnkel, J.Zhang , and Y.Zhang explored long-range correlations in various human writings [Schenkel 1993]. The Bible turned out the closest to ideal power law.

## 4. The Wundt Curve

Arousal response curve well known as the Wundt curve looks like an inverted 'U' shaped curve, is sketched in Figure 5. Berlyne refers to the Wundt curve as a hedonic function" [Berlyne, 1971].

The relationship between "arousal potential" of artworks and "hedonic value" can be explained from the theory of criticality standpoint, more exactly, the theory of second order phase transition. Order parameter equation for the second order phase transition can be written as a "cusp" catastrophe

$\Phi^3 - (K_c - K)\Phi - C = 0$

The "arousal potential" (novelty, unexpetedness and so on) $\Phi$ in this model is the order parameter. Sophisticated measures of novelty have been developed by several researchers (see, for instance,[Kohonen 1993].

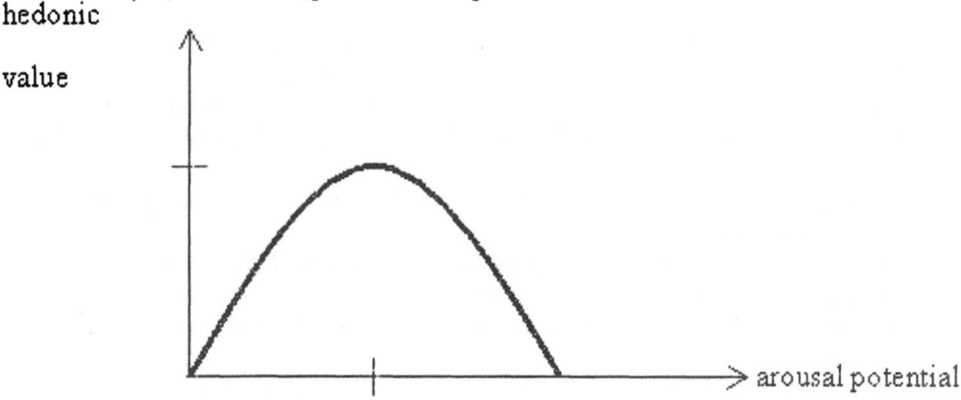


**Figure 3.** The Wundt curve.

"Hedonic value" K plays the role of control parameter (like temperature in Ginzburg-Landau model of phase transition).

## 1.5. Mutual Information between the Brain and Art

Mutual information I(A,B) between the artwork and the brain could be written using well-known Shannon expression:

$$I(A,B) = H(A) + H(B) - H(A,B)$$

H(B) - entropy of the brain is very high at critical state
H(A) -entropy of the artwork is very high if artwork is at critical state
H(A,B) -mutual entropy of the brain and the artwork is very small when both are at critical state because mutual correlations between the brain and the artworks are very high.

Therefore mutual information I(A,B) reaches the maximum magnitude when both the artwork and the brain are at critical state.

## 6. Conclusion

All natural complex system, such as biological, ecological, economical exist at the edge of chaos and order. In the theory of complexity has been able to show that human brain and behaviour also are at critical point. Now we can declare that any masterpiece and utstanding artwork exists near critical point and art as a whole exhibits all properties of self-organization: non-equilibrium phase transitions [Yevin 2001], multystability under perception of ambiguity [Yevin 2000], etc. Perceiving critical phenomena in artworks human brain have to operate near criticality. Therefore, the art should be regarded as a tool for supporting human brain at critical state.

## References

Haken, H., 1996, *Principles of Brain Functioning*, Springer, (Berlin)
Bak, P., 1996, *How Nature Works*, Copernicus (New York)
Kinouchi, O., & Copell, M., 2006, Physics of Psychophysics: Dynamics Range of Excitable Networks is Optimized at Criticality. ArXiv:q-bio. NC/0601037 v1 23Jan. 2006
Arnheim, R., 1964, *Art and Visual Perception*, University of California Press( Berkeley and Los Angeles)
Chek Ling Ngo, D., Seng Ten L. Byrne J.G. *A Mathematical Theory of Interface Aesthetics* Visual Mathematics N 1, 2001
Nodine, C. and McGinnis,J., 1982, *Artistic Style, Compositional Design, and Visual Scanning. Visual Art Research*, 12, pp.1-9,
Toulouse, G., 1979, *Theory of the Frustrational Effect in Spin Glasses. Communications on Physics*, 2, 115-119
Vygotsky, L.S., 1971, *The psychology of the Art*, (Cambridge. Massachusetts. London)
Mandelbrot, B., 1982,*The Fractal Geometry of Nature*, Freeman (San Francisco)
Voss, R., Wyatt, J., 1993, *Multifractals and the Local Connected Fractal Dimension: Classification of Early Chinese Landscape Paintings*. In: A.J.Crilly, R.A.Earnshaw, H.Jones, eds., *Applications of Fractals and Chaos*, Springer (Berlin)
Taylor, R., 2003, *Fractal Expressionism – Where Art Meets Science*, In: J.Casti and A.Karlqvist (Eds). *Art and Complexity*. Elsevier, pp. 117-144
Orlov, Yu., 1980, *Unvisible Harmony."Chislo i Mysl"*, No3, Znanie (Moscow) (in Russian)
Schenkel, A., Zhang , J., Yi-Cheng Zhang, 1993, *Long-Range Correlation in Human Writings, Symmetry: Culture and Science*. 4(3) 229-241.
Berlyne, D., 1971, *Aesthetics and Psychobiology*, Appleton-Century-Crofts (New York).
Kohonen, T.., 1993, *Self-Organization and Associative Memory*, 3rd ed., Springer (Berlin).
Yevin, I., 2000, *Ambiguity and Art*, Visual Mathematics, 2, No. 1
Yevin, I., 2001, *Complexity theory of art: Recent investigations*, InterJournal

# Chapter 17

# The Visual Complexity of Pollock's Dripped Fractals


**R.P. Taylor,[1] B. Spehar,[2] C.W.G. Clifford[3] and B.R. Newell[4]**
[1]Physics Department, University of Oregon, Eugene, USA
[2]School of Psychology, University of New South Wales, Australia
[3]School of Psychology, Sydney University, Australia
[4]Department of Psychology, University College London, UK


## 1. Introduction

Fractals have experienced considerable success in quantifying the complex structure exhibited by many natural patterns and have captured the imagination of scientists and artists alike [Mandelbrot]. With ever widening appeal, they have been referred to both as "fingerprints of nature" [Taylor et al 1999] and "the new aesthetics" [Richards]. Recently, we showed that the drip patterns of the American abstract painter Jackson Pollock are fractal [Taylor et al 1999]. In this paper, we describe visual perception tests that investigate whether fractal images generated by mathematical, natural and human processes possess a shared aesthetic quality based on visual complexity.

## 2. Dripped Complexity

The art world changed forever in 1945, the year that Jackson Pollock moved from downtown Manhattan to Springs, a quiet country town at the tip of Long Island, New York. Friends recall the many hours that Pollock spent on the back porch of his new house, staring out at the countryside as if assimilating the natural shapes surrounding him (see Figure One) [Potter]. Using an old barn as his studio, he started to perfect a radically new approach to painting that he had briefly experimented with in previous years. The procedure appeared basic. Purchasing yachting canvas from his local hardware store, he simply rolled the large canvases (sometimes spanning five meters) out across the floor of the barn. Even the traditional painting tool - the brush - was not used in its expected capacity: abandoning physical contact with the canvas, he dipped the stubby, paint-encrusted brush in and out of a can and dripped the fluid paint from the brush onto the canvas below. The uniquely continuous paint trajectories served as 'fingerprints' of his motions through the air.

These deceptively simple acts fuelled unprecedented controversy and polarized public opinion of his work. Was this painting 'style' driven by raw genius or was he simply mocking artistic traditions? Sixty years on, Pollock's brash and energetic works continue to grab public attention and command staggering prices of up to

$40M. Art theorists now recognize his patterns as a revolutionary approach to aesthetics. However, despite the millions of words written about Pollock, the real meaning behind his infamous swirls of paint has remained the source of fierce debate in the art world [Varnedoe et al].

One issue agreed upon early in the Pollock story was that his paintings represent one extreme of the spectrum of abstract art, with the paintings of his contemporary, Piet Mondrian, representing the other. Mondrian's so-called "Abstract Plasticism" generated paintings that seem as far removed from nature as they possibly could be. They consist of elements - primary colors and straight lines - that never occur in a pure form in the natural world. In contrast to Mondrian's simplicity, Pollock's "Abstract Expressionism" speaks of complexity – a tangled web of intricate paint splatters. Whereas Mondrain's patterns are traditionally described as "artificial" and "geometric", Pollock's are "natural" and "organic" [Taylor 2002]. But if Pollock's patterns are a celebration of nature's organic shapes, what shapes would these be?

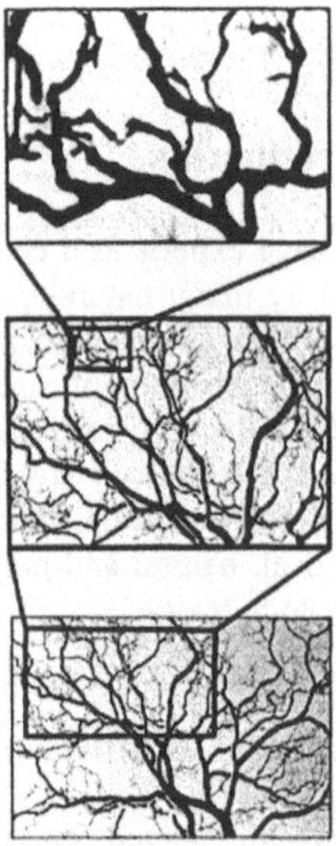

**Figure 1.** Left: Pollock's house on Long Island. In contrast to his previous life in Manhattan, Pollock perfected his drip technique surrounded by the complex patterns of nature. Right: Trees are an example of a natural fractal object. Although the patterns observed at different magnifications don't repeat exactly, analysis shows them to have the same statistical qualities (photographs by R.P. Taylor).

## 3. Nature's Fractals

Since the 1970s many of nature's patterns have been shown to be fractal [Mandelbrot]. In contrast to the smoothness of artificial lines, fractals consist of patterns that recur on finer and finer scales, building up shapes of immense complexity. Even the most common fractal objects, such as the tree shown in Figure One, contrast sharply with the simplicity of artificial shapes.

An important parameter for quantifying a fractal pattern's visual complexity is the fractal dimension, D. This parameter describes how the patterns occurring at different magnifications combine to build the resulting fractal shape. For Euclidean shapes, dimension is described by familiar integer values - for a smooth line (containing no fractal structure) D has a value of one, whilst for a completely filled area (again containing no fractal structure) its value is two. However, the repeating structure of a fractal pattern causes the line to begin to occupy area. D then lies between one and two and, as the complexity and richness of the repeating structure increases, its value moves closer to two [Mandelbrot]. For fractals described by a low D value, the patterns observed at different magnifications repeat in a way that builds a very smooth, sparse shape. However, for fractals with a D value closer to two, the repeating patterns build a shape full of intricate, detailed structure. Figure Two (left column) demonstrates how a pattern's D value has a profound effect on the visual appearance. The three natural scenes from top to bottom have D values of 1.0, 1.3 and 1.9 respectively. Table One shows D values for various natural forms:

| Natural pattern | Fractal dimension | Source |
|---|---|---|
| Coastlines:<br>South Africa, Australia, Britain<br>Norway | <br>1.05-1.25<br>1.52 | <br>Mandelbrot<br>Feder |
| Galaxies (modeled) | 1.23 | Mandelbrot |
| Cracks in ductile materials | 1.25 | Louis et al. |
| Geothermal rock patterns | 1.25-1.55 | Campbel |
| Woody plants and trees | 1.28-1.90 | Morse et al. |
| Waves | 1.3 | Werner |
| Clouds | 1.30-1.33 | Lovejoy |
| Sea Anemone | 1.6 | Burrough |
| Cracks in non-ductile materials | 1.68 | Skejltorp |
| Snowflakes (modelled) | 1.7 | Nittman et al. |
| Retinal blood vessels | 1.7 | Family et al. |
| Bacteria growth pattern | 1.7 | Matsushita et al. |
| Electrical discharges | 1.75 | Niemyer et al. |
| Mineral patterns | 1.78 | Chopard et al. |

**Table 1.** D values for various natural fractal patterns

## 4. Pollock's Fractals

In 1999, we published an analysis of twenty of Pollock's dripped patterns showing that they are fractal [Taylor et al 1999]. To do this we employed the well-established 'box-counting' method, in which digitized images of Pollock paintings were covered with a computer-generated mesh of identical squares. The number of squares N(L)

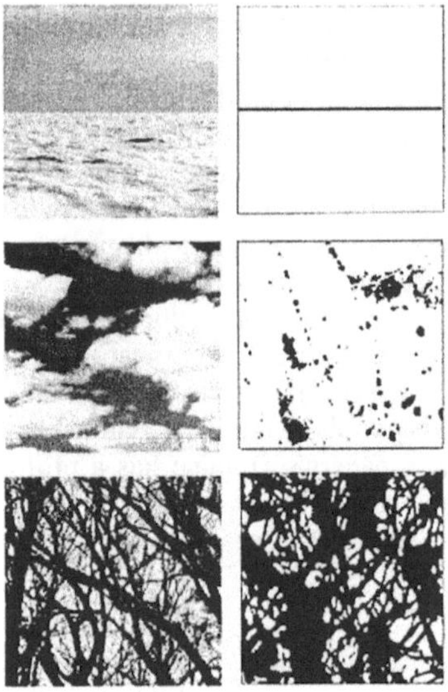

**Figure Two.** Examples of natural scenery (left column) and drip paintings (right column). Top: the horizon forms a line with D=1. Middle: Clouds and Pollock's painting *Untitled* (1945) are fractal patterns with D=1.3. Bottom: A forest and Pollock's painting *Untitled* (1950) are fractal patterns with D=1.9. (Photographs by R.P. Taylor).

that contained part of the painted pattern were then counted and this was repeated as the size, L, of the squares in the mesh was reduced. The largest size of square was chosen to match the canvas size (L~2.5m) and the smallest was chosen to match the finest paint work (L~1mm). For fractal behavior, N(L) scales according to $N(L) \sim L^{-D}$, where $1 < D < 2$. The D values were extracted from the gradient of a graph of log N(L) plotted against log L. Details of the procedure are presented elsewhere [Taylor et al 1999].

Recently, we described Pollock's style as 'Fractal Expressionism' [Taylor et al, Physics World, 1999] to distinguish it from computer-generated fractal art. Fractal Expressionism indicates an ability to generate and manipulate fractal patterns *directly*. In many ways, this ability to paint such complex patterns represents the limits of human capabilities. Our analysis of film footage taken at his peak in 1950 reveal a remarkably systematic process [Taylor et al, Leonardo, 2002]. He started by painting localized islands of trajectories distributed across the canvas, followed by longer extended trajectories that joined the islands, gradually submerging them in a dense fractal web of paint. This process was very swift with the fractal dimension rising sharply from D=1.52 at 20 seconds to D=1.89 at 47 seconds. He would then break off and later return to the painting over a period of several days, depositing extra layers on top of this initial layer. In this final stage he appeared to be fine-tuning the D value, with its value rising by less than 0.05. Pollock's multi-stage painting technique was clearly aimed at generating high D fractal paintings [Taylor et al, Leonardo, 2002].

As shown in Figure Three, he perfected this technique over ten years. Art theorists categorize the evolution of Pollock's drip technique into three phases [Varnedoe]. In the 'preliminary' phase of 1943-45, his initial efforts were characterized by low D values. An example is the fractal pattern of the painting *Untitled* from 1945 which has a D value of 1.3 (see Figure Two). During his 'transitional phase' from 1945-1947, he started to experiment with the drip technique and his D values rose sharply (as indicated by the first dashed gradient in Figure Three). In his 'classic' period of 1948-52, he perfected his technique and D rose more gradually (second dashed gradient in Figure Three) to the value of D = 1.7-1.9. An example is *Untitled* from 1950 (see Figure Two) which has a D value of 1.9. Whereas this distinct evolution has been proposed as a way of authenticating and dating Pollock's work [Taylor et al, Scientific American, 2002] it also raises a crucial question for visual scientists - do high D value fractal patterns have a special aesthetic quality?

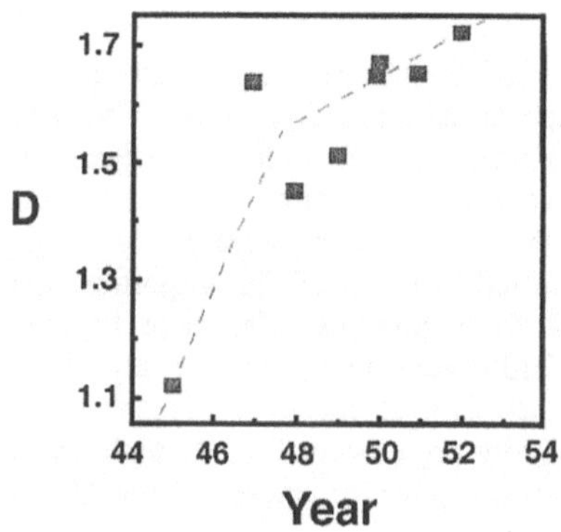


**Figure Three.** The fractal dimension D of Pollock paintings plotted against the year in which they were painted (1944 to 1954). See text for details.

## 5. The Aesthetics of Fractals

The prevalence of fractals in our natural environment has motivated a number of studies to investigate the relationship between a pattern's fractal character and its visual properties [Cutting et al, Geake et al, Gilden et al, Knill et al, Pentland, and Rogowitz et al]. Whereas these studies have concentrated on such aspects as perceived roughness, only recently has the 'visual appeal' of fractal patterns been quantified [Aks et al, Pickover]. The discovery of Pollock's fractals re-invigorates this question of fractal aesthetics. In addition to fractal patterns generated by mathematical and by natural processes, there now exists a third family of fractals - those generated by humans [Taylor 2001].

Previous ground-breaking studies have concentrated on computer-generated fractals. In 1995, Pickover used a computer to generate fractal patterns with different D values and found that people expressed a preference for fractal patterns with a high value of 1.8 [Pickover], similar to Pollock's paintings. However, a subsequent survey by Aks and Sprott also used a computer but with a different mathematical method for generating the fractals. This survey reported much lower preferred values of 1.3 [Aks et al]. Aks and Sprott noted that the preferred value of 1.3 revealed by their survey corresponds to prevalent patterns in the natural environment (for example, clouds and coastlines have this value) and suggested that perhaps people's preference is actually 'set' at 1.3 through a continuous visual exposure to patterns characterized by this D value. However, the discrepancy between the two surveys seemed to suggest that there isn't a universally preferred D value but that the aesthetic qualities of fractals instead depend specifically on how the fractals are generated.

To determine if there are any 'universal' aesthetic qualities of fractals, we carried out a survey incorporating all three categories of fractal pattern: fractals formed by nature's processes, by mathematics and by humans. We used 15 computer-generated images of simulated coastlines, 5 each with D values of 1.33, 1.50 and 1.66; 40 cropped images from Jackson Pollock's paintings, 10 each with D values of 1.12, 1.50, 1.66 and 1.89; and 11 images of natural scenes with D values ranging from 1.1 to 1.9. Figure Two shows some of the images used in the survey.

Within each category of fractals (i.e. mathematical, natural and human), we investigated the visual appeal as a function of D. This was done using a 'forced choice' visual preference technique, in which participants were shown a pair of images with different D values on a monitor and asked to chose the most "visually appealing". In the surveys, all the images were paired in all possible combinations and preference was quantified in terms of the proportion of times each image was chosen. Although details will be presented elsewhere, Figure Four shows the results from a survey involving 220 participants [Spehar et al]. Taken together, the results indicate that we can establish three categories with respect to aesthetic preference for fractal dimension: 1.1-1.2 low preference, 1.3-1.5, high preference and 1.6-1.9 low preference. (Note: a set of computer generated random dot patterns with no fractal content but matched in terms of density (area covered) to the low, medium and high fractal patterns were used to demonstrate that aesthetic preference is indeed a function of D and not simply density).

## 6. Future Studies

Is Jackson Pollock an artistic enigma? According to our surveys, the low D patterns painted in his earlier years should have more 'visual appeal' than his later *classic* drip paintings. What was motivating Pollock to paint high D fractals? Should we conclude that he wanted his work to be aesthetically challenging to the gallery audience? It is possible that he regarded the visually restful experience of a low D pattern as being too bland for an artwork and wanted to keep the viewer alert by engaging their eyes in a constant search through the dense structure of a high D pattern. We plan to investigate this intriguing possibility by performing eye-tracking experiments on Pollock's paintings, which will assess the way people visually assimilate fractal patterns with different D values.

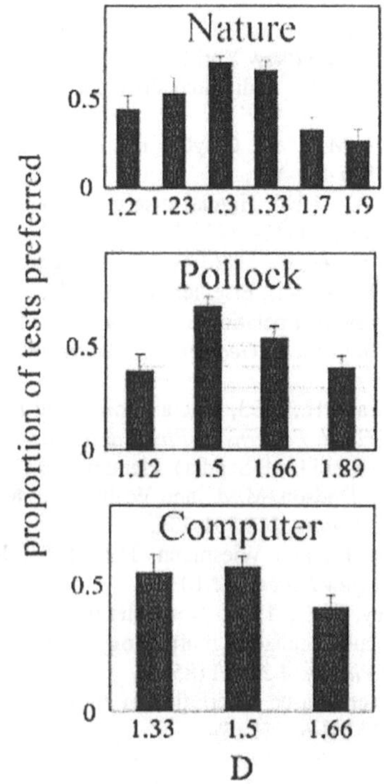


**Figure Four.** Visual preference tests for natural fractals (top), Pollock's fractals (middle) and computer fractals (bottom). In each case, the y axis corresponds to the proportion of trails for which patterns of a given D value were chosen over patterns with other D values.

## Acknowledgments

We thank Adam Micolich and David Jonas (Physics Department, University of New South Wales, Australia) for their help with the fractal analysis and Peter Namuth for permission to use the drip patterns.

## References

Aks, D, and Sprott, J, 1996, Quantifying aesthetic preference for chaotic patterns, *Empirical Studies of the Arts*, **14**, 1.

Burrough, P.A., 1981, Fractal dimensions of landscapes and other environmental data, *Nature,* **295** 240-242.

Cambel, A.B., 1993, *Applied Chaos Theory: A Paradigm for Complexity,* Academic Press, (London).

Chopard, B., Hermann, H.J., and Vicsek, T., 1991, Structure and growth mechanism of mineral dendrites, Nature, **309** 409.

Cutting, J. E. and Garvin, J.J, 1987, Fractal curves and complexity, *Perception & Psychophysics*, **42** 365-370.

Family, F., Masters, B.R. and Platt. D.E., 1989, Fractal pattern formation in human retinal vessels. *Physica D*, **38** 98.

Feder, J. 1988, *Fractals* Plenum (New York).

Geake, J., and Landini, G., 1997, Individual differences in the perception of fractal curves, *Fractals*, **5**, 129-143.

Gilden, D.L., Schmuckler, M.A. and Clayton K., 1993, The perception of natural contour, *Psychological Review*, **100** 460-478.

Knill, D.C., and Kersten, D., 1990, Human discrimination of fractal images, *Journal of the Optical Society of America*, **7** 1113-1123.

Louis, E., Guinea F. and Flores, F., 1986, The fractal nature of fracture, *Fractals In Physics* (eds. L. Pietronero and E. Tossati), Elsevier Science 177.

Lovejoy, S., 1982, Area-perimeter pelation for pain and cloud areas, *Science*, **216** 185.

Mandelbrot, B.B., 1977, *The Fractal Geometry of Nature*, W.H. Freeman and Company, (New York).

Matsushita, M. and Fukiwara. H., 1993, Fractal growth and morphological change in bacterial colony formation, In *Growth Patterns in Physical Sciences and Biology*, (eds. J.M. Garcia-Ruiz, E. louis, P. Meaken and L.M. Sander) Plenum Press (New York).

Morse, D.R., Larson, J.H., Dodson, M.M., and Williamson, M.H., 1985, Fractal dimension of anthropod body lengths, *Nature*, **315** 731-733.

Niemeyer, L., Pietronero, L., and Wiesmann, H.J., 1984, Fractal dimension of dielectric breakdown, *Physical Review Letters*, **52** 1033.

Nittmann, J., and Stanley, H.E., 1987, Non-deterministic approach to anisotropic growth patterns with continuously tunable morphology: the fractal properties of some real snowflakes, *Journal of Physics A* **20**, L1185.

Pentland, A. P., 1984, Fractal-based description of natural scenes, *IEEE Pattern Analysis and Machine Intelligence*, **PAMI-6**, 661-674.

Pickover, C., 1995, *Keys to Infinity*, Wiley (New York) 206.

Potter, J., 1985, To a Violent Grave: An Oral Biography of Jackson Pollock, G.P. Putman and Sons (New York).

Richards, R., 2001, A new aesthetic for environmental awareness: Chaos theory, the beauty of nature, and our broader humanistic identity, *Journal of Humanistic Psychology*, **41** 59-95.

Rogowitz, R.E., and Voss, R.F., 1990, Shape perception and low dimensional fractal boundary contours, Proceedings of the conference on Human vision: Methods, Models and Applications, S.P.I.E., **1249**, 387.

Skjeltorp. P., 1988, Fracture Experiments on Monolayers of Microspheres, *Random Fluctuations and Pattern Growth* (ed. H.E. Stanley and N. Ostrowsky) Kluwer Academic (Dordrecht).

Spehar, B., Clifford, C., Newell, B. and Taylor, R.P., 2002, Universal aesthetic of fractals, submitted to Leonardo.

Taylor, R.P., 2001, Architect reaches for the clouds, *Nature*, **410**, 18.

Taylor, R.P., 2002, Spotlight on a visual language, *Nature*, **415**, 961.

Taylor, R.P., Micolich, A.P., and Jonas, D., 1999, Fractal analysis of Pollock's drip paintings *Nature*, **399**, 422.

Taylor, R.P., Micolich, A.P. , and Jonas, D, 1999, Fractal expressionism, *Physics World*, **12**, 25-28.

Taylor, R.P. , Micolich, A.P., and Jonas, D., 2002, The construction of Pollock's fractal drip paintings, to be published in *Leonardo*, MIT press.

Taylor, R.P., Micolich, A.P., and Jonas, D., 2002, Using nature's geometry to authenticate art, to be published in *Scientific American*.

Varnedoe, K., and Karmel, K., 1998, *Jackson Pollock*, Abrams (New York).

Werner, B.T., 1999, Complexity in natural landform patterns, *Science*, **102** 284.

# Chapter 18

# Emergent Patterns in Dance Improvisation and Choreography

**Ivar Hagendoorn**
ivar@ivarhagendoorn.com
http://www.ivarhagendoorn.com

In a traditional choreography a choreographer determines the motions of a dancer or a group of dancers. Information theory shows that there is a limit to the complexity that can be created in any given amount of time. This is true even when building on previous work, since movements and their interactions have to be communicated to the dancers. When creating a group work, choreographers circumvent this problem by focusing either on the movements of individual dancers (giving rise to intricate movements but within a simple spatiotemporal organization) or on the overall structure (intricate patterns but simple movements) or by creating room for the dancers to fill in part of the movements. Complexity theory offers a different paradigm towards the generation of enticing patterns. Flocks of birds or schools of fish for instance are considered 'beautiful' but lack a central governing agent. Computer simulations show that a few simple rules can give rise to the emergence of the kind of patterns seen in flocks or swarms. In these models individual agents are represented by dots or equivalent shapes. To be of use to choreography and to be implemented on or rather with dancers, some additional rules will therefore have to be introduced. A number of possible rules are presented, which were extracted from 'real life' experiments with dancers. The current framework for modeling flocking behavior, based on local interactions between single agents, will be extended to include more general forms of interaction. Dancers may for instance perceive the global structure they form, e.g. a line or a cluster, and then put that knowledge to creative use according to some pre-established rules, e.g. if there is a line, form a circle or if there is a cluster spread out in all directions. Some of these rules may be applied back to other complex systems. The present paper is also an invitation to complexity theorists working in different fields to contribute additional rules and ideas.

# 1 Introduction

A choreography is a set of instructions for the organization and reconfiguration of one or several bodies in space and time. In practice there is always a 'residual term' $\varepsilon$, in the sense of performance = choreography + $\varepsilon$, which is not covered by the explicit instructions and which is left to the dancer(s) to fill in. It follows that the more complex a choreography, the longer the set of instructions. This means that there is a limit to the degree of complexity that can be created by a choreographer in any given amount of time. In dance there are various ways of dealing with increasing complexity. A choreographer can concentrate on each dancer's individual movements and organize the dancers in simple patterns, such as lines, circles and clusters, s/he can create intricate global patterns while keeping the individual movements simple or increase what I called the residual term, leaving more of the individual movements to the dancers to fill in.

Complexity theory offers a different paradigm towards the generation of complex structures. Flocks of birds and schools of fish for instance, exhibit intricate patterns that emerge from the interaction of individual agents, in the absence of a 'master mind' or central governing agent. Computer simulations have shown that a few simple rules give rise to the kind of patterns seen in swarms [e.g. Reynolds 1987, Tone & Tu 1998, Csahók & Vicsek 1995, Czirok et al 1997], pedestrian [Helbing & Molnár 1995] and traffic flow [Helbing 2001] and large crowds of people [Helbing et al. 2000]. Another intriguing finding is that large groups of interacting agents or particles not only exhibit aggregation and swarming but also spontaneous synchronization [e.g. during the applause after a performance, Néda et al. 2000], an important aspect of the aesthetic appeal of choreographed dance. It therefore seems natural to try and translate these rules to dance.

A quick glance at the principles that have been incorporated in the models underlying these computer simulations shows that some can be made to depend on other, possibly hidden, variables. For instance, we could make distance to the nearest neighbor into a dependent variable, instead of a constant and invent our own criteria for relating inter-agent direction and velocity.

To make the transition from dots on a screen to multi-limbed dancers moving in three-dimensional space requires the design of additional rules and principles. After all we would like the dancers to do more than just walk around. In the present paper I will describe some of the rules I have designed to this end, for which I transformed a dance studio into a laboratory for studying complex systems.

# 2 What I Mean by Complexity

The present paper differs from other studies of complex systems in that I do not attempt to model an existing system, but rather try to *produce* a complex system. In doing so I loosely draw on the methodology sketched out by John Holland [Holland 1995 and 1998]. The work by Eric Bonabeau and Guy Theraulaz on ant behavior may also serve as an analogy. Having analyzed how ants collectively find the shortest path to a food source, they played around with the variables in their model and were thus able to derive a more optimal solution to a problem not yet discovered by nature [Bonabeau et al. 1999 and 2000]. This is what I do with models for flocking. The present work is therefore *inspired* by research into complex systems and its scientific claims are modest.

It will be clear that an *emergent choreography* as will be described here can be regarded as a complex system in the sense that it consists of multiple interacting components (dancers), the properties of which are not fully described by those of the individual components (dancers). Analyzing the isolated movements of an individual dancer will not bring us closer to an understanding of the choreography.

In information theory the complexity of an object is defined as the length of the shortest program, which generates the object. This alternative notion of complexity, originally proposed by the Russian mathematicians Ray Solomonoff and Andrei Kolmogorov and independently by the American mathematician Gregory Chaitin, is known as Kolmogorov complexity or algorithmic complexity. A problem with Kolmogorov complexity is that it cannot be computed, as it is based on considering all possible programs for generating the object. However, limiting the number of admissible programs to a certain class makes it possible to approximate an object's algorithmic complexity [Li & Vitanyi 1997].

In *The Quark and the Jaguar* Murray Gell-Mann gives a similar definition of complexity [Gell-Mann 1994]. Take the following three pictures [adapted from Gell-Mann 1994].

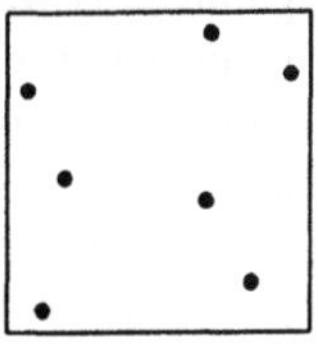

Figure 1a

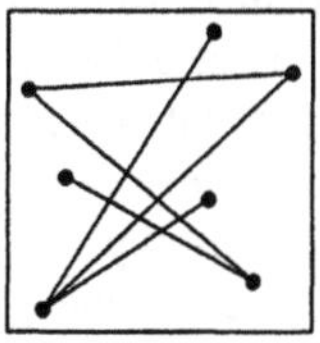

Figure 1b

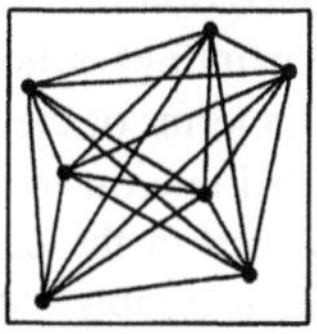

Figure 1c

Now which one is the most complex? None of the 7 dots in figure 1a are connected, so instructing someone over the phone to draw this image would require no more information than this, if we abstract from the position of the dots. In figure 1c all 7 dots are connected, and thus all we'd have to say would be 'draw 7 dots and connect all dots'. Now, if we wanted to transmit figure 1b we would have to specify exactly which dots are connected, and thus by the above definition, figure 1b is the most complex. This appears counterintuitive, because at first sight figure 1c looks the most complex. However, with 11 or 23 instead of 7 dots it would begin to look dense and from a distance there would seem to be little internal difference. Indeed it would matter little if there were 21 or 23 dots.

It's the same with movements of one or several bodies: constantly moving every conceivable part of the body may at first look complex, but is in fact quite simple: the task move every part of the body will produce a series of sequences that all look more or less the same.

The above example also illustrates the difficulty of quantifying the complexity of a choreography, whether a solo or a group work. We therefore have to rely on our senses and an estimate of what it would take to reproduce the ballet from a set of instructions. In principle one of the existing systems for dance notation, Labanotation or Benesh notation, could be used to this end. However, the fact that neither achieved popularity shows the difficulty of capturing dance in symbolic sequences. Perhaps the techniques that have been proposed to measure the complexity of RNA sequences may at some point be extended to dynamic sequences of body configurations [Adami & Cerf 2000].

# 3 Why Complexity Matters

The careful reader will have noticed that I use the term 'complex' not only as a way of characterizing a system, but also as a value or an objective, "complex=beautiful" or "complexity=great" The psychologist Mihalyi Cszikszentmihalyi has argued that the human brain actively searches for difference and more complex scenes or events, once it gets used to whatever it is currently doing or perceiving [Cszikszentmihalyi 1992]. If no such perceptual or cognitive challenges are at hand the brain will get bored. This may be why as they mature the work of many choreographers shows an increase in complexity, in a weak sense of algorithmic complexity.

In the 1930's the mathematician George Birkhoff proposed a measure of beauty defined as M=O/C, whereby M stands for 'aesthetic measure' (or beauty), O for order and C for complexity [Birkhoff 1933, 1956]. In formulating this measure Birkhoff was inspired by the observation that beauty, whether in art or in nature, has something to do with order and complexity. Order shows in such compositional elements as repetition, contrast, symmetry, balance, synchronization etc. and is found in flowers, sculptures, figurative as well as abstract paintings, schools of fish, classical ballet and so on. Complexity Birkhoff defined in terms of the effort that goes into perceiving the object or event, an idea of renewed interest in the light of contemporary cognitive neuroscience. Birkhoff also referred to the definition of beauty by the 18th century Dutch philosopher Frans Hemsterhuis, "that which gives us the greatest number of ideas in the shortest space of time". A perhaps unwanted consequence of his formula is that, as the degree of complexity goes to zero, and the degree of order is high, as in some abstract art, the measure of beauty goes to infinity. Its verbal formulation as "the density of order relations in the aesthetic object" may therefore be a more appropriate description. In a recent paper Misha Koshelev and Vladik Kreinovich formalized Birkhoff's idea by defining the complexity of an object in terms of its Kolmogorov complexity [Koshelev 1998, Kreinovich et al. 1998].

The various 20th century avant-gardes have shown the limitations of any approach to art that is purely based on an analysis of form. Recent progress in cognitive neuroscience shows that with respect to some art Birkhoff may have been on the right track [Zeki 1999, Ramachandran & Hirstein 1999] and it would be interesting to experimentally test his hypothesis. Not only scientific evidence supports Birkhoff's claims. In a recent review of a mixed program by William Forsythe and the Frankfurt Ballet [1], a Dutch dance critic wrote that "Forsythe's dance language is crystal clear, but of such a complex virtuosity [2] that the images and thoughts it evokes can hardly fall into place within the duration of a performance" [Rietstap 2002]. This almost reads as a reformulation of Frans Hemsterhuis' and George Birkhoff's definition of beauty. The critic concluded that "Forsythe's dance language [is] too complex for [the] audience", which would be explained by the gap between a choreographer with a career spanning over 20 years and an audience which sees a few dance performances a year.

# 4 Methods

In principle patterns also arise when particles or agents of whatever kind move randomly. Pattern perception is further enhanced by the tendency of the human brain to actively search for patterns and regularities [Ramachandran & Hirstein 1999]. Nonetheless in a purely random environment patterns are rare. One way of characterizing the present approach to dance and choreography is that it attempts to increase the likelihood of the emergence of patterns. We could, in mathematical terms, change the process from which the movements are drawn. This we could do by severely restraining the state space of possible movements. If all the dancers can do is move their arms up and down and walk back and forth, the dancers' movements and positions in space are more likely to momentarily synchronize.

Another possibility would be to correlate the dancers' movements, body configurations and positions in space. That is the approach described here. The idea is to describe the situations or stimuli a dancer can encounter and list the responses available to a dancer in a given situation [Holland 1995 and 1998]. As a meta-rule a dancer can either ignore a stimulus or use it as a cue for a motor response.

The present approach to dance and choreography relies on the dancers' ability to read each other's intentions. It also requires the dancers to understand the different rules and the patterns they give rise to. While being 'inside' the group they have to know how it will look from the outside. That is, the group has to develop a form of collective intelligence, with my own aesthetic preferences or more general principles of aesthetic experience, as a utility function that the group as a whole seeks to maximize [Wolpert & Tumer 1999]. This is why I work a lot with video feedback so the dancers can observe and learn from the local and global effects of their decisions. Technically speaking this is an example of reinforcement learning whereby my verdict and that of the dancers serve as reward signals.

The method I use is to extract rules from the observation of existing choreographed dance performances and other forms of human interaction and to re-apply those in a setting based on improvisation. I then observe the resulting behavior over a series of 'trials' to evaluate the emerging patterns and to determine where a conflict or a 'decision void' arises, that would require the introduction of additional rules. The work is therefore inherently inexact. Dancers may not stretch a rule to its limits, the way a computer model does, they may implicitly incorporate another rule and the number of trials may be too small to show all possible configurations. The present work may therefore serve as inspiration for more formal modeling approaches in collective robotics. It is also inevitable that my own preferences as to what I would like to see emerge on stage make themselves felt in the choice of rules. This is not per se a compromise as long as it is acknowledged.

# 5 A Selection of Rules

## 5.1 Spatial organization

Dancers need a motivation to go from A to B and to do one thing rather another. One of the simplest rules for moving through space is a random walk. This translates into the following instructions:

1. just walk around, at any moment you can turn into another direction
2. if you bump into a wall, another person or leave the stage, turn and continue in any direction
3. at any moment you can stop and stand still

From a distance and with a little imagination applying these rules gives the impression of a room filled with randomly moving particles. However, these rules are too general. Should a dancer only move forwards or is she allowed to also move backwards and sideways? This too has to be made explicit.

4. you can move forward, backward or sideways.

The above is just a basic setting to build on. The following rules add some interaction.

5. you can decide to trail another dancer (by walking behind or next to the other person)
6. if somebody is trailing you, you can try to escape by changing direction or speed
7. if you are trailing another person you can overtake that person
8. you can block another person by standing in front of that person

While these simple rules give rise to interesting patterns in terms of spatial organization they turn out to be insufficient for two reasons. On the one hand dancers tend to find them too unspecific, they don't say anything about the use of the arms or ways of walking, on the other hand it may take too long before certain desired patterns emerge. I therefore decided to introduce some specific rules for spatial organization, making use of the fact that human beings are able to look at a situation from different degrees of locality (from nearest neighbor to the entire field of view) and act accordingly.

It should be noted that most individual based models are based on local interactions: each agent only responds to the agents "within some fixed, finite distance which is assumed to be much less than the size of the 'flock'" [Toner and Tu 1998, p. 4828]. This rule is self-reinforcing: within a flock only the nearest neighbors are visible. If the distance between agents increases, other agents become visible, thus enabling long ranged interaction. As more agents become visible the potential for interaction in terms of the number of candidates also grows, but the likelihood of communication or mutual interaction, whereby dancer A interacts with dancer B and vice versa, decreases.

## 5.2 Alignment

Various species of caterpillars form queues when traversing an open space. Human beings too self organize into queues in front of check outs, box offices etc. While this is a cultural phenomenon, the phenomenon itself is no less real. One could think of various structures into which a group of people can self-organize, the simplest being alignment: standing side by side in a row, face to back in a queue, a circle, a square etc. (see figure 2).

## 5.3 Clustering

In computer simulations of complex systems (schools, herds, swarms etc.) the rules are chosen so that they give rise to spontaneous clustering (e.g. separation, alignment and cohesion, Reynolds 1987). Effectively when choreographing a group or 'block' of dancers, the dancers implicitly apply similar rules, because a choreographer will only indicate how and where s/he wants the *group* to move, and not each individual dancer. In the present context I have adopted the now classic rules for group formation to control *the movement of* a cluster. A problem from the point of view of dance improvisation is that with these rules and in the absence of external perturbations, dancers would tend to remain in a cluster. I therefore introduced specific rules to have a cluster emerge and dissolve. The first rule tells the dancers to cluster together if they see the core of a cluster forming (see figure 3a). The rule 'expand' tells the dancers to expand in all or one particular direction if a cluster has formed (see figure 3b). A cluster can also dissolve if dancers stop adhering to the group cohesion rule.

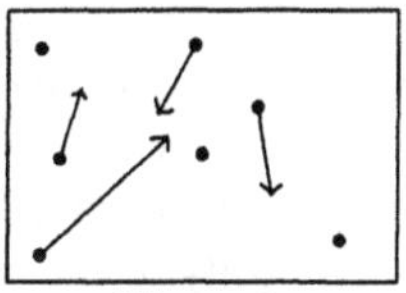
Figure 2

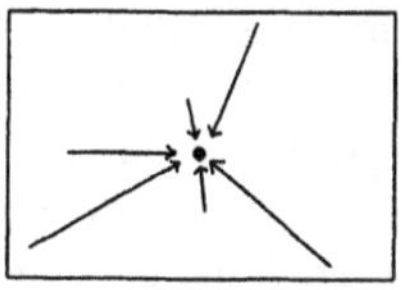
Figure 3a.

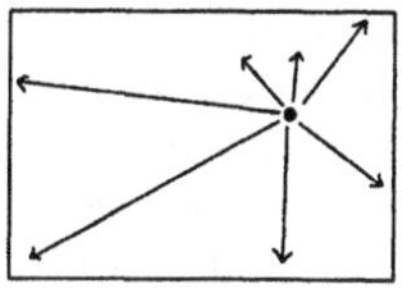
Figure 3b.

## 5.4 Levels in space

The above rules concern the layout or the map in space. To also control the organization of levels in space I designed the following rule. For simplicity I will assume there are three basic positions standing, sitting and lying on the floor. With three dancers, and not distinguishing between individual dancers, we then have 10 possible configurations (all three dancers standing, two sitting, one lying on the floor, one sitting, two standing etc). The idea is that each dancer knows all three positions and can adjust his position according to the position of the other dancers. Thus if all three dancers are standing up one dancer can decide to either sit or lie down on the floor. If two or three dancers have the same idea at the same moment in time we have spontaneous synchronization. If one dancer is lying on the floor and two others are standing, one can decide to also lie down on the floor whereupon the dancer already lying on the floor can decide to sit up, get up or keep lying on the floor.

## 5.5 Divisions of space

The above principle of adjusting the spatial configuration in relation to what the other dancers are doing is a simple and efficient means of creating spatial dynamics. It can be extended to the more general use of levels and regions (left, right, front, back) in space. With an understanding of spatial organization dancers can self-organize on the far left of the stage so as not to be constantly all over the stage. When observing a cluster a dancer can also create a contrast between group and individual by suddenly moving away. In general the space can be divided into various sections and according to different principles (a checker board, a soccer field etc.). It is then possible to devise all kinds of rules or 'games' for going from one section to another, all in relation to what another dancer is doing.

**Example: Imagine You Are in a Labyrinth**

For one particular piece I asked the dancers to imagine they are in a labyrinth. The dancers can only move back and forth and have to make rectangular turns. Because the corridors are narrow, to move the arms sideways they have to make a 90° turn. If two dancers meet they have to pass while adhering to the constraints imposed by the narrow corridors and the opportunities offered by corners [3].

## 5.6 Dynamics

The emphasis in the rules introduced so far is on spatial organization. The rules that govern the behavior of swarms also include velocity alignment. This we can translate to dance, but to make things a little more interesting and to avoid everybody dancing at the same speed, instead of adopting the same velocity we can introduce a range of options. If dancer A moves with a certain velocity, dancer B can decide to move slower, faster or with the same velocity or she can decide to freeze into a pose. Again the idea is that once the range of options available to a generic dancer have been established, they can be made interdependent.

## 5.7 Copy

It is also possible to directly link the movements of two or more dancers. If dancer A sees dancer B perform a movement ($x$) she can instantly copy the movement. Thus if dancer B raises her left arm above her head, so does dancer A, whatever the position she is in. She may be kneeling, sitting or squatting on the floor whereas dancer B is standing up. It is interesting to observe that this rule works best in non-neutral positions because it is easier to infer what the movement is going to be. If dancer A and B face each other, copying can have two meanings: if dancer B raises her left arm, dancer A either raises her left arm as well, putting herself in the shoes of dancer B, or she raises her right arm, to form a 'mirror image' of dancer B's movement. We could thus distinguish between 'copying' and 'mirroring'.

**Corollary 1**

If dancer A sees dancer B perform movement $x_1$ of a previously rehearsed movement sequence $(x_1,..., x_n)$ she will know how the movement sequence will unfold and thus be able to better synchronize her movements. If she sees dancer B perform movement $x_{n-m}$ she can perform the series $(x_{n-m+1},..., x_n)$.

**Corollary 2**

Obviously movements can only be copied if they are visible. If dancer A happens to be standing behind dancer B, dancer B won't be able to copy dancer A's movements. Unless that is, we introduce a third dancer C, who is able to see and copy dancer A's movements and is visible to dancer B (see figure 4).

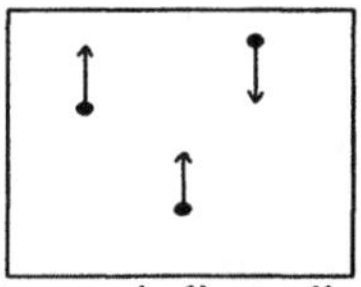

Figure 4. Arrows indicate direction of gaze

**Corollary 3**

If dancer A sees dancer B perform movement $x_i$ of a previously rehearsed sequence $(x_1,..., x_n)$ she can deliberately perform the movement sequence with a delay to create a 'ripple' effect, which becomes even stronger if dancer C copies dancer A with a delay. Dancer C may already have observed 'dancer A copying dancer B', making it easier for dancer C to copy the movement sequence.

**Corollary 4**

Instead of copying the movement of dancer B, dancer A can also use it as a cue for another movement, which can be either a previously rehearsed movement, a transformation of the original movement or in principle, 'any movement whatsoever', in which case the visible relationship with the original is lost. The difference between 'copying' and 'mirroring' when two dancers are facing each other can be regarded as a transformation of the original movement: left becomes right and vice versa. Such a left/right reversal can also be applied deliberately e.g. if dancer A is standing behind dancer B.

We thus have the following three possibilities:

1. IF $x$ THEN $x$ *(same movement)*
2. IF $x$ THEN $f(x)$ *(different movement which relates to original movement)*
3. IF $x$ THEN $y$ *(different movement)*

The effect of this principle is the spontaneous synchronization of 2 or more dancers across space. It looks 'choreographed' but it isn't. The synchronization emerges from the interaction of the dancers.

## 5.8 Internal differentiation

One of the great discoveries in the modeling of crowds is that in large enough groups, individuals can be treated as identical agents. In other words flocks, herds and crowds form a homogeneous mass. A *small* group of agents on the other hand, can be differentiated into a heterogeneous group, for instance by having the agents wear different colored outfits. Division of a group into two or more subgroups introduces novel opportunities for interaction. The fact that the 22 players in soccer are divided into two teams of 11 players each, in combination with the spatial layout, determines the behavior of the agents and the patterns that emerge. An interesting phenomenon in soccer is that children all around the world always flock around the ball. In professional soccer field players are differentiated into (left/right) defenders, strikers and midfielders. Each player has a designated area and role within the field and the team's strategy. In choreography we could again invent our own rules.

## 5.9 Equivalence relationships

An equivalence class is a set of elements defined by an equivalence relation. Examples of equivalence relations are "is parallel to", "has the same color", and perhaps a bit more daring, "has the same meaning" and "has the same perceptual effect". In my work I am trying to find different equivalence relations and equivalence classes as they apply to dance. For instance it may matter little if a dancer turns left or right or if a position is performed in a left or a right orientation. The question the dancers and I then try to address is when it *does* matter. Similarly in a duet it may matter little which person performs which role, although again the question is when it does matter.

The more interesting question is what dancer A has to do to maintain an equivalence relation relative to another movement taken as the original movement, if dancer B performs a movement in a different orientation. To give a more concrete example, the combination of dancer A standing up and dancer B lying on the floor can be equivalent to dancer A lying on the floor and dancer B standing up. So, if two dancers have rehearsed a duet, then knowing various equivalence relations pertaining to that duet, they can recognize a situation and then perform an equivalent duet based on that situation. In the previous example dancer B may notice that dancer A is lying on the floor and then enter into a duet which had been rehearsed with dancer B lying on the floor and dancer A standing up.

Now the following two constellations may constitute a perceptual equivalence class in the sense that they have a similar perceptual effect on an observer:

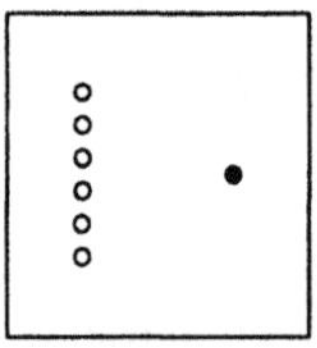

Figure 5a

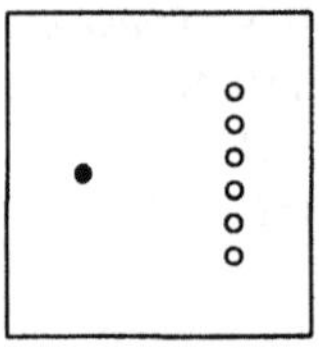

Figure 5b

In practice this means that if the dancer represented by the black dot is on the right side, all other dancers should go to the left to constitute constellation 5a, if however the dancer in black is on the left side of stage, the other dancers should align on the right side. Now, the next step is that any dancer can be the black dot. The dancers may notice that they can self-organize into either one constellation or a dancer may isolate him or herself and try to get the other dancers to align into figure 5b.

By defining a number of constellations in combination with a set of equivalence relations it is possible to direct the patterns that emerge on stage, without setting the piece as a whole. This approach has the additional benefit of creating a form of consistency across different performances while simultaneously allowing for novelty. One has, in other words, structure with variations, one of the characteristics of a complex system.

This form of interaction whereby the patterns stay the same, but the agents take turns, one finds not only in team sports such as soccer or basketball, but also in social animals that hunt as groups such as hyenas and wolves. Both species are known to chase their prey in a relay hunt and to drive an animal into an ambush formed by a (sub) group. This requires the predators to interact in response to what the others are doing. The chase itself is of course also a form of interaction, with both predator and prey performing their genetically determined escape and pursuit strategies (e.g. the zig zag of hare), which can also be implemented in dance.

# 6 Conclusion and Directions for Future Research

The present article has described some of the rules I have designed to govern the interaction of a group of dancers. The primary goal of my research is artistic: to have fascinating patterns emerge within a group of dancers. Although I myself have the dancers improvise on stage as well, the above rules can also be used as a choreographic rapid prototyping engine. When recorded on video, desired patterns can be singled out and fixed to be included in a 'standard' choreography. The equivalence classes described above are an example of this approach. Some of the rules proposed here may already have been discovered by nature. For instance prairie dogs feed in groups and always have at least one animal standing on guard on its hind legs. It appears that the moment this animal joins the group another animal takes over its role as guard. This is equivalent to the rule for levels in space introduced above (5.4). In addition to its artistic merits the present approach to dance improvisation and choreography may therefore serve as an experimental setting for investigating forms of group interaction, such as the behavior of social animals, that at present are hard to implement in a computer simulation.

*Acknowledgements: I would like to thank the dancers I have worked with in exploring and bringing to life the ideas presented here.*

## Notes

[1] The program included: *7 To 10 Passages* (2000), *The room as it was* (2002), *Double/Single* (2002) and *One flat thing, reproduced* (2000).

[2] The critic may have meant virtuoso complexity.

[3] The "labyrinth" section in Communications from the Lab (2004), see http://www.ivarhagendoorn.com for a video excerpt.

## Bibliography


[1] Adami, C., & Cerf, N.J. 2000, Physical complexity of symbolic sequences. Physica D 137, 62-69.

[2] Birkhoff, G.D., 1956, Mathematics of Aesthetics, in The World of Mathematics Vol. 4 edited by J.R. Newman, Simon and Schuster (New York), 2185-2195.

[3] Birkhoff, G.D., 1933, Aesthetic Measure, Harvard University Press (Cambridge, MA).

[4] Bonabeau, E., Dorigo, M., & Theraulaz, G., 2000, Inspiration for optimization from social insect behaviour, Nature 406, 39-42.

[5] Bonabeau, E., Dorigo, M., & Theraulaz, G., 1999, Swarm Intelligence: From Natural to Artificial Systems, Oxford University Press (Oxford).

[6] Csahók, Z., & Vicsek, T., 1995, Lattice gas model for collective biological motion, Physical Review E 52, 5297.

[7] Cszikszentmihalyi, M., 1992, Flow. The Psychology of Happines, Rider (London).

[8] Czirok, A., Stanley, H.E., & Vicsek, T., 1997, Spontaneously ordered motion of self-propelled particles, Journal of Physics A 30, 1375.

[9] Gell-Mann, M., 1994, The Quark and the Jaguar. Adventures in the Simple and Complex, Abacus (London).

[10] Helbing, D., & Molnár, P., 1995, Social force model for pedestrian dynamics, Physical Review E 51, 4282-4286.

[11] Helbing, D., 2001, Traffic and related self-driven many-particle systems, Reviews of Modern Physics 73, 1067-1141.

[12] Helbing, D., Farkas, I., & Vicsek, T., 2000, Simulating dynamical features of escape panic, Nature 407, 487-490.

[13] Holland, J.H., 1995, Hidden Order. How adaptation builds complexity, Helix Books (Reading, MA).

[14] Holland, J.H., 1998, Emergence. From chaos to order, Perseus Books (Reading, MA).

[15] Koshelev, M., 1998, Towards The Use of Aesthetics in Decision Making: Kolmogorov Complexity Formalizes Birkhoff's Idea, Bulletin of the European Association for Theoretical Computer Science, 66, 166-170.

[16] Kreinovich, V., Longpre, L., & Koshelev, M., 1998, Kolmogorov complexity, statistical regularization of inverse problems, and Birkhoff's formalization of beauty, in Bayesian Inference for Inverse Problems, Proceedings of the SPIE/International Society for Optical Engineering edited by A. Mohamad-Djafari, 3459, 159-170.

[17] Li, M., & Vitanyi, P.M.B., 1997, An introduction to Kolmogorov complexity and its applications 2nd edition, Springer (New York).

[18] Néda, Z., Ravasz, E., Brechet, Y., Vicsek, T., & Barabasi, A.-L., 2000, The sound of many hands clapping, Nature 403, 849-850.

[19] Ramachandran, V.S., & Hirstein, W., 1999, The science of art. A neurological theory of aesthetic experience, Journal of Consciousness Studies 6, 6-7, 15-51.

[20] Reynolds, C.W., 1987, Flocks, herds, and schools: A distributed behavioral model, Computer Graphics 21, 4, 25-34.

[21] Rietstap, I. 2002, Danstaal Forsythe te complex voor publiek [transl.: Dance language Forsythe too complex for audience], NRC Handelsblad, 15 April.

[22] Toner, J., & Tu, Y., 1998 Flocks, herds and schools: a quantitative theory of flocking, Physical Review E 58, 4828-4858.

[23] Wolpert, D.H., & Tumer, K., 1999, An introduction to collective intelligence, in Handbook of Agent Technology edited by J.M. Bradshaw, AAAI Press/MIT Press (Cambridge, MA).

[24] Zeki, S., 1999, Inner vision. An exploration of art and the brain, Oxford University Press (Oxford).

# Chapter 19

# A Stochastic Dynamics for the Popularity of Websites

**Chang-Yong Lee**
Department of Industrial Information
Kongju National University
Chungnam, 340-702, South Korea
clee@kongju.ac.kr

In this paper, we have studied a dynamic model to explain the observed power law distribution for the popularity of websites in the WWW. The dynamic model includes the self growth for each website and the external force acting on the website. With numerical simulations of the model, we can explain most of the important characteristics of websites, such as a power law distribution of the number of visitors to websites and fluctuation in the fractional growth of individual websites.

## 1 Introduction

As the Internet and the World Wide Web (the web, for short) plays an important role in our present society, research on these becomes more and more active. In particular, study of the characteristics of websites and their dynamical behavior has become recognized as a new field of research. Aside from the technical understandings of the Internet and the web, within this new field, the Internet can be regarded as an "artificial complex system" of which many interacting agents, or websites are composed. As is true for most complex systems, size and dynamic variations make it impractical to develop characteristics of the web deterministically.

Despite the fact that the web is a very complex system, seemingly an unstructured collection of electronic information, it is found that there exists a simple and comprehensible law: the power law distribution. According to the research [1], the number of visitors to websites exhibits a power law distribution. This finding implies that most of data traffic in the web is diverted to a few popular websites. This power law distribution of the popularity for websites is one of the characteristics of the Internet web market and contrasts with the traditional equal share markets in which the transaction cost and geological factors play important roles.

In addition to the power law distribution, the analysis of empirical data [1] shows a few additional characteristics for the number of visitors to websites: first, the number of visitors follows a power law with different exponents depending on the category of websites. More specifically, for the ".edu" domain sites the exponent $\beta = 1.45$, and for all websites, $\beta = 2.07$. This shows that the exponent of all categories is greater than that of a specific category. Second, the fluctuation of the growth rate in the number of visitors for each site is uncorrelated.

In this paper, we investigate the power law distribution of the number of visitors to websites and the dynamic properties among competing websites. In particular, we focus on the result of empirical data analysis in Ref. [1]. For this end, we first build up a stochastic model for the number of visitors to the websites, and then carry out both numerical and analytic calculations.

## 2 A Dynamic Model

In general, a dynamic system can be described schematically as

$$\frac{dX_i(t)}{dt} = f_i(\vec{X}), \tag{1}$$

where $\vec{X}$ represents the state of the system and takes values in the state or phase space. In the present case, $X = \{X_i(t)\}, i = 1, 2, \cdots, N(t)$, and $X_i(t)$ is the number of visitors to the website $i$ at time $t$. Expanding $f_i(X)$ in the powers of $X_i$ and keeping the lowest order term in $X_i$, Eq. (1) can be rewritten, after absorbing the constant term into $X_i$, as

$$\frac{dX_i(t)}{dt} = A_{ii} X_i + \sum_{j \neq i} A_{ij} X_j . \tag{2}$$

There is one more ingredient that should be taken into account: an exponential growth in the number of websites. It is known that the number of websites connected to the Internet is not constant but increases exponentially [2, 3]. Thus the number of websites $N(t)$ at time $t$ satisfies, in the continuous time limit,

$$\frac{dN(t)}{N(t)} = \lambda\, dt\ , \tag{3}$$

where $\lambda$ is the growth rate of the number of websites. To implement this exponential growth, we discretize time and take $\Delta t$ as the time step such that within $\Delta t$ a new website can be added into the Internet with the probability $N(t)\lambda\Delta t$. That is, in each time step $\Delta t$, on the average, the number of the websites will be increased at time $t$ by an amount

$$\Delta N(t) = N(t+\Delta t) - N(t) = N(t)\lambda\Delta t\,. \tag{4}$$

We further assume that no two websites can be created within $\Delta t$.

Now, let us determine the coefficients $A_{ii}$ and $A_{ij}$. The first term on the right hand side of Eq. (2) is the self growth term of the website $i$ with the growth rate $A_{ii}$. The implication of this term is that in two successive time periods the increase in the number of visitors is proportional to the number of visitors to that site. This is reasonable because once a website is created, the website will be known to more users. In consequence, more users visit the website as time progresses. We also assume that each website would grow with an equal rate so that the coefficient $A_{ii}$ could be set to the same irrespective of the website. This assumption is valid if there are no other factors affecting on the growth of a website. Furthermore, the coefficient $A_{ii}$ can be absorbed with an appropriate re-scaling of $X_i$ so that one can set $A_{ii} = 1$ for all $i$.

The second term on the right hand side of Eq. (2) can be regarded as an "external force" acting on the website $i$. The coefficient $A_{ij}$ should satisfy the following. First, the force has to be global, that is, the website $i$ experiences a force from all the others. Since websites distributed over the Internet can be accessed by a few clicks of a button [4], accessing a website does not depend on the geographical degree of freedom. Second, the force term should include environmental changes in the Internet, such as the bandwidth, Internet technologies, and topology. Since it is difficult to take these changes into account explicitly, we describe the influence of the environmental changes via a stochastic process. The environmental fluctuations in essence can be modeled as a random process, thus it is convenient to express these as a Gaussian white noise process.

More specifically, during $\Delta t$, all factors for the environmental fluctuation are absorbed into a stochastic noise, which leads to a stochastic differential equation in time step $\Delta t$. That is, we lump all environmental influence on websites during $\Delta t$ into a stochastic variable. Thus, we can write

$$A_{ij} \rightarrow \langle A\rangle + \kappa\,\eta_{ij}(t)\ , \tag{5}$$

where $\kappa$ is a time independent parameter representing the noise amplitude (or force strength) and $\eta_{ij}(t)$ is a Gaussian white noise characterized by

$$\langle \eta_{ij}(t) \rangle = 0 \text{ and } \langle \eta_{ij}(t)\eta_{kl}(t) \rangle = \delta(t-s)\delta_{ik}\,\delta_{jl}\,. \tag{6}$$

Note that we set $\langle A \rangle = 0$ for simplicity. We also assume that the external force strength acting on the website $i$ depends on the number of websites influencing the website $i$. The physical implication of this assumption is that as the number of websites increases, the effective force strength from each website onto the website $i$ decreases. Thus we take $\kappa \rightarrow \kappa / N(t)$ for the normalization.

With this stochastic nature of the external force term together with the exponential growth of the number of websites, the dynamics of the number of visitors to the website $i$ can be expressed as

$$\frac{\Delta X_i}{\Delta t} = X_i + \frac{\kappa}{N(t)} \sum_{j \neq i}^{N(t)} \eta_{ij}(t) X_j\,, \tag{7}$$

where $\Delta X_i(t) = X_i(t+\Delta t) - X_i(t)$, and $N(t)$ satisfies Eq. (3). From the model, one finds that there are two parameters, $\kappa$ and $\lambda$: $\kappa$ being the noise strength and $\lambda$ being the growth rate of the number of websites. Since the number of websites in the model is not constant but increases in time, it is not easy to solve the coupled dynamic equation analytically.

## 3 Simulation Results

With the stochastic dynamic equation of Eq. (7) and the exponential growth of the number of websites, we perform numerical simulations. In the simulation, we start with a small number of the websites (say, $N(0) = 10$) and at every time step $\Delta t$, a new website is added to the system with the probability .

Figure 1 shows cumulative distribution functions (CDF) of the number of visitors to websites with different number of total websites $N_{total}$ at the end of each simulation. In the simulation, the growth rate and the force strength are held fixed as $\lambda = 0.5$ and $\kappa = 2.0$. From Fig. 1, one obtains a power law distribution as CDF, $C(x) \approx x^{-\alpha}$, and finds $\alpha \approx 0.5$. In terms of the probability density function (PDF), $P(x)$ of the number of visitors to the websites, we get $p(x) \propto x^{-\beta}$ with an exponent $\beta = 1 + \alpha \approx 1.5$. One can also see that the distribution of the number of visitors to websites follows a universal power law with the same exponent $\beta$ irrespective of the total number of websites, $N_{total}$.

To see the effect of the force strength, we carried out simulations with different force strengths $\kappa$ while keeping the other parameters fixed ($N_{total} = 2000$ and $\lambda = 0.5$). As can be seen in Fig. 2, the results for different $\kappa$ fall into the same distribution, thus one can infer that the exponent does not depend on the force strength $\kappa$.

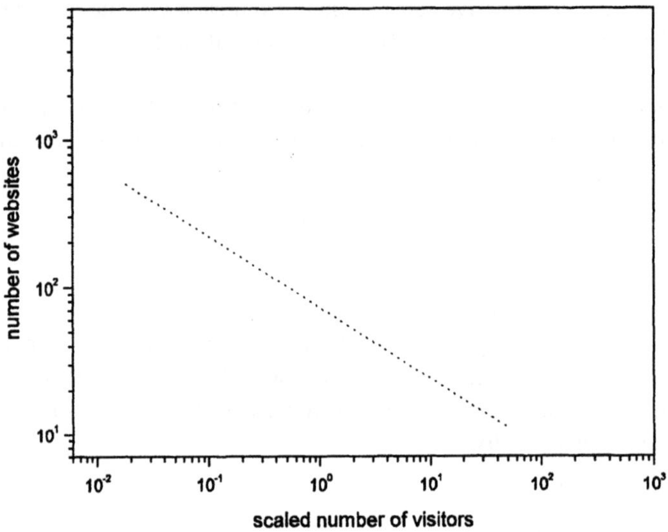


Figure 1 : Log-log scale plots of cumulative distribution functions of the number of visitors for the total number of websites at the end of simulations $N_{total} = 1000$ (_), $N_{total} = 2000$ (_), and $N_{total} = 5000$ (_). The dotted line has slope -0.5.

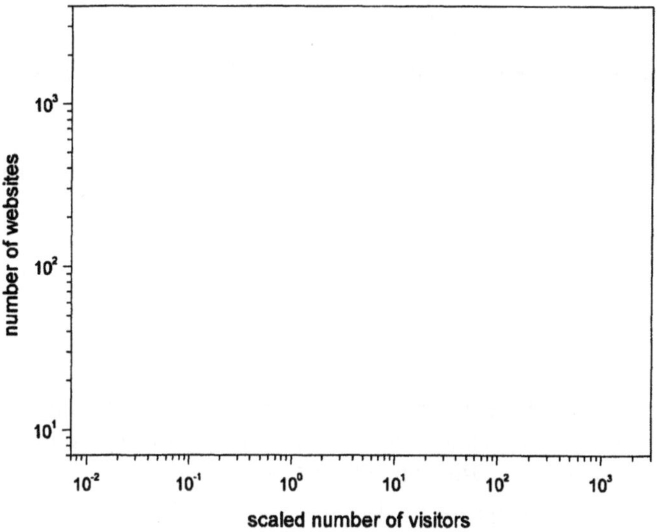


Figure 2 : Log-log scale plots of cumulative distribution functions of the number of visitors for $\kappa = 0.5$ (_), $\kappa = 1.0$ (_), and $\kappa = 2.0$ (_).

The force term in the model is responsible for the fluctuation of the number of visitors to websites. It is found in Ref. [1] that the fractional fluctuations in the number of visitors for a given website are uncorrelated to each other. To show this, we calculate the quantity

$$g(t) \equiv \frac{X(t+\Delta t) - X(t)}{X(t)}, \tag{8}$$

as a function of time. This random fluctuation of the fractional growth can be verified in terms of the auto-correlation. The calculation of the auto-correlation function shows that the fractional fluctuation is linearly uncorrelated. It should be also stressed that this uncorrelated fluctuation is independent of the force strength $\kappa$ as well as the growth rate $\lambda$.

The power law distribution observed in Fig. 1 and Fig. 2 can be derived analytically within an appropriate approximation. Following the procedure similar to Ref. [5], we plot in Fig.3 the number of visitors $X_i(t)$ to various websites as a function of time with parameters $N_{total} = 2000$, $\kappa = 1.0$, and $\lambda = 0.5$. From Fig. 3 one can obtain an approximate differential equation for $X_i$ as

$$\frac{\partial \ln X_i(t)}{\partial t} \approx \alpha, \tag{9}$$

where $\alpha$ is estimated from Fig. 3 as $\alpha \approx 1$.

The solution to Eq. (9) is given as

$$X_i(t) = m_0\, e^{(t-t_i)}, \tag{10}$$

where $m_0 = X_i(0)$, and $t_i$ is the time at which the website $i$ is added to the system. Equation (10) implies that older websites (smaller $t_i$) increase their visitors at the expense of younger ones (larger $t_i$); « rich-get-richer » phenomenon that was observed in the dynamics of the various networks [5].

With the above result, we can get the probability distribution analytically. The probability that a website $i$ has visitors smaller than $x$, $P(X_i(t) \le x)$, can be written as $P(t_i \ge \tau)$, where $\tau = t - \ln(x/m_0)$. Note that $P(t_i \ge \tau)$ is the probability that the website $i$ can be found in the system up to time $\tau$. Therefore, the desired probability is just a fraction of the number of added websites up to time $\tau$ to the total number of websites up to time $t$. Thus we have

$$P(t_i \ge \tau) = 1 - P(t_i \le \tau) = 1 - e^{\lambda(t-\tau)}, \tag{11}$$

where $\lambda$ is the growth rate of the number of websites. With the above, we get,

$$P(X_i(t) \le x) = 1 - (m_0/x)^{\lambda}, \tag{12}$$

which yields

$$P(x) = \frac{\partial P(X_i(t) < x)}{\partial x} \propto x^{-(1+\lambda)}, \tag{13}$$

from which the exponent of the power law distribution can be obtained as $\beta = 1 + \lambda$. This result is consistent with the simulation results that are shown in Fig. 1 in which we obtained $\beta \approx 1.5$ with $\lambda = 0.5$.

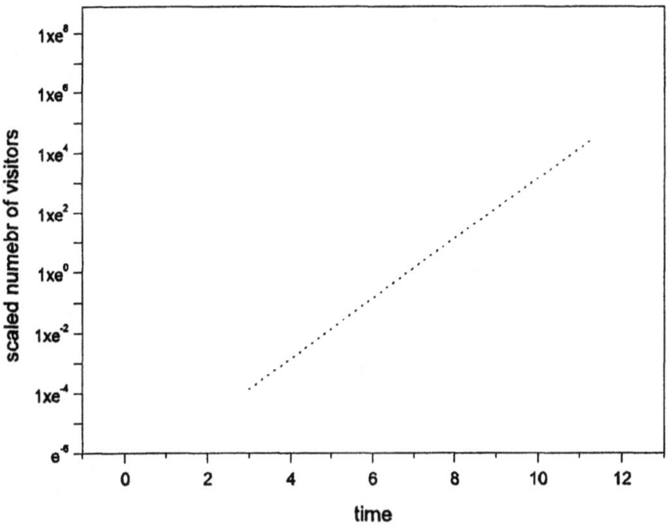


Figure 3 : Time evolution of the number of visitors $X_i(t)$ for websites $i = 10$ (_), 50(_), and 100(_) added to the system. The ordinate is in the logarithmic scale with the natural base and the dotted line has slope 1.0

From the above result, we infer that the exponent in the power law distribution depends only on the growth rate $\lambda$: the higher is the growth rate, the greater the exponent. This relationship between $\beta$ and $\lambda$ also explains dependence of the exponent on the category of the websites that are observed in the empirical study [1]. In Ref. [1] it was found that for the .edu category the power law exponent $\beta = 1.45$, while for all categories the exponent $\beta = 2.07$. Since the growth rate $\lambda$ of all categories is greater than that of one specific category (.edu category for instance), the exponent for overall websites should be greater. This explains why the exponent for overall websites is bigger than that for .edu sites.

# 4 Summary and Conclusion

In this paper we investigated the origin of the empirically observed power law distribution of the number of visitors to websites. In order to explain the characteristics of the websites, we established a stochastic dynamic model, which includes the following: the growth of an individual website, the external forces acting on each website, and the exponential growth of the number of websites. With the model, we were able to show most of the characteristics of the dynamics of the websites, such as power law distributions of the number of visitors to websites and the fluctuation in the individual website's growth. Moreover, we found that the exponential growth rate $\lambda$ of the number of websites determines the exponent $\beta$ in the power law distribution: the higher the growth rate, the bigger the exponent. We also performed an analytic calculation and compared the result with that of the numerical simulations. Within the approximation we formulated the exponent in terms of the growth rate $\lambda$ and confirmed the simulation results.

Thus the key ingredients in the dynamics of the websites are the following. First, there is a global interaction in terms of the stochastic force strength among websites with which one can view the web ecology as a competitive complex system. Second, the web ecological system stays in non-equilibrium in the sense that the number of the websites in the system is not fixed but exponentially increased. These two ingredients in the web ecological system lead to the characteristics of the system.

Needless to say, this approach is not the unique way to explain the power law nature of the dynamics of the websites. Other approaches that lead to the same characteristics of the dynamics of the websites are possible and one candidate model might be the one in which the interaction among websites are included. This could be one of the further directions of research in this field. This work was supported by Grant No. R02-2000-00292 from the Korea Science & Engineering Foundation (KOSEF).

# Bibliography


[1] L. Adamic and B. Huberman, 2000, Quarterly Journal of Electronic Commerce, 1, 5.

[2] Source for the exponential growth of the websites are from the World Wide Web Consortium, Mark Gray, Netcraft Server Survey and can be obtained at http://www.w3.org/Talks/1998/10/WAP-NG-Overview/slide10-3.html.

[3] It is known in Ref. [2] that between August of 1992 and August 1995, the number of web servers increases 100 times for every 18 months, and between August 1995 and February 1998, 10 times every 30 months.

[4] R. Albert, H. Jeong, and A.-L. Barabasi, 1999, Nature, 401, 130.

[5] A.-L. Barabasi and R. Albert, 1999, Science, 286, 509.

# Chapter 20

# Design Patterns for the Generation and the Analysis of Chemical Reaction Networks

**Hugues Bersini**
IRIDIA – ULB –CP 194/6
50, av. Franklin Roosevelt
1050 Bruxelles - Belgium
bersini@ulb.ac.be

## 1. Introduction

Chemical reactions among molecules rapidly create a complex web of interactions where it is nearly impossible by analytical means to predict which molecule will emerge in high concentration and which will not « survive » the network interactions. Among others, a lot of factors make this prediction quite delicate: The description of the molecules and the reaction mechanisms, the calculation of the molecular internal energy and accordingly the establishment of the reaction rate, the non-linearity of the reaction dynamics (made even more complicated by the order of the reaction). The reaction network evolves according to two levels of change called dynamics and metadynamics. The dynamics is the evolution in time of the concentration of the units currently present in the network. Their concentration changes as a function of their network interaction with the other units. The metadynamics amounts to the generation of new molecules by recombining chemical materials constituting the molecules existing so far in the network. An analytical approach is made hard by the continuous appearance of new variables in the set of the kinetic differential equations. Computer simulations could help a lot in facilitating the prediction of the structure of the emerging network as well as the nature of the "winning" and "loosing" molecules.

The model presented here is trying to capture in software, and in a very preliminary attempt, some scholar chemistry like the molecular composition and combinatorial structure, basic reaction mechanisms as chemical crossover, bonds opening/closing, ionic reactions, and simple kinetics aspects as first or second order reactions. A lot of finer chemical aspects are still however left aside (which reaction takes place, which active group and which bonds are really involved, what is the value of the reaction rate,...) but in such a way that it should be easy for a more informed chemist to parameterize and precise the simulation in order to account for the influence of these aspects.

To make this parameterization and the interface with the chemist possible, Object Oriented (00) type of computation appears to be the natural way to follow, at least today. The choice of object-oriented programming results from the attempt to reduce the gap between the computer modeling and its "real-world" counterpart. It is intrinsic to OO programming that by thinking about the problem in real-world terms, you naturally discover the computational objects and the way they interact. The Unified Modeling Language (UML) is a major opportunity for our complex system community to improve the deployment, the better diffusion and better understanding of the computer models, and especially when proposed to researchers not feeling comfortable reading programs.

The next section will present, through the UML class diagram, the basic architecture of the chemical reactor software. The main classes are « Atom », « Molecule », « AtomInMolecule », « Link », « Reaction » and all the sub-classes of possible reactions. Classical problems like the « canonization » of molecules will be just sketched in the following section. In the fourth section, some simulated reaction mechanisms will be presented while the fifth will sketch the complete chemical simulator and some preliminary results obtained in mineral and organic chemistry, departing from very simple molecules. These results will testify for the hardness in predicting the emerging molecules of the reaction network, and the benefits gained in resorting to this type of computer simulations.

## 2. The Chemical OO Class Diagram

UML proposes a set of well-defined diagrams (transcending any specific OO programming language) to naturally describe and resolve problems with the high level concepts inherent in the formulation of the problem. It is enough to discover the main actors of the problem and how they mutually relate and interact in time to build the algorithmic solution of this problem. A simple and introductory overview of the UML language can be found in (Eriksson and Penker, 1998). However, by deliberately restricting our use of UML to the only class diagram, reader familiar enough with OO programming should not have any understanding problem. The main class diagram of the chemical computer platform is shown in figure 1.

The main classes necessary to represent the molecular structure as a computational graph are: « Molecule », « Atom », « Group », and their three ionic counterparts, « MolecularIon », « AtomicIon », « IonicGroup ». Additional classes are « Link » and

« AtomInMolecule ». The Atom is the first basic class of the whole system. A fundamental attribute is the *valence*, which indicates in which proportion this atom will connect with another one to form a molecule. For instance, an atom "1" (this is the value of the identity) with valence 4 (such as carbon) will connect with four atoms "4" of valence 1 (such as hydrogen) to form the molecule: *1(4 4 4 4)*. The second major attribute is the *identity*, inherited from the super-class chemical component, which, in our simulation, relates to the value of the variance. Atom with a high variance will be given a small identity value. This identity simply needs to be an ordered index ("1", "2") for the canonical rules shaping the molecular graph to be possible. Since this identity takes a unique value for each atom object, the way it is defined depends on what we take to be unique to any atom. Actually, in chemistry this identity is given by the atomic mass.

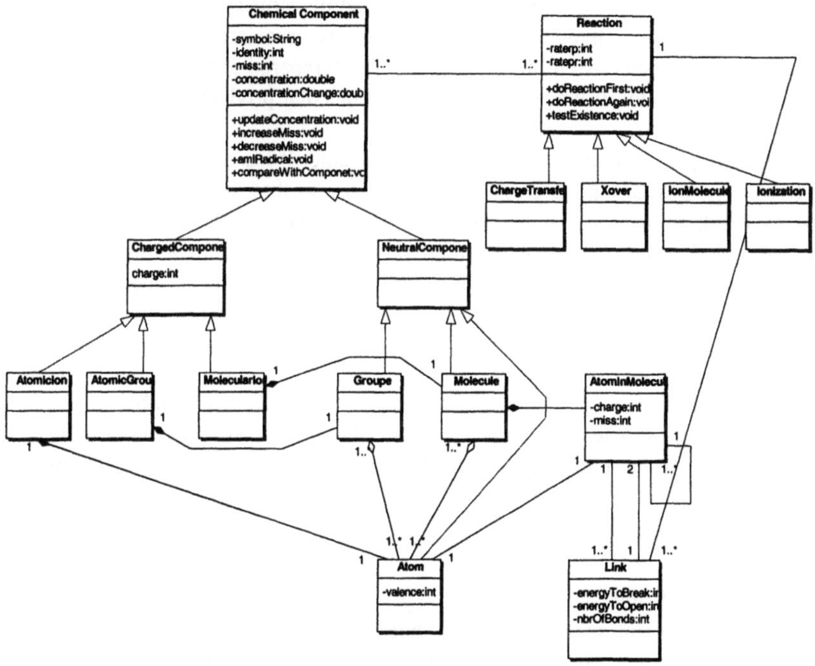


Figure 1: The class diagram of the OO chemical computational platform

The second basic class is the Molecule. As shown in the class diagram, molecule objects are compounds of atom objects. An attribute called *numberOfInstances* is a vector of integers whose elements are the number of times one specific atom appears in the molecule (i.e. four "4" and one "1" in the molecule *1(4 4 4 4)*). Molecules can be represented in a string linear form (like *1(4 4 4 4)* for the methane), called their symbol (inherited from the chemical component super-class). Concentration is a very essential property of any molecule because, for obvious memory size constraints, an object of the class molecule is not a single chemical element but rather the set (whose cardinality is the concentration) of all identical chemical elements. A molecular object remains one and only one specific object, with its concentration evolving as a result of chemical reactions and their specific rate. Molecules can be compared, and a set of methods is defined to describe the molecule or to retrieve part of it, such as atomic sub-groups, weak or strong links. Molecules, groups and atoms have their ionic counterparts with, as only difference, the additional presence of a "*charge*" attribute.

Molecules are graphs that are computationally structured (in a canonical form to be discussed in the next section) with nodes of class AtomInMolecule pointing to other atomInMolecule nodes. Each molecule possesses one and only one AtomInMolecule attribute called the *headAtom* and which can be seen as its "front door" (it would be the "1" in the molecule *1(4 4 4 4))*. As soon as an atom enters into a molecule, it is transformed into an AtomInMolecule object. AtomInMolecule relates to atom since the identity of such an object is the same as its associated atom. Using natural recursive mechanisms, AtomInMolecules can be compared and duplicated to be part of new molecules (for instance to compose the molecular products of some reactions).

Finally, an object Link connects two atomInMolecule. It has a *number of bonds* and a given *energyToBreak* and a given *energyToOpen*, so that the weakest links are the first to break or to open in the reaction mechanism. For instance, two atoms of valence 4 will connect (to form a diatomic molecule *4(4)*) with a link containing 4 bonds, and one atom of valence 4 will connect with four atoms of valence 1 (*1(4 4 4 4)*), each link now containing one bond. Link objects intervene in the unfolding and the coding of the reaction mechanisms. For instance, one major method associated with the class Link is "*exchangeLink*" involved in crossover molecular reactions. Additionally, links can be opened and increase or decrease their number of bonds. In the model here, the energy of any link will only depend on the identity of the two atomic poles and the number of bonds. These energies are given at the beginning of the simulation

*The Reactions*

Reaction is an abstract class that can only pave the way to derived or sub classes representing different specific reaction mechanisms like *crossover*, *openBond* and *closeBond*, *ionization*, *ionMolecule*, that will be illustrated next. Actually in our simulation, 20 sub-classes of Reactions inherit from the super-class. Common to all reactions is that they associate chemical reactants (molecule, molecularIon, Atom, AtomicIon) with chemical products (the same). In the simulations to be presented, chemical reactants and chemical products are at most two. Also reactions involve links. For instance, two links are exchanged in the crossover reaction and one link is just open to free some bonds in the openBond reaction. Among the several methods to be "concretized" in the derived classes, the "*doTheReactionFirst*" realizes the reaction for the first time, namely generates the molecular products, test their existence (to verify is they are new or already existing), calculates the rate (in a way to be explained next) and executes the reaction just once. The "*doTheReactionAgain*" method just executes the reaction one time and boils down to modify the concentration of the molecular reactants and products according to the reaction rate. For obvious reasons of computational economy, an object reaction is first created and calibrated by the doTheReactionFirst method, then it is repeatedly executed by the doTheReactionAgain method

### 1.3 The Molecular Unique Computational Structure

Whatever chemical notation you adopt, for instance the "line-bond" or Kékulé, one way of reproducing the connectivity pattern of a molecule is by means of a computational graph. Here a symbolic linear notation perfectly equivalent to the graph, is adopted to describe the molecular computational graph. One example will be enough to understand it. Take the following molecule:

*1 ( 1 ( 4 4 4 ) 2 ( 1 ( 3 3 3 ) ) 2 ( 2 ( 3 ) ) 2 ( 4 ) )*

"1" is an atom with valence 4, "2" is an atom with valence 2, and "3" and "4" are atoms with valence 1. The graphic tree version is given in fig.2.

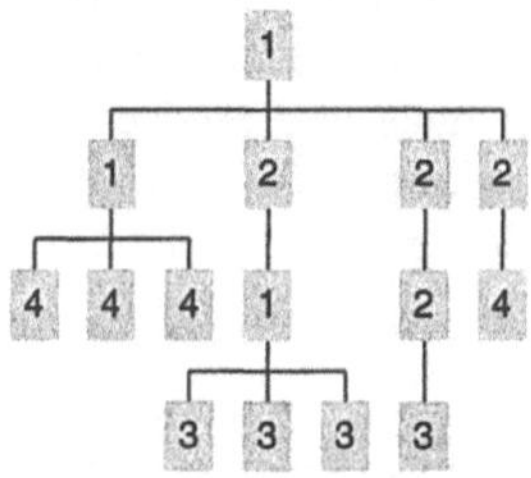

Figure 2: The computational tree structure for a molecule

In our linear notation, a "butane" molecule become: 1( 1 (1 (4 4 4) 4 4) 1 (4 4 4) 4 4) This string notation is automatically generated from the graph structure by a method being recursively executed through the graph. Cycles need to be detected by the method and are indicated by adding brackets at the two atomInMolecule which close the graph, like in the example below representing the benzene.

1[1]( 1( 1( 1( 1( 1[1]( 4( 4 ( 4( 4( 4( 4)))))))))))

In a first approximation, the connectivity shows symmetry both vertically and horizontally. The following rules need to be respected in order to shape the molecular graph in a unique canonical way:

**Vertically**: The highest node, i.e. the front door of the molecule (the initial "1" in our example of fig.2) must be the smallest of all the AtomInMolecule objects composing the tree.

**Horizontally**: Below any node (i.e. any AtomInMolecule) the sub-nodes are arranged from left to right in an increasing order, the smallest to the left, the greatest to the right.

Clearly these two rules depend on the definition of "smaller" between two AtomInMolecule nodes. It is precisely defined in a way discussed in (Bersini 2000a and b). Organizing the graph is such a canonical way allows differentiating two structural isomers, i.e. molecules that contain the same number of the same atoms but in a different arrangement (like the chemical examples of the butane and the methylpropane molecules, both $C_4H_{10}$ but with distinct connectivity patterns). However such a re-organisation can miss the differences still remaining between molecules showing a same connectivity pattern but with different spatial organisations i.e. the geometrical isomers.

## 4 The different reaction mechanisms

Several reaction mechanisms called "decomposition", "combination", "replacement" are repeatedly described in the chemical literature. Based on our syntactical definition of what is the molecular identity, a new nomenclature will be used here for the possible chemical reactions, with a simple illustration for a few of these reaction mechanisms. Every time a new molecular product is created as a result of the combination of two molecular reactants, this new molecule needs to be reshaped according to the canonical rules previously defined. Among the 20 reactions implemented in the code, there are:

*Single-link crossover*: the weakest links of each molecule are exchanged (suppose that "1" has valence 4, "2" has valence 2 and "3" and "4" have valence 1, and suppose the link "2-3" is weaker than the link "2-4").

*1(1) +* ***[4]*** *2 (3 4) → 1(3 3 3 3) + 1(2(4) 2(4) 2(4) 2(4))*

The bold values between brackets are *stoichiometric coefficients* needed in order to balance the chemical equations

*Open-bond reactions*: in the first molecule, a link containing i bonds opens itself to make j links of i/j bonds (here the first link "1-2" of the first molecule opens itself (i.e. frees one bond) and the first link of the second molecule breaks)

*1(2 2) + 2(3 3) → 1(2(3) 2(3) 2)*

*Charge Transfer reaction*: it just amounts to the transfer of charges between two components

$1^+ + 4^- \rightarrow 1^- + 4^+$

Each reaction is an object with (as attributes) at most two chemical reactants and two chemical products. Another important attribute to derive is the reaction rate. Suppose the following reaction: A+B->C+D, the reaction rate $K_{ABCD}$ is calculated as follows: $K_{ABCD} = \alpha exp(-E_{ABCD}/\beta T)$ where α and β are two parametric attributes of any reaction. Their chemical meaning derive from the orientation and the shape of the molecules and influence the likelihood of the reaction. T is the temperature. $E_{ABCD}$ is the activation energy of the reaction and, in a very naïve preliminary attempt, is calculated as follows. Every link of the molecule has a certain energy associated with it, which can be seen as the energy required to break the link:

$E_{ABCD}$ = *(Σ links energy broken in the reactant – Σ links energy created in the product) + Δ* *if the first difference is positive*

$E_{ABCD} = \Delta$ *otherwise*

A reaction leading to a more stable molecule (i.e. when the difference is negative) needs to absorb much less energy to take place (just the energetic barrier Δ). This reaction is exothermic and much more likely and faster. Indeed the chemical kinetics (just first order reactions are considered) drives the concentration evolutions:

$$d[A]/dt = d[B]/dt = -K_{ABCD}[A][B]$$
$$d[C]/dt = d[D]/dt = K_{ABDC}[A][B]$$

Whenever the Reaction method "*doTheReactionAgain()*" is executed, a one time step integration of the four concentrations is achieved.

## 5 Results

In the final simulator, three vectors of objects are important: one containing all molecules, one containing all new molecules appearing at each time step of the simulation (and which will participate to create new reaction objects) and one containing all the reaction objects. In order to avoid the computational explosion that will rapidly occur as a consequence of the new molecules appearing in the simulation, molecules will enter a reaction only if their concentration exceeds a certain threshold. For similar reasons, only the weakest links of any molecules will be opened or broken for the reaction to occur.

Let's now present two simulation results obtained, one for mineral and the other for organic chemistry. In the first one, the simulation is released with the following two molecules: 2(2) and 4(4), 2 is an atom with valence 3 (like N). After a great number of simulation steps the following molecules appear in the simulator: 2 ( 4 4 4 ) , 2 ( 2 ( 4 ) 4 ), 2 ( 2 ( 4 4 ) 2 ( 4 4 ) 2 ( 4 4 ) ) , 2 ( 2 ( 2 ( 4 ) ) 4 4 ) , 2 ( 2 ( 4 4 ) 4 4 ), 2 ( 2 ( 4 4 ) 2 ( 4 4 ) 4 ), 2 ( 2 ( 2 ( 4 ) ) 2 ( 2 ( 4 ) ) 2 ( 2 ( 4 ) ) ), 2 ( 2 ( 2 ( 4 ) ) 2 ( 4 ) ), 2 ( 2 ( 2 ( 4 4 ) 4 ) 2 ( 4 4 ) 4 ), 2 ( 2 ( 2 ( 2 ( 4 4 ) ) ) 4 4 ) with their respective final concentration. The second simulation departs with one organic molecule: the benzene, and a second diatomic molecule 5(5) (it could be $Cl_2$). An explosion of new organic molecules appear like: 1[1] ( 1 ( 1 ( 1 ( 1 ( 1 ( 1 ( 1 ( 1 ( 1 ( 1[1] ( 4 ) 4 ) 4 ) 4 ) 4 ) 4 ) 4 ) 4 ) 4 ) 4 ) 4 ) 4 ), 1 ( 1 ( 1 ( 1 ( 4 5 ) 4 ) 4 ) 1 ( 1 ( 4 5 ) 4 ) 4 ),

1[1](1(1(1(1(1[1](4)4)4)4)45)45),1[1](1(1(1[1](4)4)4)4),1(1(4)4),1[1](1
(1(1(1(1(1(1(1(1(1(1(1(1(1(1(1(1[1](4)4)4)4)4)4)4)4)4)4)4)4)4)4)
4)4)4), 1(1(1(1(1(1(1(45)4)4)4)4)4)1(1(1(1(1(45)4)4)4)4)4),1[1](1(1
(1(1(1(1(1(1(1(1(1[1](4)4)4)4)4)4)4)4)4)4)45)45),1[1](1(1(1(1(1(1(1
(1(1(1(1(1(1(1(1(1(1(1(1(1(1(1(1[1](4)4)4)4)4)4)4)4)4)4)4)4)4)4)
4)4)4)4)4)4)4)4)4),1[1](1(1(1(1(1(1(1(1(1[1](4)4)4)4)4)4)4)4)4),
1[1](1(1(1(1(1(1(1[1](4)4)4)4)4)4)4),1[1](1(1(1(1(1[1](4)4)4)4)45)1(
1(1(1(1(1(1(1(1(1(1(1(45)4)4)4)4)4)4)4)4)4)4)4),1[1](1(1(1(1(1[1]
(45)4)4)45)45)45),1[2](1[1](1(1(1(1(1[1](4)4)4)4)45)4)1(1(1(1(1[2](4)
4)4)4)45)4).

One can verify the way all molecules are canonized. In both cases, an additional result, not shown here, is the evolution in time of the concentration of all these molecules, so that it is easy to rank them by their concentration and to pinpoint which of them emerge in high concentration and which tends to disappear.

## 6 Conclusions

Without doubt, this paper has a richer computational than chemical content. All the chemistry addressed in the simulations is rather naïve. The motivation is indeed to invite chemists to bring all the necessary chemical complements to improve the realism of the computational platforms. The design patterns presented in the UML form should be more attractive for them than pure code. Using UML to describe molecules and reactions is an effort to be compared with projects like CML or SMILES, to express chemistry by understandable and semi-formal computer tools. It puts pressure on chemists who find some interest in these tools to clarify in computational terms some still fuzzy chemical notions, like the definition of the isomerism or the precise description of the reaction mechanisms. However the tools presented here must be seen as complementary since the emphasis is not on the coding of these computational objects but on the simulation of their behavior and interactions. Making prediction of chemical reactions, like for instance which molecules will be presented in high concentration once the reactions reach its equilibrium, is the challenge. Not only the software, once achieved a certain chemical realism, could help to predict the reaction outcomes, but like the "tracing" of an expert system reasoning, the reaction network graphs could help to explain why these molecules indeed emerge. Also this reaction network could be used to anticipate the effect of increasing the concentration of some of the molecules, and in general to anticipate the displacement of the equilibrium resulting from any perturbation.

## References

Bersini, H. (2000): Chemical crossover – In Proceedings of the GECCO-2000 conference – pp. 825-832.

Bersini, H. (2000): Reaction mechanisms in the OO Chemistry – In Proceedings of the Artificial Life 7 Conference.

Eriksson, H-E, Penker, M. (1998): UML Toolkit – John Wiley and Sons

# Chapter 21

# On the adaptive value of Sex

**Klaus Jaffe**
Departamento de Biología de Organismos
Universidad Simón Bolívar
kjaffe@usb.ve

Using computer simulations I studied the conditions under which sex was evolutionary stable. The parameters that showed relevance to the stability of sex were: variable environments, mutation rates, ploidy, number of loci subject to evolution, mate selection strategy and reproductive systems. The simulations showed that mutants for sex and recombination are evolutionarily stable, displacing alleles for monosexuality in diploid populations mating assortatively when four conditions were fulfilled simultaneously: selection pressure was variable, mate selection was not random, ploydy was two or the reproductive strategy was haplo-dipoid or hermaphroditic, and the complexity of the genome was large (more than 4 loci suffered adaptation). The results suggest that at least three phenomena, related to sex, have convergent adaptive values: Diploidy, sexual reproduction (recombination) and the segregation of sexes. The results suggest that the emergence of sex had to be preceded by the emergence of diploid monosexual organisms and provide an explanation for the emergence and maintenance of sex among diploids and for the scarcity of sex among haploid organisms. The divergence of the evolutionary adaptation of the sexes is a derived consequence of the emergence of sex. A corollary of these simulations is that gene mixing, achieved by sex, is advantageous if the degree of mixing is not very great, suggesting that an optimal degree of gene mixing should exist for each species.

# 1 Introduction

What selective forces maintain sexual reproduction and genetic recombination in nature? The answer to this question has been an elusive mystery (Maynard-Smith 1978, Judson and Normak 1996, Hurst and Peck 1996). Asexual reproduction is theoretically much more likely to occur than sexual one due to at least three inherent advantages: parthenogenic females do not need to find mates; they produce twice as many daughters and four times as many granddaughters compared to the average sexual ones; and natural selection drives adaptation and thus selection of relevant genetic traits much faster in asexual organisms compared to sexual ones (Maynard-Smith 1978, Jaffe 1996). Despite these relative theoretical advantages of asexuality, most higher organisms are sexual. The various hypotheses put forward to explain this mystery can be grouped into three broad categories:

1- The ecological genetic models and the Red Queen Hypothesis which postulate that sex is adaptive in variable environments or variable parasite pressure because it enables genetic variation and the rapid spread and creation of advantageous traits (Bell and Maynard-Smith 1987, Hamilton et al 1990, Ebert and Hamilton 1996, Howard and Lively 1994). This model has been shown to be incomplete in explaining the emergence and maintenance of sex (Ochoa and Jaffe, 1999 for example)

2- The mutation-accumulation models (Muller 1964, Hill and Robertson 1966, Kondrashov 1984, 1988, 1994, Taylor and Williams 1982, Heisler 1984), which suggest that sex is adaptive because it performs the efficient removal of deleterious mutations or DNA repair. Experimental results have shown that this model can not explain the genetic dynamics of extant sexual organisms (Cutter and Payseur 2002 for example).

3- The mate selection models, which assume that sex allows for the selection of 'good genes' by orientating the evolutionary process towards the fixation of beneficial traits (Kodric-Brown and Brown 1987, Jaffe 1996, 1999). Specifically, assortative mating has been shown to be very successful in increasing the fitness of sexual species (Davis 1995, Jaffe 1998, 2000). Here I want to explore this last model further.

The model Biodynamica used here (Jaffe 1996, 1998, 1999, 2000, 2001, 2004), has been shown to have heuristic properties in explaining or predicting experimental data. It explains many aspects of the emergence of genetic resistance to antibiotics and pesticides (Jaffe et al 1997), it predicted divergent behavior of production of males in facultative sexual nematodes (Rincones et al. 2001), the importance of economic aspects in the evolution of social behavior (Silva and Jaffe 2002), it predicted the existence of homophily among humans, such as the physical similarity between faces of married couples (Alvarez and Jaffe 2004) and the similarities dog pets and their human owners (Payne and Jaffe 2005), and it predicted sperm selection by sperm-plasma in humans (Jaffe et al. 2006). Thus it seemed promising in uncovering some remaining mysteries of sex.

# 2 Methods

In this multi-agent, adaptive model, each individual was simulated as an autonomous agent who interacted with the environment and with other individuals according to five evolutionary steps (see below) and to the alleles it carried in its set of up to 8 loci as given in Table 1. Simulations were competitions between agents with alleles coding for different strategies. The population of agents (organisms) after being created with a given random seed, suffered a 5 step evolutionary process which mathematically speaking (the program was built in visual-basic) is equivalent to the following:

**2.1 Mate selection:** Females of bisexual species choose a male of the same species, whereas hermaphrodites mated with a conspecific individual. When random mating was simulated, females and hermaphrodites mated with a randomly chosen mate, whereas in assortative mating females and hermaphrodites mated with the genetically most similar mate among 20 randomly chosen individuals (This number had been shown to be close to the optimal for asortative mating to work under the present set of parameters, see Jaffe 1999). Genetic similarity was estimated by comparing the phenotypes of both individuals. In some rare moments of some simulations no sexually mature mate was found by some females or hermaphrodites. Then the individual did not reproduce during that time step if bisexual, or reproduced monosexually if hermaphrodite. The simulations did not distinguish between mate selection and gamete selection, as the model simulated the transmission of only one gamete in each mating act.

**2.2 Reproduction:** The reproductive strategy could be for haploid (**H**) or diploid (**D**) organisms. If sexual (i.e. not monosexual) organisms could mate randomly (**RM**) or assortatively (**AM**). Thus, ten different reproductive strategies were simulated. Monosexuals simulated parthenogenesis or thelytoky. That is, monosexual organisms did not mate. In **H-Monosex** (monosexual haploids), the individual transmitted all its genes to the offspring (cloning) with no variance except that allowed by mutations, simulating asexuality. **D-Monosex** (monosexual diploids) did not mate and produced offspring by uniform random crossovers of the alleles in each loci of the parent. Bisexuals (either H- or D- and -RM or -AM) produced equal numbers of males and females randomly (**Bisexual-r**) or produced a biased ratio of 60 % more females (**Bisexual-b**). Males could mate several times each reproductive step. Hermaphrodites (either H- or D- and -RM or -AM) produced only females and reproduced similar to bisexuals if finding another hermaphroditic female (**Herma1**) or any female (**Herma2**), or else reproduced as the corresponding H- or D-monosexuals. Herma1-RM, thus, mated assortatively with females having the same disposition for sex, even when mating randomly regarding all other loci.

Females produced offspring according to their phenotypically determined clutch size (see below), transmitting their genes following Mendelian rules of reproduction (free recombination). If sexual, each parent provided half of its alleles to the newborn, so that for each locus, one allele came from each parent if diploid, or each parent had a probability of 0.5 to transmit its allele to each locus if haploid.

**2.3 Variation:** In each offspring, randomly selected genes mutated, changing their allelic value randomly in those loci which allowed for allelic variance, with a probability determined by their allele in gene 2 (Table 1).

**2.4 Phenotypic expression:** As commonly done with genetic algorithms and as it is known to occur frequently in real organisms, total allelic dominance was simulated. That is, in diploid organisms, only one allele per loci was expressed phenotypically during the lifetime of each organism, which was selected randomly at birth. In simulations comparing the relative evolutionary success of two alleles and which in addition aimed to assess the effect of allelic dominance on the competition between the two alleles, the dominant allele, defined by the experimenter, was phenotipically expressed if present in the diploid genome. For example, in experiments assessing the relative evolutionary between diploid hermaphrodites and diploid monosexuals (Fig 1) when the allele for hermaphroditism was programmed as dominant, then, if the corresponding allele was present in the organism, the individual would behave as a diploid hermaphrodite.

**2.5 Selection:** The model did not assume any simplified expression of fitness but reproduction and individual survival were decomposed into different aspects for selection to act. Individuals were excluded from the population at the end of each time step when any of the following criteria applied:

1- Their age exceeded their genetically prefixed life span.

2- When randomly selected with a probability which increased with population density as given by the formula:

$$\text{survival of individual i at time step t} = \begin{cases} 0 & \text{if } r_1 * N_t \geq ops * r_2 \\ 1 & \text{if } r_1 * N_t < ops * r_2 \end{cases}$$

where ops is the optimal population size, $N_t$ the population size at time-step t and $r_1$ and $r_2$ are random numbers between 0 and 1

3- Individuals not possessing the resistant phenotype of genes 6 to 8 in Table 1 were killed randomly each time step with probabilities which varied randomly each time step from 0 to 0.6, simulating an environment in which two different biocides or parasites trimmed the population by killing non resistant individuals.

Optimal size of populations was 400 and the initial size of the populations was 200 individuals.

**Table 1:** Genes and their possible alleles defining the agents-organisms. Simulations of genes with allelic variance allowed mutant alleles to appear in the range given below. Initial populations had individuals possessing any of the alleles indicated in that range. In simulations in which some genes had no allelic variance, the default allele, indicated in parenthesis, was assigned to all the corresponding loci in all organisms.

| Gene | Range | Effect on phenotype for alleles |
|---|---|---|
| 0 | 1-2 | Ploidy. Either haploid or diploid. |
| 1 | 1-6 | Reproductive strategy |
| 2 | 0-10 | Mutation rate: from 0.2 to $10^{-7}$ mutations per gene in logarithmic decrements (0.008) |
| 3 | 0-10 | Maximum life span coding for life spans from 0 to 10 time steps (5) |
| 4 | 0-10 | Clutch size from 0 to 10 offspring (5) |
| 5 | 0-5 | Minimum age for initiating reproduction of females in t-steps (0) |
| 6 | 0-10 | Resistance to biocide 1: Only allele 0 was resistant to that biocide (0) |
| 7 | 0-10 | Resistance to biocide 2: Idem as gene 6 but for biocide 2 (0) |
| 8 | 0-10 | Resistance to biocide 3: Idem as gene 6 but for biocide 3 (0) |

# 3 Results

Mutant alleles coding for sexuality displaced the corresponding alleles coding for monosexual strategies when simulating hermaphrodites or diploid bisexual organisms which mated assortatively and produced more females than males. Mutant alleles coding for sexual diploid hermaphrodites with assortative mating displaced alleles coding for monosexual diploid alleles from the populations. Bisexuality seemed unable to displace monosexuality in haploid populations (see Figure 1).

When making tournaments, in which two alleles coding for different sexual phenotypes had to compete with ach other in invading a population of agents, we find that the most successive sexual strategies in displacing asexuality in the evolutionary game are a combination of bisexuality with assortative mating, with a large variance in males reproductive success, and with sperm selection acting on spermatozoa prior to the production o a new offspring (see Figure 2).

More results can be obtained independently by downloading the program Biodynamica at http://atta.labb.usb.ve/Klaus/klaus.htm

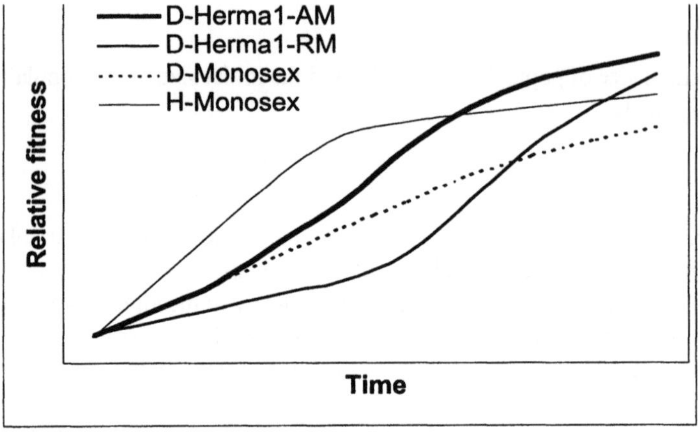


**Figure 1**: Approximate time course of mean fitness of organisms in an evolving population using different reproductive strategies.

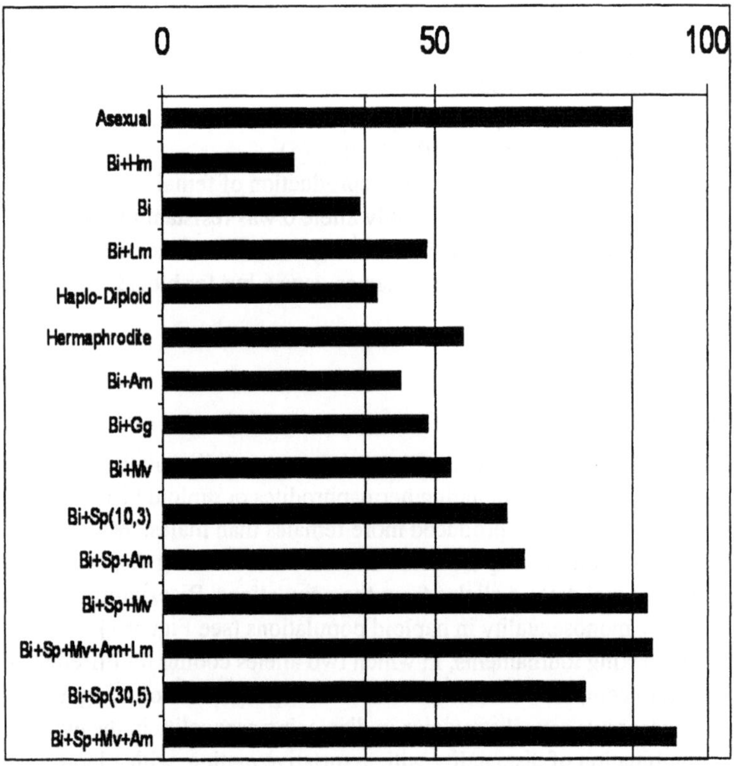


**Figure 2:** Percent of alleles of the given reproductive strategy present in a population of 800 agents in relation to alleles coding for monosexual diploids (thelytoky), when alleles coding for both strategies are competing between them in a single population. Bars smaller than 50% indicate strategies less efficient than monosexual diploids; bars larger than 50% indicate

strategies that are more successful than the monosexual diploids. Each bar is the average of 2000 simulations using the agent based computer simulation Biodynamica (Jaffe, 1995). Tournaments were run, starting with populations having a 50/50 proportion of alleles for monosexuality and for the ones indicated in the y axis. Abbreviations indicate; Bi: bisexual, Haplo-Diploid: females are diploid and males haploid, Am: assortative mating, Gg: females select males for good genes, Mv: reproductive variance in males was twice the variance for females, Lm: low mutation rate was 0.008 mutations per loci, Hm: high mutation rate was 0.2 mutations per loci. Else 0.04 random mutations per loci were simulated. Sp(x,y) indicate simulation of x spermatozoa per male expressing y genes.

# 4 Discussion

I postulated earlier (Jaffe 2000) that the genetic variance produced by sex differs from that produced by random mutations in that sex with assortative mating produces a better blend of variation, allowing faster adaptation in scenarios with very large genetic combinatorial possibilities, if compared to random mating. That is, sex slows the speed of evolution (Jaffe 1996) as advantageous mutations are not always transmitted to the offspring and are often mixed with disadvantageous alleles in other loci during recombination. Assortative mating reduces the extent to which this "dilution effect" of advantageous mutations occurs (Jaffe 1999), by reducing the variance of allelic composition between mates and thus producing offspring which have a greater likelihood of possessing the advantageous genes of their parents. Thus, assortative mating accelerates the fixation of advantageous alleles in the population canceling the effect of sex in slowing evolution. On the other hand, the long term advantage of sex is that it can produce advantageous blends of alleles faster than asexual reproduction does, but only if the number of loci is large (Jaffe 1998). For genomes with low genetic complexity (number of loci), mutations together with asexual reproduction is faster than sex in achieving optimal allelic combinations in the genome. Thus, the advantage of sex will be evidenced only if organisms do not mate randomly and the simulated genome has sufficient complexity (Ochoa and Jaffe 1999). Most studies on the emergence and maintenance of sex have focused on models using random mating, failing to find this phenomenon.

Yet, other simulation models have not obtained these results. The careful comparative analysis of the various models used allows us to establish a list of critical features in models that may explain these differences:

- Modeling diploid organisms versus haploids
- Modeling true random mating versus truncated fitness determining reproduction
- Modeling agents with many genes versus agents with up to 3 genes
- Modeling various selection criteria versus assessing a fitness value

Thus, I suggest that the more realistic the agent based simulations become, the closer we get to understand the mysteries of biological evolution.

# Bibliography

[1] Alvarez L., Jaffe K. 2004. Narcissism guides mate selection: Humans mate assortatively, as revealed by facial resemblance, following an algorithm of "self seeking like". Evolutionary Psychology, 2: 177-194.

[2] Bell G. and J. Maynard Smith 1987. Short-term selection for recombination among mutually antagonistic species. Nature, 328 : 66-68.

[3] Cutter, A.D. and B.A. Payseur 2002. Implications of the deleterious mutation rate in three *Caenorhabditis* species for the evolution of sex. http://www.u.arizona.edu/~acutter/abstracts.htm

[4] Ebert D. and W.D. Hamilton 1996. Sex against virulence: The coevolution of parasitic diseases. Trend. Ecol. Evol, 11 : 79-82.

[5] Hamilton W.D., R. Axelrod and R. Tanese, 1990. Sexual reproduction as an adaptation to resist parasites (A review). Proc. Nat. Acad. Sci. (USA), 87: 3566-3573

[6] Heisler, I.L. 1984. A quantitative genetic model for the origin of mating preferences. Evolution 36: 1283.

[7] Hill W.G. and A. Robertson 1966. The effect of linkage on limits to artificial selection. Genet. Res. 8: 269-294.

[8] Howard R.S. and C.M. Lively 1994. Parasitism, mutation accumulation and the maintenance of sex. Nature, 367 : 554-556 .

[9] Hurst L.D. and J.R. Peck 1996. Recent advances in the understanding of the evolution and maintenance of sex. Trend. Ecol. Evol. 11: 46-52.

[10] Jaffe K. 1996. The dynamics of the evolution of sex: Why the sexes are, in fact, always two? Interciencia 21: 259-267 and *errata* in 22: 48.

[11] Jaffe, K. 1998. Sex, mate selection and evolution. In : Lecture Notes in Computer Science 1447 : Evolutionary Programming VII, Springer Verlag, V.W. Porto, N. Saravanan, D. Waagen and A.E. Eiben (Eds.), pp. 483-492.

[12] Jaffe, K. 1999. On the adaptive value of some mate selection strategies. Acta Biotheoretica 47: 29-40.

[13] Jaffe, K. 2000. Emergence and maintenance of sex among diploid organisms aided by assortative mating. Acta Biotheoretica 48: 137-147.

[14] Jaffe K. 2001. On the relative importance of Haplo-Diploidy, Assortative Mating and Social Synergy on the Evolutionary Emergence of Social Behavior. Acta Biotheoretica 49: 29-42.

[15] Jaffe K. 2004. Sex promotes gamete selection: A quantitative comparative study of features favoring the evolution of sex. Complexity 9: 43-51.

[16] Jaffe K., S. Issa, E. Daniels and D. Haile 1997. Dynamics of the emergence of genetic resistance to pesticides among asexual and sexual organisms. J. Theor. Biol. 188: 289-299.

[17] Jaffe K., Camjo M.I., T.E. Carrillo, M. Weffer, M.G. Muñoz. 2006. Evidence favoring sperm selection over sperm competition in the interaction between human seminal plasma and sperm motility *in vitro*. Achieves of Andrology 52: 45-50.

[18] Judson O.P. and B.B. Normak 1996. Ancient asexual scandals. Trend. Ecol. Evol. 11: 41-46.

[19] Kodric-Brown, A. and J.H. Brown 1987. Anisogamy, sexual selection, and the evolution and maintenance of sex. Evolut. Ecol. 1: 95-105.

[20] Kondrashov, A.S. 1984. Deleterious mutations as an evolutionary factor. I. the advantage of recombinations. Genet. Res. 44: 199.

[21] Kondrashov, A.S. 1988. Deleterious mutations as an evolutionary factor. III. Mating preferences and some general remarks. J. Theor. Biol. 131: 487-496.

[22] Kondrashov, A.S. 1994. The asexual ploidy cycle and the origin of sex. Nature, 370 : 213-216.

[23] Maynard-Smith J.M. 1978. The Evolution of Sex. Cambridge University Press, U.K.

[24] Muller H.J. 1964. The relation of recombination to mutational change. Mut. Res., 1:2-9.

[25] Ochoa G. and Jaffe K. 1999. On sex, mate selection and the Red Queen. J. Theor. Biol 199: 1-9.

[26] Payne C., Jaffe K. 2005. Self seeks like: Many humans choose their dog-pets following rules used for assortative mating. J. Ethol. 23: 15-18.

[27]Rincones J., Mauleon H., Jaffe K. 2001. Bacteria modulate the degree of ampimix of their symbiotic entomopathogenic nematodes in response to nutritional stress. Naturwissenschaften 88: 310-312.

[28] Silva E.R. and Jaffe K. 2002. Expanded food choice as a possible factor in the evolution of eusociality in Vespidae Sociobiology 39:25-36.

[29] Taylor P.D. and Williams G.C. 1982. The lek paradox is not resolved. Theor. Pop. Biol. 22: 392

# Chapter 22
# Optimal Sampling for Complexity in Soil Ecosystems

**Arturo H. Ariño**
Department of Zoology and Ecology
University of Navarra
artarip@unav.es

**Carlos Belascoáin**
TB-Solutions S.A.
belascoc@tb-solutions.com

**Rafael Jordana**
Department of Zoology and Ecology
University of Navarra
rjordana@unav.es

## 1. Introduction

### 1.1. Complexity in Soil Biology

Complexity in soil biology is a multi-level concept. Soil itself is the result of multiple interactions between physical structure, interface phenomena, soil biota activity, population dynamics, chemical composition, time, and environmental conditions. In turn, the resulting system (the soil) influences all those factors except time. Soil complexity can thus be observed at different physical levels (i.e., frequency distribution of aggregates' sizes, order of strata, etc.), biological levels (i.e., taxocoenoses, oxidable organic matter availability, population distribution, etc.), interaction levels (i.e. mineral paths between compartments, food web, etc.), or evolutionary levels (short-term variations on water availability, long-term erosion, etc.).

Dealing with all the complexity of soil structure at all these levels, and its evolution, may be beyond our current ability. However, we might try to capture some measure of at least the complexity of some of these levels. Should we be able to measure that at different levels, we could use these measures to construct a model that could relate the underlying processes from whence the complexity emerges. In doing so, it should be taken into account that simple models of complex ecological systems tend to be closer to the truth than more complex models of the same systems (Mikkelson, 2001).

Soil ecology studies whose main purpose is to serve another goal, i.e. community characterisation or biodiversity determination (which, in turn, may be used as tools for yet another purpose, i.e. "biotope quality" assessment, as indicators - Kurtz et al., 2001) can be regarded as a way to obtain a simplified view of a naturally complex system. Their goal is often to produce a "black box" that ensnares a given set of complex factors, and then concentrate on the inputs and outputs of that subsystem as related to the rest of the system. How these relations affect the system often represents "a more useful kind of truth" (Mikkelson, 2001) than the sheer number of factors that actuate within the black box.

Complexity in soil systems is, naturally, more than simply the number of these black boxes and their interactions. However, one can look for operators that can encode a great deal of information, yet convey clearly some of the essence of what makes the system complex (Paul K. Davis, personal communication). Biodiversity estimates can fit within this category, as they are related to how the soil has come to be as a result of the interactions depicted above. It has already been empirically demonstrated that there is a cross-scale influence of complexity levels when considering different aspects of the soil system, such as the close relationship between animal biodiversity and physical soil structure (Jordana et al., 2000) or even chemical characteristics (Peltier et al., 2001).

### 1.2. Soil Sampling Issues

These soil ecology studies usually require processing a series of samples, that must meet the basic conditions for ecological studies: appropriateness, homogeneity, objectiveness, standardisation and efficiency (Gauch, 1982, p. 44-8). The fulfilment of these conditions would make data from several isochronal samples comparable (it is another matter to sample along time). Ideally, a soil fauna sample should be large enough so as to allow the characteristic(s) under study to be measured under no more error than acceptable statistical sampling error. However, when it comes to quantitatively determine the soil zoocoenosis (which is necessary for diversity or richness measurements), the spatial distribution of many soil fauna populations, and very frequently their aggregated nature, must also be taken into account. Thus, one may choose between taking a very large sample that encompasses a number of clumps, or taking a set of smaller, random and independent (Kasprzak, 1993) subsamples that are spread over a larger area and accumulate the information drawn from these.

A sufficiently large number of subsamples should statistically gather both the species richness information, or purely alpha diversity, and the beta diversity, that is, the portion of diversity that comes from the spatial distribution of the individuals of the populations (Margalef, 1980; Caswell and Cohen, 1993). It is well known that the number of species usually increases with the sample size, and so does the diversity (Margalef, 1980; Magurran, 1989; Huston, 1994; Roszenweig, 1995). Aggregating the results of several subsamples would thus yield a value of diversity usually larger than the value derived from one single subsample. If it is necessary to determine the diversity of the whole sample, then the size of each subsample, as well as the number of subsamples that make up one sample, have to be determined. Very small subsamples will result in a large variance between subsamples; but very large subsamples may mean an unnecessary amount of material to process.

In spite of the importance of ensuring a sufficiently representative sample, ecologists have not always appreciated the true effect of the sampling effort for the assessment of biodiversity (Gotelly and Colwell, 2001). Extracting and classifying soil fauna, which are necessary for determining diversity, is not an easy task. If the study deals with large areas from which many samples are taken, i.e. transects across ecotones or heterogeneous regions, the number of samples to process may rapidly reach the "unmanageable" level. It is thus of paramount importance to determine which is the minimal sample size for these type of studies, that will yield representative results without imposing excessive (and unnecessary) work (Kasprzak, 1993).

The problem of minimal sampling has been constantly attacked over time. There are several "rules of the thumb", as well as formal methods, regarding the increase of the number of species in the sample vs. the size of the sample, or number of subsamples for soil diversity studies. Optimisation of the sample size would be ideally achieved by sequential sampling (Krebs, 1989, p. 238) where only the strictly necessary number of subsamples would be taken; but this is usually not feasible for soil studies as the process of the subsamples is a lengthy one (Krebs, 1989, p.196) and may include even the determination, to species level, of the zoocoenosis. It is a common strategy, thus, that studies dealing with comparison of soil fauna diversity across several biotopes may need a systematic sampling where the minimal sample size is to be determined beforehand in a pilot study, as it is done in other, perhaps less complex systems where a single ecological indicator is sought (Kurtz et al., 2001).

## 1.3. Accumulation Curves

Most usual methods of prior minimal sample size determination would use some derivative of the species–accumulation curve. The sample size is plotted against the number of species, and minimal sample size correspond to one of several "stopping criterion" that in general mean that, beyond a certain sample size, the increase on species richness measured as species number is low enough so as to justify not taking a larger sample.

When "diversity" is meant to be just "richness", i.e. number of species, there are a number of parametric and non-parametric estimators that will give the number of species of standardised, smaller samples (such as rarefaction) or extrapolate to standardised, larger samples (such as Michaelis-Menten curve or Chao estimator). Most of them rely on randomisation of existing data, such as collected individuals, to get the necessary variance estimates. A number of computer programs are routinely used for these estimations; the most popular ones being EstimateS (Colwell, 2000), WS2M (Turner et al., 2001), and EcoSim (Gotelli and Enstminger, 2001). It should be noted that these methods will yield and estimate of the total richness of the biotope sampled, much better than can be obtained from the direct reading of these parameters from the subsamples themselves; but they do not ensure that the asymptotic (total) richness has been reached (Gotelli and Colwell, 2001).

A minimal sample size for diversity studies of soil fauna has to ascertain that the measured diversity, and not just the species richness, is representative of the biotope being studied. Therefore, it is the accumulation of diversity what should be used in order to determine that critical size (Magurran, 1989, p. 59). A diversity-area curve would measure both alpha and beta diversity, and would ideally flatten at the point where the total diversity of the biotope is measured, i.e., the structure of the community under study is caught. (We do not take into account now the gamma diversity, *sensu* Margalef, or diversity through time).

However, to construct such a curve from the aggregation of results of smaller subsamples, as it would be the general case, or by adding specimens one by one, has also a major drawback. The order in which subsamples, or specimens, are added to the curve significantly influences the value of diversity reached at each point. The joint list of species and their abundances is different depending on which subsamples from a larger sample are added up. Thus, one may obtain different diversity curves depending upon the particular choice of subsamples and the order in which they are added to obtain the progressively larger area's species list. Therefore, the selection of minimal sample size by any of the usual subjective criteria is also affected, yielding different possibilities. This effect parallels the one observed when constructing species accumulation curves, and Monte Carlo randomisations of subsample aggregates are the choice methods for the construction of smooth species accumulation curves that can be treated analytically or statistically (Christen and Nakamura, 2000). Again, these methods may permit to infer the richness of the biotope being sampled from the bootstrap samples, but cannot ascertain that the diversity measured statistically correspond to that of a complete enumeration of the biocoenosis being sampled. This could be accomplished by somehow ensuring that the samples being analysed are true minimal samples of the biotope *for* the parameter being measured.

Accumulation curves can be used for this purpose. An asymptotic curve could, by definition, yield the maximum value of the parameter once the asymptote has been reached, and that would in turn give the value of the parameter and no estimate would be necessary. However, this is not usually the case, and a stopping criterion must be chosen.

Contrasting to the analytical and statistical treatment of inference from accumulation curves, the stopping criterion for an accumulation curve has been somewhat left to the researcher's discretion. Most scientific literature that resorts to a species-area curve (a species accumulation curve where the x-axis is not individuals but sampled area which prevents it from analytical description and requires bootstrap sampling and randomisations –Gotelli and Colwell, 2001-) for minimal sampling assessment do not explicitly develop a formal stopping criterion other than "seeing the flattening" of the accumulation curve or other arbitrary criteria.

## 2. Minimal Sample Size for Diversity Measurement

We propose a method for determining the minimal sample size by calculating the accumulation of diversity from standardised subsamples, that deals with the problems explained above. Our method is based on the accumulation curve for the parameter of interest, which can be a measure of the complexity of the system. As a stopping criterion, we resort to the slope analysis, a family of methods used for two-sample comparisons (Dawson, 1998). We construct a diversity accumulation curve from subsamples, where each point represents the average diversity of all possible combinations of subsamples, or of a bootstrap sample of these combinations. This allows us to determine the point of non-significant diversity variation, and thus minimal sample size, by a simple statistical test on the slopes of the accumulation curve points. The method can be used to assess the number of subsamples that will be necessary to process in order to estimate diversity from soil fauna samples on large studies.

### 2.1. Method

Let it be *N* subsamples of equal dimensions (be them surface, volume, or mass) coming from a pilot study. We intend to determine the ***M*** number of subsamples that conform the minimal sample for calculating the total diversity. Each subsample is processed, and their faunistic contents are extracted and classified. The total extracted faunal categories (taxa) are ***T***. The abundance results can be represented as the usual ***NT*** array, where columns are subsamples and rows are the taxa.

One will first calculate the chosen diversity parameter ***D*** of each subsample, which is assumed to be subminimal. We thus obtain a series of *N* individual values of diversity. Next, we combine the first and the second subsample, and obtain the diversity of this new double-sized, pooled sample. We do the same for all possible pairs of subsamples, that is, all possible combinations of *N* vectors taken by twos, obtaining a new series of diversity values for all possible pairs. The algorithm thus continues by combining all possible trios, quartets, and so forth, of the subsamples and always calculating a set of diversity values for each series. The number of possible combinations of ***x*** subsamples within the sample drawn from the *N* total samples, *N!/[x!(N-x)!]*, can be a rather large figure; in that case, the algorithm will select a random set of those combinations.

In the next step, all series of diversities for the combinations of 1, 2, 3 .... ***x*** ... ***N*** subsamples are averaged. We thus obtain a single value of average diversity, along with its variance, for each sample size. Plotting the average diversity for each sample size against the sample size, which is the number of subsamples ***x***, we obtain the familiar diversity/area curve.

Now we select the stopping criterion. For the purpose of measuring complexity in the form of diversity, we assume that we have captured the diversity of the system when additional samples do not add additional value to the diversity index that we have chosen, that is, when the slope of the curve between a pair of consecutive points is not significantly different from zero (i.e., the curve is statistically 'flat'). Thus, we can perform a simple ***t***-test for the mean of the slopes between each averaged diversity measure and all possible values of the following one, and stop when we cannot reject the null hypothesis of these slopes averaging zero at the desired significance level.

### 2.2. Algorithm Implementation

The algorithm can be summarised as textual metacode in the following steps:

1. Read the data file, which will contain records in a suitable format that will express the abundance of each taxon in each subsample.
2. Create a vector **S[1...*N*]** containing the subsample codes.
3. Obtain *N* vectors **X[1...** $\binom{N}{x}$ **]** of all possible combinations of *N* subsample codes from **S** taken by ***x*** where ***x*** (1•*N*).
4. For each of the *N* vectors **X** do:

4.1. Create a new empty vector $\mathbf{D}_n$**[1...** $\binom{N}{n}$ ] for the $D$ values.

4.2. For each element in **X** do:

4.2.1. Merge the faunistic lists of each subsample contained in the element, adding the abundances for equal taxa.

4.2.2. Add a new $\overline{D}_n$ element to $\mathbf{D}_n$ with the $D$ value obtained.

4.3. Calculate $\overline{D}$ for $\mathbf{D}_n$.

4.4. Compute one-sided *t*-test for $H_0=0$ for the slopes between $\mathbf{D}_n$ and $\overline{D}_{n-1}$ if it exists, against $H_1$=greater mean.

4.5. Plot $\overline{D}$ against $\boldsymbol{n}$.

5. Choose sample size=$\boldsymbol{n}$-1 when $H_0$=true at the desired significance level.

## 3. Discussion

While this diversity accumulation algorithm does not avoid initial oversampling, for it is necessary to combine several small subsamples in order to be able to statistically compare the means of their diversities, it may save extra work if used as a pilot study prior to larger, systematic studies by assuring that the samples taken will have adequate size. Whereas this is the goal of any sampling optimisation technique, we provide a way to do it for repetitive diversity measurements that includes some "statistical certainty" that, on average, we are dealing with samples that will represent the true (up to beta) diversity of the biotope.

Our choice of averaging the slopes between the average diversity of all combinations of subsamples and all the possible combinations of subsamples at the next point may overshadow another possible technique, perhaps more straightforward, that would consist on drawing all possible diversity accumulation curves for a set of subsamples, calculating the average slope of all curves at each point and stopping when the average slope does not significantly differ from zero. Though this may be more intuitive, we discarded it for the sake of parsimony: There are *N!* possible accumulation curves for a given set of *N* subsamples and $2^N$-1 possible diversity values giving

$$\sum_{i=1}^{N-1} \frac{N!}{i!}$$

possible slopes between these values along the curve. However, the number of possible slopes if only the slopes between the average diversity at one point and all possible diversities at the next point are used is

$$\sum_{i=2}^{N-1} \binom{N}{i},$$

which is a much smaller figure.

The algorithm can be regarded as a null model that deliberately excludes the mechanism of diversity formation (Gotelli, 2001), as it seeks a stopping criterion that is explicitly statistical in nature and that does not depend on what the parameter means. Thus, any complexity measure could, in principle, be subject to the same algorithmic decision if they yield an accumulation curve.

As null models are usually based on real data sets of an index of community structure (Gotelli, 2001), we have used a set of 12 initial soil nematode subsamples taken from an oak forest in Navarra, North of Spain, as a part of a project on forest soil recovery after fire. Samples were taken by a cylindrical corer, 25 mm in diameter. The original TxN matrix, species level, is 104 x 12. The algorithm was programmed as a set of pure C routines, debugged and tested with a series of benchmark files, and the program was applied to the real data matrix. The accumulation curve for species for Shannon's diversity ***H'*** (fig. 1) shows no distinct flattening, whereas the slope of the diversity curve becomes not significantly different from zero (95% confidence level) after the tenth sample. The total number of ***H'*** values calculated were 4,095. The number of necessary slopes to test if all individual curves had been plotted to look for non-significant slope change would have been over 823 million.

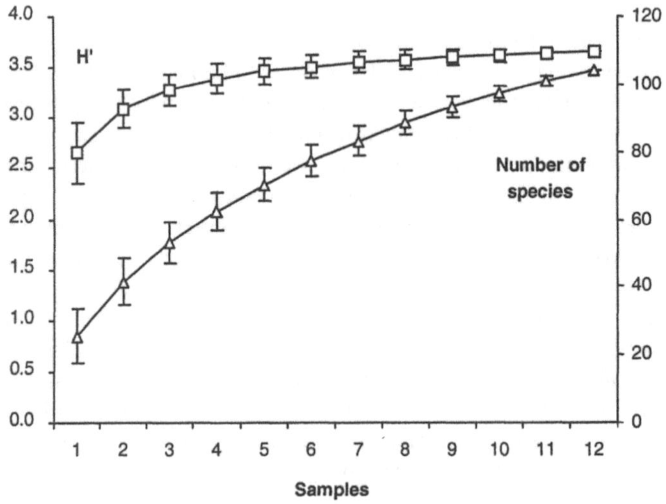


***Figure 1.*** *Species/area and* **H'** *diversity/area curves when combining samples as described in the text. Values are averages, rounded to integer in the case of species. Error bars represent standard deviations. The mean of the slopes between the mean diversity at point 10 and the diversity values at point 11 become not significantly (p<0.05) different from zero, whereas the species numbers are always significantly different from each other.*

## 4. Acknowledgements

Samples used in the example provided were taken by the Department of Zoology and Ecology of the University of Navarra during the Project 'Study of the impact of forest fires and their recovery: soil science, fauna, flora, and erosion", project no. FOR 90-0982-CO2 of the National Institute of Agronomic Research, Spain. C. Belascoáin was successively a grantee of the Friends of the University of Navarra Association and of the above Institute.

## 5. References

Caswell, H., & Cohen, J.E., 1993. Local and Regional Regulation of Species-Area Relations - A Patch-Occupancy Model. In: R.E. Ricklefs and D. Schluter (Editors), *Species Diversity in Ecological Communities: historical and geographical perspectives.* University of Chicago Press, Chicago, pp 99-107.

Christen, J.A., & Nakamura, M., 2000. On the Analysis of Accumulation Curves. *Biometrics*, **56,** 748-754.

Colwell, R.K., 2000. *EstimateS: Statistical Estimation of Species Richness and Shared Species from Samples (Software and User's Guide)*, Version 6. http://viceroy.eeb.uconn.edu/estimates

Dawson, J.D., 1998. Sample Size Calculations Based on Slopes and Other Summary Statistics. *Biometrics*, **54**: 323-330.

Gauch, Jr., H.G., 1982. *Multivariate Analysis in Community Ecology.* Cambridge University Press, Cambridge.

Gotelli, N.J., 2001. Research frontiers in null model analysis. *Global Ecology & Biogeography*, **10**, 337-343.

Gotelli, N.J. & Colwell, R.K., 2001. Quantifying biodiversity: procedures and pitfalls in the measurement and comparison of species richness. *Ecology Letters*, **4,** 379-391

Gotelli, N.J. & Entsminger, G.L., 2001. *EcoSim: Null models software for ecology.* Version 7.0. Acquired Intelligence Inc. & Kesey-Bear. http://homepages.together.net/~gentsmin/ecosim.htm.

Jordana R, Arpin P., Vinciguerra M.T., Gonzalez S., Aramburu M.P., Ariño A.H., Armendariz I., Belascoain C., Cifuentes P., Clausi M., Escribano R., Garcia Abril A., Garcia-Mina J.M., Hernandez M., Imaz A., Moraza M.L., Ponge J.F., Puig J. & Ramos A., 2000. Biodiversity across ecotones in desertificable Mediterranean areas. In: Balabanis P., Peter D., Ghazi A., Tsogas M. (Eds).: *Mediterranean Desertification Research results and policy implications*, Vol 2.: pp 497-505. European Comission EUR 19303.

Huston, M.A., 1994. *Biological Diversity. The coexistence of species on changing landscapes.* Cambridge University Press.

Kasprzak, K., 1993. Selected aspects of mathematical statistics. In: M. Górny and L. Grüm (Editors), *Methods in Soil Zoology.* Elsevier, Amsterdam, pp. 16-69.

Krebs, C.J., 1989. *Ecological Methodology.* Harper & Row, New York.

Kurtz, J.C., Jackson, L.E. & Fisher, W.S., 2001. Strategies for evaluating indicators based on guidelines from the Environmental Protection Agency's Office of Research and Development. *Ecological Indicators*, **1**: 49-60.

Magurran, A., 1989. *Diversidad ecológica y su medición.* Ediciones Vedrà, Barcelona.

Mikkelson, G.M., 2001. Complexity and Verisimilitude: Realism for Ecology. *Biology and Philosophy*, **16**: 533-546.

Margalef, R., 1980. *Ecología.* Omega, Barcelona.

Peltier A., Ponge J.F., Jordana R. & Ariño A., 2001. Humus forms in Mediterranean scrublands with aleppo pine. *Soil Science Society of America Journal*, **65** (3): 884-896

Roszenweig, M.L., 1995. *Species diversity in space and time.* Cambridge University Press.

Turner, W., Leitner, W. & Rosenzweig, M., 2001. WS2M. Software for the measurement and analysis of species diversity (Software and User's Manual). http:// eebweb.arizona.edu/diversity

# Chapter 23

# Predator-Prey Dynamics for Rabbits, Trees, and Romance

**Julien Clinton Sprott**
Department of Physics
University of Wisconsin - Madison
sprott@physics.wisc.edu

## 1. Introduction

The Lotka-Volterra equations represent a simple nonlinear model for the dynamic interaction between two biological species in which one species (the predator) benefits at the expense of the other (the prey). With a change in signs, the same model can apply to two species that compete for resources or that symbiotically interact. However, the model is not structurally stable, since persistent time-dependent (oscillatory) solutions occur for only a single value of the parameters.

This paper considers structurally stable variants of the Lotka-Volterra equations with arbitrarily many species solved on a homogeneous two-dimensional grid with coupling between neighboring cells. Interesting, biologically-realistic, spatio-temporal patterns are produced. These patterns emerge from random initial conditions and thus exhibit self-organization. The extent to which the patterns are self-organized critical (spatially and temporally scale-invariant) and chaotic (positive Lyapunov exponent) will be examined.

The same equations, without the spatial interactions, can be used to model romantic relationships between individuals. Different romantic styles lead to different dynamics and ultimate fates. Love affairs involving more than two individuals can lead to chaos. The likely fate of couples with different romantic styles will be described.

## 2. Lotka-Volterra Equations

One variant of the Lotka-Volterra equations (Murray 1993) for two species (such as rabbits and foxes) is

$$\begin{aligned} dR/dt &= r_1 R(1-R-a_1 F) \\ dF/dt &= r_2 F(1-F-a_2 R) \end{aligned} \qquad (1)$$

where $R$ is the number of rabbits and $F$ is the number of foxes, both positive and each normalized to its respective carrying capacity (the maximum allowed in the absence of the other), $r_1$ and $r_2$ are the respective initial growth rates in the absence of competition, and $a_1$ and $a_2$ determine the interspecies interaction. In a predator-prey model, the predator (foxes) would have $r_2 < 0$ (since they gradually die in the absence of rabbits to eat) and the other constants would be positive. However, the same equations with all positive constants could model competition, or with both growth rates negative could model cooperation or symbiosis (chickens and eggs, plants and seeds, bees and flowers, etc.).

## 3. Equilibrium and Stability

The system in Eq. (1) has four equilibria, one with no rabbits, one with no foxes, one with neither, and a coexisting one with

$$
\begin{aligned}
R &= \frac{1-a_1}{1-a_1 a_2} \\
F &= \frac{1-a_2}{1-a_1 a_2}
\end{aligned}
\tag{2}
$$

which is of primary interest.

For the predator-prey case ($a_1 r_1 > 0$, $a_1 r_2 < 0$), the coexisting equilibrium is a stable focus for $r_1(1 - a_1) < -r_2(1 - a_2)$, at which point it undergoes a Hopf bifurcation, after which the trajectory spirals outward without bound from the unstable focus. Hence there are no structurally stable oscillatory solutions. For the competition case ($a_1 r_1 > 0$, $a_1 r_2 > 0$), the coexisting equilibrium is a stable node for $a_1 < 1$ and $a_2 < 1$, at which point it undergoes a saddle-node bifurcation, after which one of the species dies while the other goes to its carrying capacity ($R = 1$ or $F = 1$). Thus there are no oscillatory solutions, and stability requires that the intraspecies competition dominates. When the interspecies competition dominates, the weaker species is extinguished by the principle of 'competitive exclusion' or 'survival of the fittest' (Gause 1971). For the cooperation case ($a_1 r_1 < 0$, $a_1 r_2 < 0$), both species either die or grow without bound in what May (1981) calls an 'orgy of mutual benefaction.'

With $N$ species there are $2^N$ equilibria, only one of which represents coexistence, and it is unlikely that this equilibrium is stable since all of its eigenvalues must have only negative real parts. Ecological systems exhibit diversity presumably because there are so many species from which to choose, because they are able to spread out over the landscape to minimize competition, and because species evolve to fill stable niches (Chesson 2000). If many arbitrary species are introduced into a highly interacting environment, most would probably die.

## 4. Spatio-temporal Generalization

Now assume there are $N$ species with population $S_i$ for $i = 1$ to $N$ and that they are spread out over a two-dimensional landscape $S_i(x, y)$. The species could be plants or animals or both. For convenience, take the landscape to be a square of size $L$ with periodic boundary conditions (a torus), so that $S_i(L, y) = S_i(0, y)$ and $S_i(x, L) = S_i(x, 0)$. One commonly assumes that each species obeys a reaction-diffusion equation of the form

$$\frac{\partial S_i}{\partial t} = r_i S_i (1 - \sum_{j=1}^{N} a_{ij} S_j) + D_i \nabla^2 S_i \tag{3}$$

This system is usually solved on a finite spatial grid of cells each of size $d$ so that $\nabla^2 S_i(x, y) = [S_i(x+d, y) + S_i(x-d, y) + S_i(x, y+d) + S_i(x, y-d) - 4S_i(x, y)] / d^2$.

Diffusion is perhaps not the best model for biology, however, and if too large, it tends to produce spatially homogeneity. Instead, assume that each species interacts not just with the other species in its own cell but also in its four nearest-neighbor cells (a von Neumann neighborhood), giving

$$\frac{\partial S_i}{\partial t} = r_i S_i (1 - S_i - \sum_{j=1, j \neq i}^{N} a_{ij} \overline{S}_j) \tag{4}$$

where $\overline{S}_j(x, y) = S_j(x+d, y) + S_j(x-d, y) + S_j(x, y+d) + S_j(x, y-d) + \alpha_j S_j(x, y)$ is a weighted average of the neighborhood. In the example of rabbits and foxes, you can think of $\alpha_j$ as the tendency for the foxes to eat at home. With $\alpha_j = 0$, the foxes never eat at home, and with $\alpha_j = 1$ they forage uniformly over a five-cell neighborhood including their home cell. In the case of trees and seeds, this term is where one would include a seed dispersion kernel. The example that follows uses $\alpha_j = 1$ for all $j$, but the results are not sensitive to the choice. Including only nearest neighbor cells normalizes space so that the cell size $d$ is the order of the mean foraging (or dispersal) distance. Note that time can also be normalized to one of the growth times, so that we can take $r_1 = 1$ without loss of generality.

## 5. Numerical Example

In Eq. (4) all the biology is contained in the vector $r_i$, the interaction matrix $a_{ij}$, and the dispersal vector $\alpha_j$ here taken as unitary. Instead of modeling realistic biology, we choose the values of $r_i$ and $a_{ij}$ from an IID random normal distribution with zero mean and unit variance and examine many instances of the model to explore a range of possible ecologies.

For brevity, Fig. 1 illustrates most of the common behaviors. It starts with six species with uniform random values $S_i(x, y)$ in the range of 0 to 0.2 on a $100 \times 100$ grid. The upper plots show the spatial structure of the six species after 100 growth times, and the lower plot shows the cumulative relative abundance of each species versus time. One species (the fourth) dies out. The first, second, and sixth, nearly die, and then recover, after which the five species coexist with aperiodic temporal fluctuations and spatial heterogeneity.

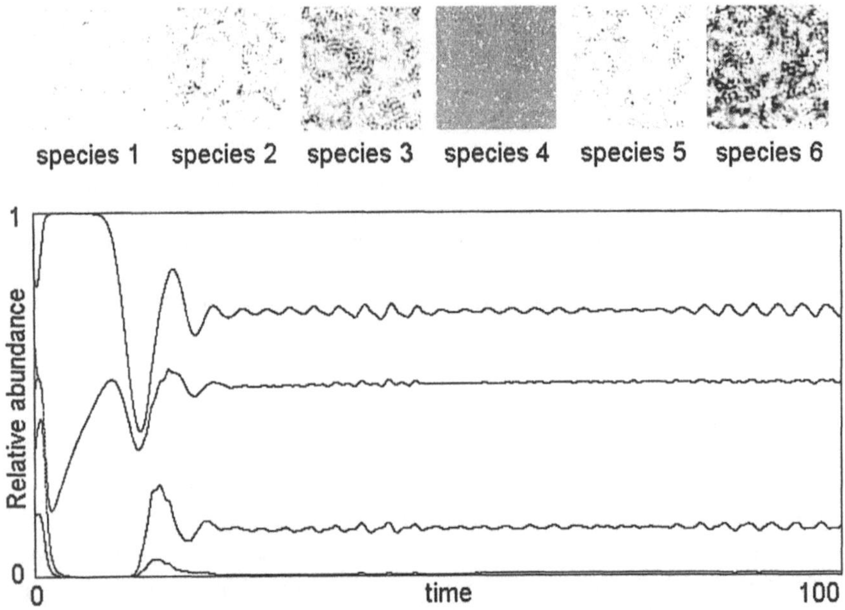


**Figure 1.** Typical example of a spatio-temporal solution Eq. (4) with six initial species, one of which died.

Figure 2 shows the dominant species in each cell after 100 growth times, each in a different shade of gray. This display facilitates comparison with real data and with the results of cellular automata models (Sprott 2002). As in earlier studies, we define the cluster probability as the fraction of cells that are the same as their four nearest neighbors and Fourier analyze the temporal fluctuations in cluster probability to obtain its power spectrum. The result in Fig. 3 shows a power law over about a decade and a half, implying temporal scale invariance as suggestive of self-organized criticality (Bak 1996). Other quantities such as the total biomass $\Sigma S_i(t)$ also have power-law spectra.

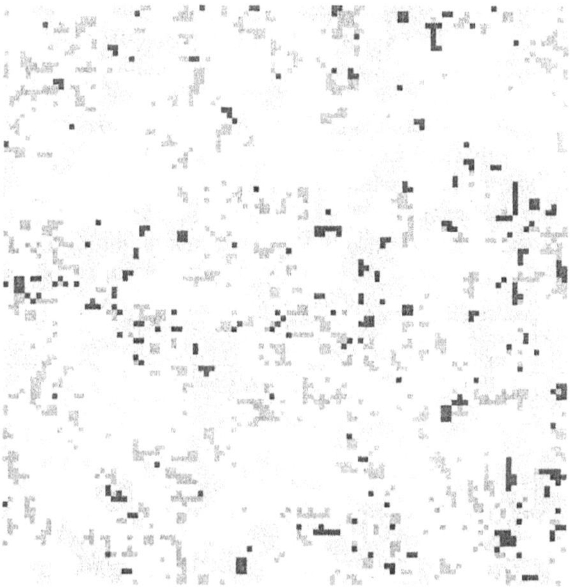

**Figure 2.** Landsccape pattern showing the dominant species in each cell.

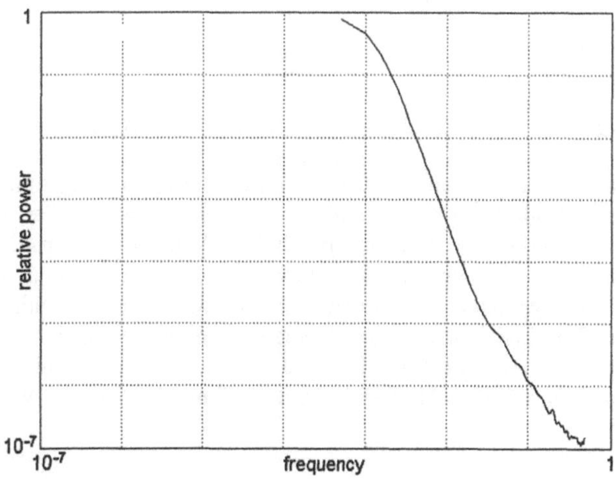


**Figure 3.** Power spectrum of fluctuations in cluster probability.

To assess whether the dynamics are chaotic, we follow Lorenz (1963) and round the values of $S_i(x, y)$ to four significant digits after the initial transient has decayed and calculate the growth of the error in the total biomass as the perturbed and unpertubed systems evolve deterministically. The result in Fig. 4 suggests an exponential growth in the error with a growth rate the order of $0.1r_1$. If the system modeled a forest with a typical $r_1$ of 50 years, the predictability time would be about 500 years. Five species is the minimum number for which such chaotic solutions were found. With four species, limit cycles are common, and with three or fewer species, all stable solutions attract to a time-independent equilibrium with no spatial structure. Spatial heterogeneity always correlates with temporal fluctuations.

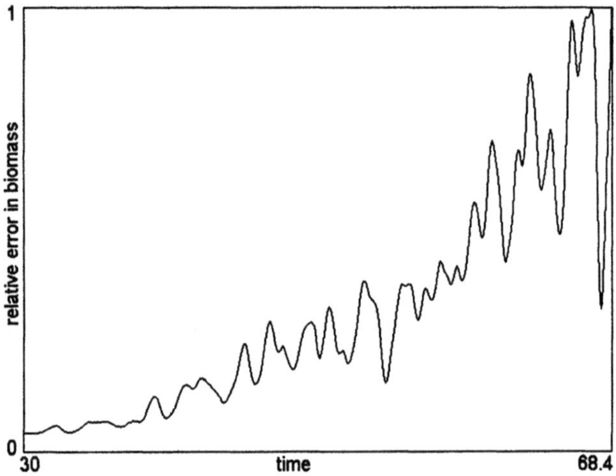


**Figure 4.** Exponential growth in total biomass error suggesting chaos.

Note that the chaos and spatial structure arise from a purely deterministic model in which the only randomness is in the initial condition. In fact, similar structures arise from highly ordered initial conditions, perturbed only by noise as small as one part in $10^6$. The model is endogenous (no external effects), homogeneous (every cell is equivalent), and egalitarian (all species obey the same equation, with only different coefficients). The spatial patterns and fluctuations are inherent in the equations whose solutions spontaneously break the imposed spatial symmetry. Similar behavior has been observed in models with a single spatial dimension (Sprott, Wildenberg, and Azizi 2005)

## 6. Application to Romantic Relationships

To stress the generality of the model, we can apply it to romantic relationships. Imagine two lovers, Romeo and Juliet, characterized by a pair of equations such as Eq. (1) in which $R$ is Romeo's love for Juliet and $F$ is Juliet's love for Romeo, both positive. Each lover can be characterized by one of four romantic styles depending on the signs of $r$ and $a$ as shown in Table I using names adapted from Strogatz (1988). The variable $r$ determines whether one's love grows or dies in the absence of a response from the other ($a = 0$), and the variable $a$ determines whether requited love enhances or suppresses the intensity of one's feelings.

**Table I.** Romantic styles

| | $a$ | |
|---|---|---|
| − + Cautious lover | | + + Narcissicist nerd |
| | | $r$ |
| − − Hermit | | + − Eager beaver |

With two interacting lovers, there are thus $2^4 = 16$ different combinations of romantic styles, the fate of which are determined by the strength of the interations. As an example, choosing $r$ and $a$ from an IID random normal distribution with zero mean and unit variance gives the results in Table II, where the percentages are the probability that a stable steady state is reached. In some sense, the best pairing is between an eager beaver and a narcissistic nerd, although two eager beavers have solutions that grow mutually without bound. Not surprisingly, the prospects are dismal for a hermit and not much better for a cautious lover. Fortunately, humans often seem capable of adapting their romantic styles to fit the situation.

**Table II.** Probability of mutual stable love for various pairings.

| | **Narcissistic nerd** | **Eager beaver** | **Cautious lover** | **Hermit** |
|---|---|---|---|---|
| **Narcissistic nerd** | 46% | 67% | 5% | 0% |
| **Eager beaver** | 67% | 39% | 0% | 0% |
| **Cautious lover** | 5% | 0% | 0% | 0% |
| **Hermit** | 0% | 0% | 0% | 0% |

It is also instructive to examine love triangles, in which case there are four variables if two of the lovers are unaware of one another, and six variables if each person has feelings for the other two. The variables need not be romatic love; the third person could be the child of a couple or perhaps an in-law. With six variables, there are $2^6 = 64$ equilibria, only one of which represents a universally happy arrangement and $4^6 = 4096$ different combinations of styles, assuming each person can adopt a different interaction style toward each of the others. Not surprisingly, the prognosis for coexisting positive feelings is very low unless the individuals exhibit strong adaptability of their styles. Perhaps humans adapt to such situtions much the way plants and animals do, by limiting their interactions, evolving to fill a stable niche, drawing sustinence from others, and maintaining spatial separation.

With three or more variables, there is the possibility of chaotic solutions. However, a search for such solutions in the system of Eq. (1) generalized to six variables failed to reveal any such solutions, although other similar models do exhibit chaos (Sprott 2004).

## Acknowledgments

I would like to thank Warren Porter and Heike Lischke whose research inspired this work and George Rowlands and Janine Bolliger whose ongoing collaboration facilitated its completion.

## References

Bak, P., 1996, *How Nature Works: The Science of Self-organized Criticality*, Corpernicus (New York).

Chesson, P., 2000, Mechanisms of Maintenance of Species Diversity. *Annu. Rev. Ecol. Syst.*, **31**, 343.

Gause, G.F., 1971, *The Struggle for Existence*, Dover (New York).

Lorenz, E.N., 1963, Deterministic Nonperiodic Flow. *J. Atmos. Sci.*, **20**, 130.

May, R.M., 1981, *Theoretical Ecology: Principles and Applications*, second ed., Blackwell Scientific (Oxford).

Murray, J. D., 1993, *Mathematical Biology*, Springer (Berlin).

Sprott, J.C., 2004, Dynamical Models of Love, *Nonlinear Dynamics, Psychology, and Life Sciences* **8**, 303.

Sprott, J.C., Bolliger, J., and Mladenoff, D.J., 2002, Self-organized Criticality in Forest-landscape Evolution. *Phys. Lett. A*, **297**, 267.

Sprott, J.C., Wildenberg, J.C., and Azizi, Y., 2005, A Simple Spatiotemporal Chaotic Lotka-Volterra Model, *Chaos, Solitons and Fractals* **26**, 1035.

Strogatz, S.H., 1988, Love Affairs and Differential Equations. *Mathematics Magazine*, **61**, 35.

# Chapter 24
# The Evolution of an Ecosystem: Pleistocene Extinctions

**Elin Whitney-Smith, Ph.D.**
Geobiology, George Washington University
elin@quaternary.net

## Abstract

*It is generally assumed that evolution is an issue of looking at how a species fits into its environment. This over-constrains our thinking; we should look at how the species and the ecosystem evolve together.*

*The current theories of the Pleistocene extinction (Climate change and Overkill by H. sapiens) are inadequate. Neither explains why: (1) browsers, mixed feeders and non-ruminant grazer species suffered most, while ruminant grazers like bison generally survived, (2) surviving mammal species, including both subspecies of bison, were sharply diminished in size; and (3) vegetative environments shifted from plaid to striped (Guthrie, 1980).*

*In addition, climate change theories do not explain why mammoths and other megaherbivores survived changes of similar magnitude. Although flawed, the simple overkill hypothesis does link the extinctions and the arrival of* H. sapiens. *However, it omits the reciprocal impact of prey decline on* H. Sapiens; *standard predator-prey models, which include this effect, demonstrate that predators cannot hunt their prey to extinction without themselves succumbing to starvation.*

*An alternate scenario and computer simulation (download at http://quaternary.net) characterized by a boom/bust population pattern is presented. It suggests* H. sapiens *reduced* ***predator*** *populations, causing a herbivore population boom, leading to overgrazing of trees and grass, resulting in environmental exhaustion and extinction of herbivores. If true, bison survival and differentiation into two sub-species, (*B. bison bison *[plains bison] and* B. bison athabascae *[woodland bison,] through the Pleistocene may be accounted for thus: environmental exhaustion selectively favors animals that could extract maximum energy from low quality forage to survive and reproduce the split into sub species is a reflection of the new vegetative environment.*

## Background

At the end of the last ice age, the Pleistocene, the New World experienced a severe extinction of large animals. Mammoths, mastodons, giant beavers and giant ground sloths went extinct. New world horses and camels though not as large also went extinct.

There are two theories put forward to explain these extinctions – climate change and overkill by humans. Neither is satisfactory. Climate change theories are unsatisfactory because the animals survived previous changes of similar magnitude. The overkill theory is unsatisfactory because predators cannot hunt their prey to extinction without starving themselves.

There are also observations about the pattern of extinction that cannot be accounted for by either theory.

A shift in the vegetation pattern from plaids – patchy mixed woodland – to stripes – closed canopy forest near the mountains, unbroken grassland in the center and tundra in the north of the continent (Guthrie, 1989).

A bias in favor of small size and for ruminants observed in many species including bison.

The extinction of cecal digesters, like horses, that should not have suffered from climate change, since they can live in the environment today.

And, the extinction of animals that were not hunted by *H. sapiens* – the Shasta ground sloth (*Nothrotheriops shastensis*). Hansen's (1978) study of sloth dung and the surrounding environment from Rampart cave shows no evidence of climate change but does show that the diet of the sloth changed showing less and less of its staple, globemallow (*Sphaeralcea ambigua*), and more and more Mormon tea (*Ephedra navadensis*), its less preferred food. Globemallow is used by other herbivores and Mormon tea is not.

## Hypothesis

Imagine the following scenario:

*Homo sapiens* enters the New World. The introduction of a new predator, *H. sapiens*, reduces the number of herbivores available to each of the existing predators.

Predators who are unable to find enough to eat and who have no experience with *H. sapiens* prey on *H. sapiens.*

### *H. sapiens* reduces predator populations

In revenge and to cut down on competition, *H. sapiens* establishes a policy of killing predators. Through this policy, predator populations are reduced below the level where they are able to control herbivore populations. Herbivore populations boom.

*H. sapiens* populations expand, but more slowly than the predators they kill because humans recruit more slowly. *H. sapiens* does not control herbivore populations as well as the now scarce predators did formerly.

### Herbivores boom and exhaust the environment

Herbivore populations overgraze the environment. Herbivores are forced to eat their less preferred foods. Mammoths and mastodons knock over trees, eventually turning mixed parkland into grassland (Wing and Buss, 1970).

Without sufficient food, herbivore populations crash (Leader-Williams, 1980; May, 1973; Scheffer, 1951).

As *H. sapiens* populations begin to experience food stress and competition with non-human predator becomes more important, predator killing increases.

In the denuded environment, herbivores that can get the most nutrition and reproduce soonest from resources which recruit quickly are selectively favored – bison. *H. sapiens* populations have been decimated. Relict groups establish new life ways. The new Holocene equilibrium is established (Whitney-Smith, 1995).

## Methods

Two system dynamics computer models were designed as a test of the hypothesis presented above. Assumptions: (1) ecosystems are generally in equilibrium, (2) vegetation continues to grow until it fills up the available area, (3) each sector of the ecosystem (predator, herbivore, *H. sapiens*) is dependent upon its food source, and (4) the food source is depleted by the populations which use it.

**The steps in the simulation are:** 1 – Establish an equilibrating ecosystem with three sectors; Plants, herbivores, and predators, 2a – Introduce *H. sapiens* as a second predator, 2b – Link the *H. sapiens* sector and the predator sector to simulate *H. sapiens* killing predators – (Second Order Overkill), 3 – Build a model with vegetation partitioned into big and small trees, high and low quality grass; herbivores partitioned into browsers, mixed feeders and ruminants and non-ruminant grazers.

**Values used in these simulations:** are based directly on those used by Whittington and Dyke (1989). Some of the values have had to be modified to fit the model. For example, Whittington and Dyke (1989) use a 0.25 recruitment rate which is represented by a birth and death rate which results in a 0.25 recruitment rate.

The hunting rate used in the simulation presented in this paper is based on food needed per pound of predator per year. Data from extant predators suggests 20 pounds of food per pound per year (International Wolf Center, 1996; Cat House, 1996; Petersen, 1977; Schaller, 1972). It is assumed that *H. sapiens* requires half the amount of meat that an obligate predator needs. A number of values were derived by running the model with approximated values until the model reached equilibrium. The initial division between trees and grass is arbitrarily set at 50%.

**Modeling paradigm:** Each of the sectors in all of the models take a similar form: The amount of stock in the sector at any given time is determined by the amount of stock in the sector at the previous time times some growth rate, discussed above, modified by the limit of that sector and minus the death rate and the hunting or consumption rate. Birth and death rates are modified by available resources for the members of the sector. Thus herbivores are limited by the amount of plants times the efficiency rate for that animal.

Hunting values are based on density of prey. If prey is sufficiently dense, predators it will kill all the prey they desire and reproduce at their biological maximum. If prey is less dense then the predator population is not able to recruit at its biological maximum. Over time the system stabilizes where predator, prey and plant populations are in balance.

## Results – Steps 1 & 2

The first step is to establish a stable ecosystem with three sectors; plants, herbivores, and carnivores (Figure 1a). Since the goal of this step is a stable ecosystem, the model was perturbed and subsequently returned to normal.

The second step is a model with a second predator, *H. sapiens*. It is broken into two parts.

**Step 2a** introduces *H. sapiens* as a second predator. *H. sapiens* enters the New World -11500 BP, 100 years after the start of the model. Figure 1b shows the impact of a second predator – the overkill hypothesis.

Herbivores decreased less than predators, and food for predators increased. At the end of the model predator populations are 59% of what they were when the model started; Herbivore populations are 63% and Plants has increased to 103% of the starting value (there are fewer herbivores therefore there are more plants). It has been assumed that *H. sapiens* needs half the food per year per pound (10 lb.) than non- human predators. *H sapiens* is considered to be as effective a hunter as are non-human predators. As in the Whittington and Dyke model *H. sapiens* is 200 individuals at year -11500 BP. Each person is 100 lb. of biomass.

**Step 2b** is the position of the scenario presented above. It links the *H. sapiens* sector and the predator sector to simulate *H. sapiens* killing predators – (second order predation – 2op) (Figure 1c)

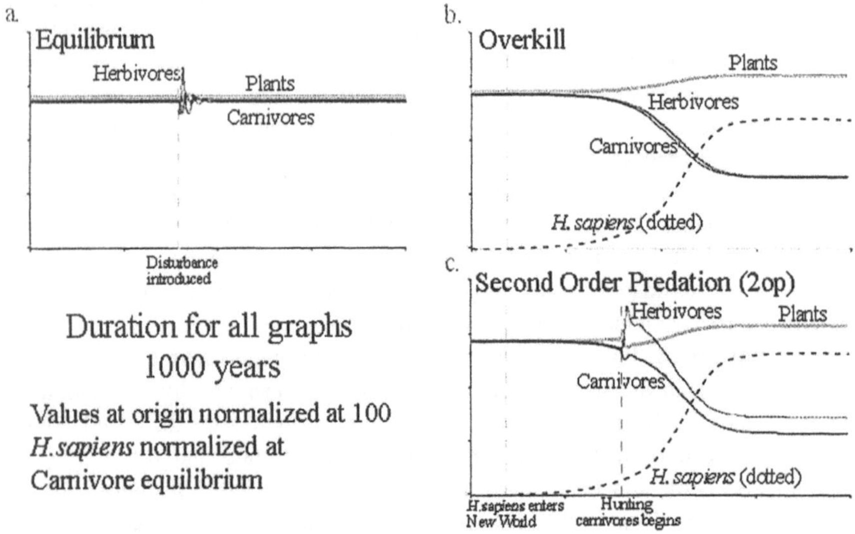


**Figure 1** – Graphs of the base model results.

In the second order predation (2op) run of the base model, 400 years after *H. sapiens* enters the New World predator populations are reduced by 1.5%. This destabilizes the system. Herbivore populations escape from predator control. This creates a population boom. Herbivores eat Plants faster than they can be replaced. Ultimately herbivores crash allowing plants to recover. The system then stabilizes. At the end there are fewer herbivores and carnivores than with the overkill run and there are more plants and *H. sapiens*.

## Discussion

The highly aggregated base model shows that second 2op leads to a boom and bust but doesn't show extinction *per se*. We may speculate that during the boom phase of the run less efficient herbivores perished in the competition for scarce resources but this is not conclusive. It is necessary to disaggregate the model to see if extinctions are actually occurring during the boom phase. In addition, this version of the model doesn't explain the pattern or extinction or the changes in the vegetation pattern identified above.

## Results – Step 3

**Step 3** is to build a model with vegetation partitioned into big and small trees, high and low quality grass; herbivores partitioned into browsers, mixed feeders and ruminant and non-ruminant grazers. Test equilibrium, overkill and 2op in this model.

The reduction in carnivore populations is based on carnivores killed per *H. sapiens* per year. The reduction in carnivore populations is a factor of carnivore density and kills per *H. sapiens*. At maximum density a 100 lb *H. sapiens* kills 2.5 lbs. of carnivore per year. As carnivore densities fall kills per unit of *H. sapiens* drop off to zero.

The stable ecosystem graph is like the stable ecosystem in the Base Model (Figure 1a) presented above. The only difference is that the increase in size and stability of a more complex system made it necessary to perturb the predator sector with a pulse reduction of 3% before any disturbance of the system was observable.

In the four herbivore model the introduction of *H. sapiens* causes very little disturbance to any of the sectors. Populations are 90% of starting values. Predator populations are reduced relatively more than Herbivores but only a fraction of a percent, and plants still increase slightly to 101% of its starting value (Figure 2a.).

In second-order predation mode, where *H. sapiens* hunts carnivores, first, as in the second order predation mode of the base model, herbivores increase, then crash; plants dip then recover; carnivores decline. With this version of the model after the first crash there is a leveling off and then a second crash of herbivores, carnivores, and *H. sapiens,* all leveling off at much lower levels and a second increase in plants with a leveling off at much higher levels. This is a new dynamic (Figure 2b).

The Herbivore graph Figure 2c shows ruminant grazers and browsers expand mixed feeders remain more or less level responding to competition from both grazers and browsers, non-ruminant grazers slump, responding more to competition from ruminants then to predation. Then mixed feeders crash followed by browsers. Ruminant grazers dip as they bear impact of predation from both carnivores and *H. sapiens*. Non-ruminant grazers rebound. A temporary equilibrium is established but then non-ruminant grazers crash.

The vegetation graph (Figure 2d) shows the reason. During the boom phase, browsers and mixed feeders eat trees faster than trees can recruit, resulting in a complete crash of small trees, followed by near extinction of large trees as well. The loss of trees frees up more land for quickly recruiting grass and so grazers are relatively favored. Grass and grazer populations are stable for a while. Then trees, freed from predation by browsers and mixed feeders recover and begin to colonize new territory. The loss of grassland to trees coupled with the decline in predators (human and non-human) sets up a competitive situation between the grazer populations. Since non-ruminants are less efficient than ruminants the competition drives non-ruminants to final extinction.

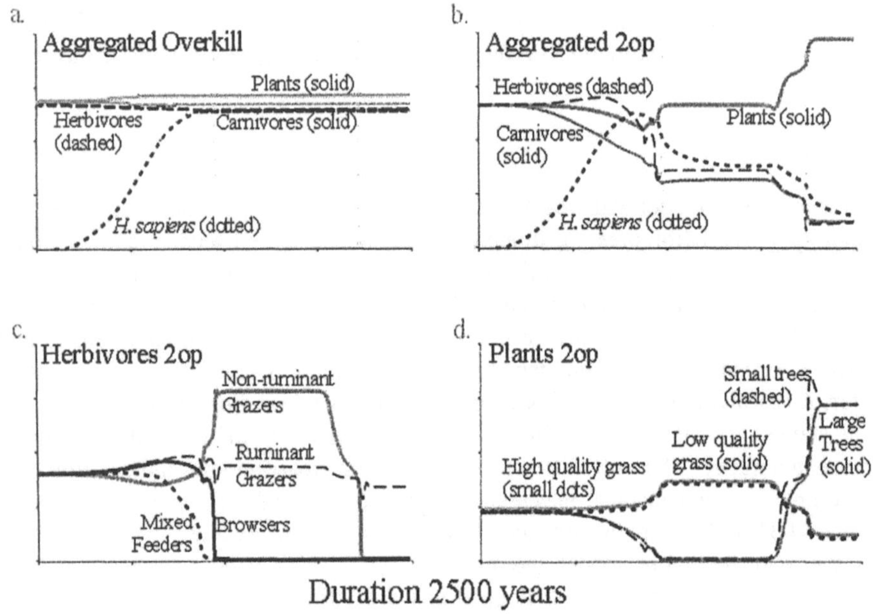


**Figure 2** – Four Herbivore model results

## Discussion

An unspoken assumption by scientists has been that population levels of the species which went extinct declined monotonically to zero. We see in the models presented here that the path of extinction is more likely to include recovery and overshoot as well as decline.

This research effort begins with the assumption that North American ecosystems were in equilibrium prior to the arrival of *H. sapiens*. This modeling paradigm forces an understanding of the web of relationships that support equilibrium. Only then are we able to come to grips with the perturbations associated with the migration of *H. sapiens*.

In this modeling effort, the more complex four herbivore model has shown the impact of *H. sapiens* in the overkill mode is much less severe than in the simple, base model. In the second order predation mode, the more complex model shows much greater impact with very little reduction in predator populations. These results are both the due to the interactions between herbivore species and their respective environments.

In the more complex model as one kind of herbivore or plant type is reduced other herbivores or plant types take their place so that he impact of predation (human or non-human) is buffered allowing the herbivore or plant type to recover – a complementary interaction. However, in the boom situation, due to loss of population control through predation, animals are in competition for scarce resources and an increase in one kind of herbivore means there is less available for others

Given the assumption of equilibrium, Overkill has been shown to be inconsistent with extinction. Second Order Predation *(H. sapiens* reducing predator populations) has been shown to be more consistent with extinctions.

The boom and bust created environmental exhaustion – a period of extreme vegetation scarcity. This accounts for some of the observations which are not accounted for by either climate change or overkill.

This model accounts for the shift from plaids to stripes (Guthrie, 1989) mentioned above. During the time of scarcity mammoths would knock down the trees in the mixed woodland to get at the leaves at the top. Bison, mammoths and all the extant herbivores ate everything in sight. Then once the number of animals had been reduced, trees were able to re-invade from mountain refugia. Bison, which by that time had evolved into the smaller, obligate grazers of today bison kept the woodland from encroaching on the plains and thus preserved the prairie. The shift, in turn, suggests why two ecophenotypes or two sub-species of bison emerged – *B. bison bison*, grazers of the prairie and *B. bison athabascae*, mixed feeders of the woodland. *B. bison*

During the boom bust period there would have been a bottleneck of vegetation. Scarcity favors smaller animals over larger animals and ruminants (bison, elk, deer) and against cecal digesters (mammoths, mastodons, horses). The bottleneck accounts for the extinction of mid-size mixed feeders – North American camels – who were in competition with both grazers and browsers.

Environmental exhaustion accounts for the extinction of ground sloth. It explains the data from Hansen's (1978) study of Rampart cave. Scarcity from herbivore population growth put the sloth in competition with more efficient herbivores and forced the ground sloth to eat its less preferred food – Mormon tea.

## References

Budyko, M. I. 1974. Climate and life. Academic Press, NY (quoted by Whittington and Dyke Simulating overkill: experiment with the Mossiman and Martin model. In, Martin, P.S. & Klein, R.G. (eds.) *Quaternary extinctions: A prehistoric revolution.* Univ. Arizona Press, Tucson.) page-451-464

Birdsell, J. B. 1957. Some population problems involving Pleistocene man. Population studies: animal ecology and demography, Cold Spring Harbor symposium on quantitative biology, 22:47-69. (quoted by Whittington and Dyke Simulating overkill: experiment with the Mossiman and Martin model. In, Martin, P.S. & Klein, R.G. (eds.) *Quaternary extinctions: A prehistoric revolution.* Univ. Arizona Press, Tucson.) page-451-464

Cat House 1996. personal communication in response to an e-mail request. electronic publication site *http://www.cathouse-fcc.org/*

Guthrie, R. D. 1989. Mosaics, allochemics, and nutrients: an ecological theory of Late Pleistocene megafaunal extinctions. In, Martin, P.S. & Klein, R.G. (eds.) *Quaternary extinctions: A prehistoric revolution.* Univ. Arizona Press, Tucson. page-259-298.

Hansen, R. M. 1978. Shasta ground sloth food habits, Rampart Cave, Arizona. *Paleobiology* vol. 4 page-302-319.

International Wolf Center 1996. Frequently asked questions (FAQ), electronic publication site *http://www.wolf.org/*

Leader-Williams, N. 1980. Population dynamics and regulation of reindeer introduced into South Georgia. *J. Wildl. Management.*, vol. 44 page-640-57.

May, R. M. 1973. *Stability and complexity in model ecosystems.* Princeton Univ. Press. Princeton, NJ.

Mossimann, J. E. & Martin, P. S. 1975. Simulating overkill by Paleoindians. *American Scientist* vol. 63 page-304-313.

Peterson, R. O. 1977. Wolf ecology and prey relationships on Isle Royale. *National Park Service Scientific Monog.* vol. 11, Washington, DC.

Schaller, G. 1972. The Serengetti lion: a study of predator prey relations. Univ. Chicago Press, IL.

Scheffer, V. B. 1951. The rise and fall of a reindeer herd. *Scientific Monthly*, vol. 75 page-356-362.

Whitney-Smith, E. 1995 Pleistocene extinctions: The case of the arboricidal megaherbivores, Canadian Quaternary Association (CANQUA) plenary presentation; electronic publication *http://quaternary.net/mstry.htm*

Whitney-Smith, E. 1996. New World Pleistocene extinctions system dynamics and carrying capacity: a critique of Whittington and Dyke (1989). American Quaternary Association (AMQUA) poster presentation; electronic publication *http://www.well.com/user/elin/w&d-txt.htm*

Whittington, S. L. and Dyke, B. 1989. Simulating overkill: experiment with the Mossiman and Martin model. In, Martin, P.S. & Klein, R.G. (eds.) *Quaternary extinctions: A prehistoric revolution.* Univ. Arizona Press, Tucson. page -451-464

Wing, L.D. & Buss, I.O. 1970. Elephants and Forests. *Wildl. Mong.* vol. 19.

# Chapter 25

# The Accuracy of Auto-adaptive Models for Estimating Romanian Firms' Cost of Equity

**Irina Manolescu**
"Alexandru Ioan Cuza" University, Iasi, Romania
Faculty of Economics and Business Administration
iciorasc@uaic.ro
**Claudia-Gabriela Ciorăscu**
"Alexandru Ioan Cuza" University, Iasi, Romania
Faculty of Computer Science
Evolutionary Computing Group
claudiag@info.uaic.ro

This paper analyzis the mutations of the Romanian firms capital structures and the relations with the cost of equity capital. The validity of leverage effect of capital structure over financial return is also tested. We consider in this analysis the public data of more than one hundred Romanian firms listed at Bucharest Stock Market and on RASDAQ, between 1997 and 2000.

Because of their adaptation to the specific of the input data, the neural models can be successfully used in real problems with large data sets. We are studying the quality of this approach in the case of estimating firms' cost of equity. This also defines the pre-requisites for considering new instances of the problem: share valuation, investment decision-making, etc.

The neural system designed for approximating the input data will be generated automatically by an evolutionary method based on genetic procedures. The supervised learning method used for training the neural network consists in a powerful combination between evolutionary techniques and back-propagation algorithm.

# 1 Introduction

Establishing the level of the financial structure is one of the most important decisions of the financial policy in a company. The estimation of this value depends on multiple factors, some of them especially volatile. Once this value is established, it will have to be modified whenever changes appear in the relevant factors, especially those that are bound to the governmental economic policy (monetary and fiscal).

We will dynamically generate a model for approximating the financial structure, by combining the paradigms specific to neural networks (NN) and evolutionary algorithms (EA). Both techniques are inspired from biological models, and, most important, both of them require little information about the problem to solve.

Different association schemes between genetic algorithms and neural networks have been proposed. The neuro-genetic system described below represents a collaborative combination, where the genetic model is used for selecting the proper topology of the feedforward multilayer perceptron network. A local optimization step is given by back-propagation procedure, assuring a better convergence rate of the algorithm.

# 2 Financial Structure

The level of indebtedness of the company represents a very suggestive index, as an estimator of the financial yield and the financial risk of the company. When the rate of indebtedness increases, it will determine a raise of the yield of the shareholders (as long as the rate of economic yield is superior to the interest rate), but the assumed risk for shareholders will be also higher.

The indices considered in the analysis are not the classic ones, but they have undergone adaptations to the conditions of the Romanian financial market. The rate of indebtedness has been determined in percentage, like ratio between the total debts and total equity capitals of the company. This ratio has been preferred because the limitation to the debts to average and length term is not unusable for the Romanian companies, the great majority of the debts being on short term (less then a year), but having a permanent character.

In addition, the yield of the equity capital considered is represented as the ratio net benefit / equity capitals and not as a financial yield, of the shareholders, in the true sense. From the 107 studied companies, near half (52) have not distributed dividends in this period. As consequence, if we considered the normal ratio for the financial yield, between the annual gains of the shareholders (in terms of dividends and gains of capital) and the equity capitals, we will obtain negative indices, or very small values.

# 3 Linear Regression Model

The instrument used in the econometric analysis of the financial data is the simple linear regression – the leverage effect is a linear function, having the financial yield as dependent variable and the rate of indebtedness as independent variable. The informatics support we used is SPSS 10.0 for Windows.

We included 9 variables in the regression and only 8 of them included explicitly (and for which there is a statistical description in Table 1):

- one *grouping variable* – the activity sector;
- four *dependent variables* – financial yield for each of the 4 years considered in the analysis, expressed in percentage (named rf00–rf97);
- four *independent variables* – the rate of indebtedness for each of the 4 years considered in the analysis, expressed in percentage (named indat00–indat97).

In order to verify the leverage effect of capital structure over financial return, the econometric models have been generated using simple linear regression. We also tested statistically the resulted models.

**Table 1**: Statistical description of the variables included in the models

| Variable (%) | Minimum | Maximum | Mean | Std. deviation |
|---|---|---|---|---|
| indat00 | 3,50 | 20000,00 | 406,12 | 2103,39 |
| indat99 | 4,00 | 2500,00 | 189,98 | 366,64 |
| indat98 | 3,46 | 9900,00 | 244,79 | 1042,90 |
| indat97 | 2,10 | 9500,00 | 222,21 | 988,14 |
| rf00 | −132,00 | 120,00 | 9,56 | 22,50 |
| rf99 | −157,00 | 120,00 | 8,97 | 26,30 |
| rf98 | −106,00 | 130,60 | 7,82 | 23,41 |
| rf97 | −50,00 | 100,00 | 12,03 | 22,25 |

In order to prepare the regression, we have tested the existence of significant differences between the values of the rate of indebtedness of 2 consecutive years. The average values of the variables differ from one period to another, the rate of indebtedness reaching its maximum in 2000 (406.12%) and minimum in 1999 (189.98%), while for the financial yield it was a maximum of 12.03% in 1997 and a minimum of 7.82% in 1998. The null hypothesis of the comparison consists in the fact that the difference between 2 consecutive years is insignificant from the statistical point of view. The results are presented in the Table 2.

From this statistical analysis we notice relatively low correlations between the consecutive series of years 2000, 1999 and 1998, and a high correlation in the case of years 1998 and 1997. All the calculated coefficients of correlation have been significant at a level of 2.5%.

Table 2: The results of comparison between the rates of indebtedness

| Paired series | Correlation coeficients | Sig. | Paired differences Mean | Paired differences Std. dev. | t | df | Sig. |
|---|---|---|---|---|---|---|---|
| indat00/99 | 0,234 | 0,025 | 216,1390 | 2048,8561 | 1,012 | 91 | ,314 |
| indat99/98 | 0,532 | 0,000 | −54,8117 | 902,9352 | −,582 | 91 | ,562 |
| indat98/97 | 0,988 | 0,000 | 22,5838 | 169,2817 | 1,280 | 91 | ,204 |

# 4 CAPM Model

The determination of financial rentability by CAPM (Capital Asset Pricing Model) as mean on each sector, can be visualised in the Table 3.

Table 3: CAPM Model

| sector | 1 | 2 | 3 | 4 | 5 | 6 | 7 | 8 | 9 |
|---|---|---|---|---|---|---|---|---|---|
| RF97 | 31.66 | 9.24 | −16.34 | −20.37 | 32.21 | −34.64 | 67.04 | 61.39 | 47.25 |
| RF98 | −9.29 | −25.58 | −44.18 | −47.10 | −8.89 | −57.48 | 16.44 | 12.33 | 2.05 |
| RF99 | 37.93 | 28.70 | 18.17 | 16.51 | 38.15 | 10.63 | 52.49 | 50.17 | 44.35 |

The differences between resulted and expected values are very high, which makes this model unusable for the Romanian stock market. Other researchers presented the same issue concerning the emergent stock markets.

# 5 Auto-adaptive Model

We are proposing a neuro-genetic system for dynamically generate an adaptive model for approximating the Romanian financial structure.

For a complete description of neural architecture, we used a direct encoding of the feedforward neural network. A chromosome is represented as a three-level tree, where the first level encodes all the layers of the networks, the second level assigns a number of neurons for each layer and the leaves are input connections to neurons. In order to introduce nonlinearity into the evolving networks, the activation function of hidden units is given by hyperbolic tangent function. The neural models have only one output assigned to financial rentability variable. The output unit uses a linear activation function.

The general scheme of the genetic algorithm used by our system could be represented as follow:

```
1. Initial population
2. Evaluate
3. While (stop condition) do
```

```
4. Select individuals for reproduction
5. Recombination of the selected individuals
6. Mutation on offspring generated by recombination
7. Evaluate new individuals
8. Create next generation
   from parents and their offspring
```

We used a constant size of the population during the evolution of the genetic algorithm. The initial population is randomly generated. Each chromosome has the number of layers and neurons on each layer limited by predefined values. The connection weights and bias are randomly selected in the range $[-0.5, 0.5]$.

The objective function of the genetic algorithm is defined by normalized squared error function computed for the whole training set. By training the phenotype with a gradient-descent algorithm locally optimizes the evaluated chromosome. The back-propagation is applied for a small number of epochs. After each evaluation, the trained network is re-encoded and preserved in the population.

This hybrid method allows the genetic model to search more intensively in the space of the permitted neural architectures and assures a good convergence rate of the algorithm. The learning rate of the back-propagation algorithm is decreased at each epoch.

## 5.1 Genetic Operators

**Selection.** The selection of parents for reproduction is made according to a linear ranking method, as suggested bellow:

$$p_i = pos_i * \frac{2 * SP}{PS^2 * (PS + 1) * NP}$$

where $p_i$ is the probability for selection of individ $i$, $pos_i$ represents the rank occupied by the chromosome in the population, considering a decreasing order (the worst individual has rank 1) and $SP$ represents the selection pressure. The constants $PS$ and $NP$ are given by the number of individuals in population and the training data set size.

After applying genetic operators, the same procedure is used for selecting $PS$ individuals into the new generation, the offspring fighting with their parents for surviving. This procedure allows keeping good individuals and also gives small chances to survive for worse chromosomes.

**Crossover.** Two types of crossover operators are proposed, each one generating two offspring.

The **weight operator** applies an arithmetical crossover over the connections weights. The first offspring will have the same topology as the first parent, but the weight corresponding to the selected connections will be modified as suggested in (1); the second offspring copies the architectures of the second parent, using the weights suggested in (2):

$$w_{ij}^{C_1} = w_{ij}^{A} + a * (w_{ij}^{B} - w_{ij}^{A}) \qquad (1)$$

$$w_{ij}^{C_2} = w_{ij}^{B} + a * (w_{ij}^{A} - w_{ij}^{B}) \qquad (2)$$

where $A$ and $B$ are the parents selected for reproduction, $w_{ij}$ represents the weight corresponding to the $i$-th incoming connection of the unit $j$ and $a$ is a random number between $(-0.25, 0.25)$.

The **connection crossover** interchanges the sub-trees of two selected neurons from the same layer of both parents. By replacing the incoming connections, it is possible that not all the connections to have an input neuron or to loose all the forwarded connections for neurons from the previous layer. In the first case, new neurons and their incoming connections are generated randomly. In the second case all the neural units that are not connected to any neural unit from the next layer are deleted.

**Mutation.** The mutation is applied on the resulted offspring. We defined three types of mutation: weight mutation, connection mutation and unit mutation, with different probabilities.

The **weight mutation** uniformly affects some connections weights. These are randomly selected from the topology. The weights are modified adding a value within $[-1.5, 1.5]$.

By **connection mutation** a new incoming connection is created between two randomly selected neural units placed in consecutive layers, or the connection weight is modified if the connection already exists.

The **unit mutation** has the highest impact over the structure of the chromosome. It interchanges the sub-trees of two neural units from the same layer.

## 5.2 Experimental results

We performed 200 runs, each one started with a different random number seed. The algorithm was allowed to continue until a maximum number of 100 generations is reached. The population size has been set to 10 and the local optimization is performed for 5 epochs with a learning rate set to 0.3. The learning rate is dynamically changed for each individual, depending on its age in the population. Long training processes assure a better accuracy, but also increase the computational time.

The data set includes 8 variables for input (including financial yield and the rate of indebtedness) and 1 variable for output (the next rate of indebtedness). All the input variables from the training data set have been standardized. For this analyze only two years have been considered (1998 and 1999).

The results obtained are indicated in the Table 4. The best individual gives the minimum error for the aproximation of the model. The differences between generated and expected values for all test cases of the best solution are presented in the Fig. 1:

Table 4: Results obtained for 200 runs

| Mean error | Standard deviation | Median error | Best error |
|---|---|---|---|
| 0.0824 | 0.0016 | 0.0817 | 0.0797 |

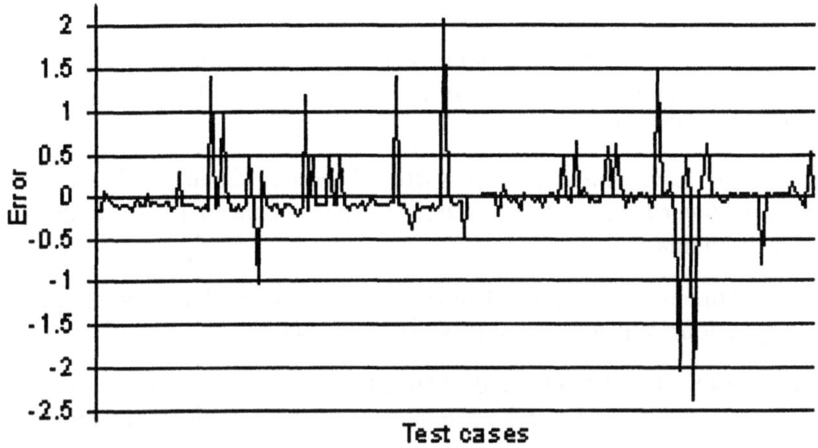


**Figure 1**: Differences between generated and expected values

# 6 Conclusions

The yield of the shareholders is, in average, constantly inferior to the interest rate, during the analyzed period. Although the cost of the credit, when it is compared with the cost of the equity capital, must be diminished as a result of the inflation, it seems that for the Romanian financial market this recommendation of the financial theory is not applicable. Even for an adaptive system it is difficult to find a model that can aproximate the correlation factors. This is because the equity capitals does not undergo annual re-evaluations, and the increase of the prices of the shares (at least proportionally with the index of the inflation) is purely accidental.

All those elements have opposed effects on the cost of capital and the resulting conclusion is the impossibility to obtain an optimal capital structure, especially on Romanian capital market. The impact of the debt is constantly negative for the Romanian firms.

The existence on Romanian stock market of the listed firms with a debt ratio (debt / equity capital) higher than 500% (but sometimes reaching incredible values like 10000%) raises serious questions over the admittance criterions on stock market quotation. These highly indebt firms have descending trends in activity and profits and the questions about financing policy can be raised both on firm and financial institutes level.

# Bibliography


[1] BISHOP, Christopher, *Neural Networks for Pattern Recognition*, Oxford University Press, ISBN 0-19-853864-2

[2] CHARREAUX, G., *Gestion financiére. Principes, études de cas, solutions*, 3eme Ed., Editions Litec, Paris, 1991

[3] DIVECHA, A. B., DRACH, J. and STEFEK, D., *Emerging Markets: A Quantitative Perspective, Readings in Investments*, John Wiley & Sons, Chichester: 457–472, 1994

[4] FERARIU, Lavinia and BÎRNOVEANU (Ciorăscu), Claudia, *Neural Genetic System Identification*, SACCS'98, *Automatic Control Proceedings*, vol. I, Iaşi, România

[5] GEN, Mitsuo and CHENG, Runwei, *Genetic Algorithms and Engineering Design*, John Wiley & Sons, ISBN 0-471-12741-8

[6] GORDON S., LEVEY D.H. and MAHONEY C., *Credit Analysis and Ratings, International Bond Portfolio Management*, Euromoney Publications: 127–148, 1989

[7] GOURIEROUX, C., SCAILLET, O. and SZAFARZ, A., *Econométrie de la finance. Analyses historiques*, Economica, Paris 1997

[8] GRIMBERT, D., MORDACQ, P. and TCHEMENI, E., *Les Marchés Emergents*, Economica, Paris, 1995

[9] LUMBY, S., *Investment Appraisal and Financial Decisions*, 5th Ed., Chapman & Hall, London, 1994

[10] MARESCHAL B. and BRANS J.P., *An Industrial Evaluation System, European Journal of Operational Research*, vol. 54: 318–324, 1991

[11] MICHALEWICZ, Zbigniew, *Genetic Algortihms + Data Structures = Evolution Programs*, 3rd Ed., Springer, ISBN 3-540-60676-9

[12] REED, Russell D. and MARKS, Robert J., *Neural Smithing – Supervised Learning in Feedforward Artificial Neural Networks*, MIT Press, ISBN 0-262-18190-8

[13] STANCU, I., *Finanţe. Teoria pieţelor financiare. Finanţele întreprinderilor. Analiza şi gestiunea financiară.*, Editura Economică, Bucureşti, 1996

[14] SUÁREZ, A. S., *Economía financiera de la empresa*, Edición Pirámide. Madrid, 1990

[15] ZAIŢ, D., *Fundamentele economice ale investiţiilor*, Editura Sanvialy, 1996

# Chapter 26

# Self-Organizing Geography

## *Scaled Objects and Regional Planning in the U.S.*

**Gary G. Nelson**
Homeland Security Institute
Arlington, VA, USA
gary.nelson@hsi.dhs.gov

**Peter M. Allen**
Cranfield University
Bedford, UK
p.m.allen@cranfield.ac.uk

The deficiencies of any kind of central planning are exposed by comparison with the paradigm of complex, evolutionary systems. Ideologically, market-democracies borrow that paradigm, subject to the distortions imposed by economic and political concentrations of power that seek control of resources. Economic space-time (geography) is a resource very much subject to such would-be concentrations of control by "top-down" forms of planning. In the U.S. particularly, the dichotomization of "land use" and "transport" planning and the ongoing contention about how such planning should be allocated and supported by models, illustrates the divergence between the ideological theory and contention for control. This paper traces some history of that contention and offers the scale hierarchy as a prescriptive framework that respects both the evolved governance hierarchy and the distributed preferences and uncertainties such governance must accommodate.

# 1 Introduction

In the 1970's, the U.S. Department of Transportation, inspired by Ilya Prigogine [1], sponsored work on self-organizing models of cities and regions. These nonequilibrium, evolutionary, models integrated land use and transport for long-range planning and have been adopted by some countries for regional and environmental planning. However, the work was never adopted by the intended market in the U.S.: The Metropolitan Planning Organizations (MPOs) that maintain regional models for 20-year long range plans as required for federal-aid transport project programming [2]. The interest in integrating transport and land use models—in short, true models of human geography—has never gone away and efforts continue to revamp the conventional (non-integrated, deterministic, quasi-equilibrium) models used by almost all of the nearly 400 MPOs. But the divergence between theory and practice was clear from the beginning, in the debate between planners and highway engineers that resulted in the 1962 Highway Act that created the MPOs.

It is often said that the highway program, now the federal-aid surface transport program, got the country out of the mud at the beginning of the 20th century, and into the muddle of congestion, pollution and sprawl by the middle of the last century. There are no political jurisdictions commensurate with urban-regional interactions: In lieu of such jurisdictions, MPOs were created by federal law to allocate federal funds via regional planning under multiple jurisdictional goals, striving toward the models and the process to deal with geographical complexity. This paper observes that a notion of scaled objects exists in the MPO process, but that these objects are mis-allocated from both a political and technical-capability perspective.

# 2 The Problem of Planning in a Market Democracy

Scale is a key parameter of planning, and in the ideological contention over what is individual, what is in "the market", and what is "public". Scale has spatial and temporal dimension. Space includes the diversity of interests concerned and the number of choices affecting a plan. The time horizon is the interval between the dissatisfaction that initiates planning and the (hopefully) satisfactory outcome of planning. Uncertainty in planning grows exponentially with scale and requires the definition of objects appropriate to each scale of planning (which is just prospective decision making).

All complex societies encounter the Tragedy of the Commons (TOTC). The TOTC is simply the generation of externalities (e.g., congestion, pollution, and sprawl) from individual activities. The term "externality" refers to the scale differentiation between a purposeful action and unintended consequences that emerge as context for those actions. Location, thrice cited by realtors as the determinant of property value, is an emergent property of geography from human interaction. Emergence scales objects above the individual level, and so justifies governmental planning that involves police (regulatory) and eminent domain (project construction) powers. But planning, that is also a code word for socialism, raises ideological questions in a market-democracy because it adds an inevitable top-down component to what is idealized as a bottom-up emergent organization. Debates about planning, including the dichotomy of land use planning—"bad" and delegated to the lowest jurisdictions—and transport planning—"good", and assumed by the highest eminent domain jurisdictions—continue to represent an ideological schism and concern the allocation of scaled objects in planning to jurisdictional levels.

## 2.1 Scale Hierarchy

Emergence, as desirable or undesirable externalities, indicates self-organization and complexity within a scale hierarchy. The scale hierarchy concept used here owes much to Stan Salthe [3] who eloquently described it in the biosphere, with applications to adaptive control systems, government and any sort of complex system. The basic scheme is in figure 1:

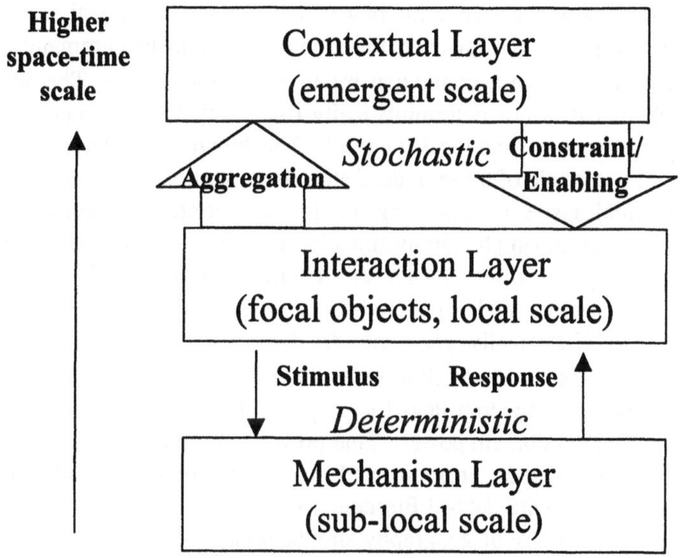


**Figure 1.** The triadic layer structure of the scale hierarchy.

The scale hierarchy is triadic and centered on a focal layer partitioned into interacting objects such as purposeful individuals or jurisdictions with local information limits that correlate with geographical space-time extent. Obviously both mobility and telecommunications give individuals information about a variety of places that can be geographically separated. However, many planning issues involve discrete and compact jurisdictions, each of whose information and interests are correspondingly discrete and compact. Below the focal scale are mechanisms that the interactors control reliably in order to act on their purposes: For instance, individuals use a mode to go places. The relation is *relatively* causal and deterministic. The stimulus-response couple can be learned from experience, or is governed by physical "law".

Above the focal scale is a context of action that in self-organizing systems also emerges from interaction. Geography is a good example. The relation between the focal and contextual scales is *relatively* stochastic: No local action determines context and context does not determine local action. Even though each actor perceives a context that is static over the scale of local action, the context is evolving non-deterministically at its scale. Context can be physical (roads, buildings), or logical (the price of housing, or a set of rules). Context both constrains and enables. It may be said that the focal-mechanism interaction represents *development*, while the focal-contextual interaction represents *evolution* within the overall self-organizing system, analogous to the "evo-devo" complementation in biology.

We can shift the focal scale layer to other space-time physical, or conceptual, scales such as jurisdictions. However, there must be a distinction between the physical and *ex officio* individual at different scales. Every person, no matter what their position of political or economic power, has the same physical scale limits and uses mechanistic physics. However, the idea of an *adaptive* command structure is that a higher official is in a contextual relation to subordinates: The scale relation must be such that it is not deterministically rigid, but allows a degree of freedom between higher "strategy" and local "tactics". The functional differentiation of scales is determined by this rule of loose, indeed stochastic, vertical coupling.

The political issue of planning is in the contention between a tighter corporate/military command hierarchy and a looser scale hierarchy. Few deny the need for a contextual layer of rules. As Hayek [4] emphasized, there are evolutionary or common laws (nomos) and legislated laws (thesis) that might also be called "developed" laws. Historically, geography has been a self-organized network mostly beyond thesis. The *right-of-ways* (links) co-evolved with common law, *rights-of-way* rules. The power to condemn land for purposefully established roads or tracks is governmental eminent domain power. That power supports thesis, in the form of the federal-aid surface transport program, and many other planning and environmental regulations in what is here dubbed Eminent Domain Planning (EDP). The problem is how the scale hierarchy of geography should relate to a set of scaled objects in EDP.

## 2.2 Scaled Objects in Transport Planning

The region is a scaled object in EDP. MPOs were federally created to program federal-aid transport projects absent regional political jurisdictions, and a region is legally defined in census terms as an urbanized area. MPO boards include county jurisdictions overlapping the urbanized area. The state department of transportation, and regional transit authorities, as the actual recipients of federal aid and the actual project builders/operators, are also voting MPO members. Conceptually, a region is a super-jurisdictional area defined by externalities, at least air pollution and congestion as related to traffic. The MPO produces long range plans (LRPs) over 20-year horizons at five-year updates, and Transportation Improvement Programs (TIPs) at 2-year updates [23 USC 134]. Nominally, the LRP is a strategy and the TIP the programming menu for the sequence of project steps (tactics). The MPO itself, but sometimes the state, operates a regional network model for the LRP to certify TIP entrants.

Eminent domain clearly scale-ranks jurisdictions (federal-state-local...) but the ambiguity is the degree of freedom allowed between political levels. The federal-aid program, in which no projects (with the exception of those on federal lands) are built or operated by the federal government, is a response to historically-strong constitutional scruples against federal involvement in "internal improvements". Table 1 shows the year 2000 proportions of highway user revenues collected (that are themselves only 63.5% of highway disbursements) and highway disbursement [5, Table HF-10]. The federal highway trust fund revenues go mostly to states, providing significant leverage for major highways, and similarly for about $5 billion per year in federal aid to transit authorities. States and localities chafe under the categorical constraints of top-down funding, and the federal planning/environmental procedure that comes with the funding.

**Table 1:** Distribution of Highway Revenues and Disbursements in 2000

| | **Federal** | **State** | **Local** | **Total** |
|---|---|---|---|---|
| **User Revenue $billions** | $29.7 B | $49.0 B | $2.3B | $81.0B |
| **Row %** | 36.7% | 60.5% | 2.8% | 100% |
| **Disbursed $billions** | $2.3B | $77.9B | $47.3B | $127.5B |
| **Row %** | 1.8% | 61.1% | 37.1% | 100% |

One ambiguity in EDP concerns accountability for projects. This is called the ping-pong planning syndrome, enunciated in Mowbray [6]. It refers to the experience of highway opponents who are referred from planners to politicians up the eminent domain chain, finally to discover that federal authorities claim no determination of what projects get built. This ambiguity of accountability destroys the political enfranchisement of citizens to affect plans politically. The MPO is nominally the local-approval authority for expenditure of significant public funds, but the MPO is not an electoral jurisdiction, and is strongly affected by state and MPO planning technicians. A second ambiguity concerns the relation of LRP to TIP. There is a strong normative contention, especially on the part of federal certifiers of MPO activity, that the LRP should not be a "staple job" of projects that simply go into the TIP: In other words the LRP and its network modeling should be a contextual strategy. But achieving this is hindered by technical limits of analyzing and predicting the evolution of geography.

# 3 Models of Geographical Complexity

The conventional MPO network model has four "steps", as shown in figure 2. The hierarchical view of the model represents a nest of contextual conditions that can be entered as sequential steps, and hence the common name of "four step" for the model. The focal step is assignment of trip paths to transport links, summing to volumes that can be compared to capacity as a level of service (LOS) measure of link performance. LOS is the primary criterion for projects of capacity expansion. Assignment is an equilibrium problem of trips trying to take shortest routes while the TOTC of congestion alters what is the shortest route, invoking the feedbacks from volume to travel time to volume. The significant issues about the four-step model concern how it handles all the other feedbacks that exist in geography. It should be noted that a new model structure called TRANSIMS [7] is being introduced. While TRANSIMS has its origins in self-organizing concepts, its use will not fundamentally alter the problem of feedbacks and scaled objects, especially with "land use" that remains an optional module under TRANSIMS.

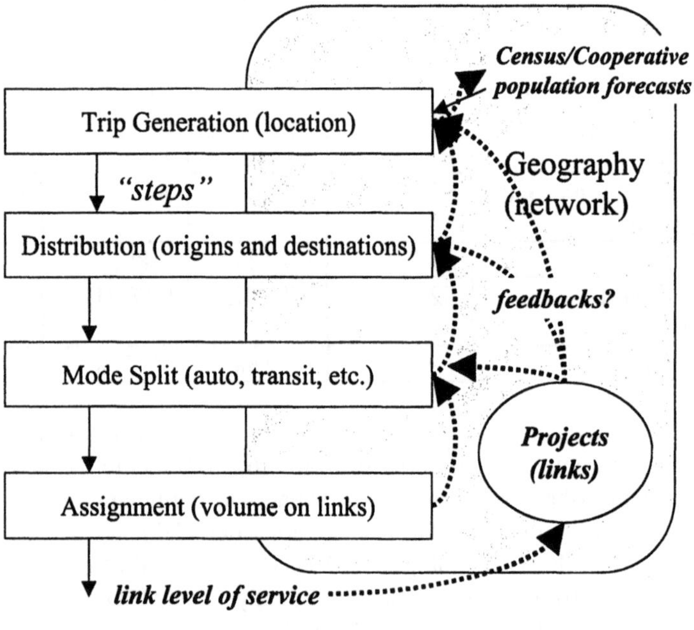


**Figure 2.** The "four step" regional network model.

Many efforts have been devoted to various levels of feedback within the four-step or TRANSIMS structures. However, the fundamental problem is really in the scaled allocation of EDP decisions versus the scaled dynamics of geography. The regional decision-object of EDP, as a 20-year LRP, needs a context, but geography itself, especially as affected by multiple 5-year cycles of LRPs and TIPs, *cannot be taken even approximately as a stable context for long-range strategic planning.*

Regional models analyze limited sets of alternatives for LRPs. The models artificially create what looks like a planning context by eliminating feedbacks and entering upper layers as static constraints. The models then deliver equilibrium predictions of geographical state, especially traffic LOS performance and its related air pollution emissions at a 20-year future. While sensitivity analyses are often done, their value as externally defined variations in improperly identified models is questionable. However, the business of the MPO in programming projects is mostly based on *current* LOS deficiencies that create the menu of projects "tested" by the LRP modeling as a *pro forma* step before TIP approval. The LOS deficiencies make projects "necessary". This inverts what is supposedly a "strategy"-driven process. It promotes the ping-pong planning syndrome because tracing accountability back up the scales only returns to local conditions. The modeling does not invite close examination by its policy customers, who are neither technicians nor interested in destabilizing the political-power enhancements of distributing federal-aid funding.

The regional modeling of the LRP is used throughout the project-scale environmental analysis to defer strategic questions such as regional air pollution (a project can increase emissions but be "OK" under aggregate emissions budgets) or the transfer of bottlenecks throughout the network by removing a bottleneck in one location. Yet project link-level predictions of the regional models are *not* made public to allow scrutiny of their relation to projects. This helps suppress the fact that the regional models *do not have* the accuracy at the assignment level to resolve link LOS within the critical project-decision thresholds. The models are expected *at best* to be within a 15% volume error against *current* volume data [8, Figure 7-4, Maximum Desirable Error for Link Volumes] and that performance is consistent with the few model results that are available for inspection. But that error encompasses the whole range between acceptable LOS and link capacity (designated LOS D to F). The 20-year forecast of such models has *unknown* prospective validity. However, comparing LRPs from the 1970's and 1990's for the DC area shows that fast growth areas, where location and transport interact intensively, have errors around 100%. There is a significant uncertainty gap between the regional and project level. This has been recognized in federal attempts to introduce an intermediate level of analysis as the "major investment" or "corridor" study that addresses some area beyond local scale (especially for modal choice, trip generation and bottleneck-shift effects) but hopefully with higher resolution and less error than regional models.

The self-organizing models [9] [10] [11] included the range of geographical feedbacks, as nonlinear morphogenetic equations, in an urban and inter-urban hierarchy, to simulate realistic geographical evolution. Computer limitations in the 1970's prevented the self-organizing models from being convincingly detailed. TRANSIMS today, with its heritage from cellular automata and its lavish use of computer power on micro-scale (individual trip) simulations represents the capability to implement the self-organizing models. The basic reason fully integrated models have not been pursued is that there is no expectation that any equilibrium exists for geography in full (viewed as a problem of network design with both competitive location and congestion). Geography modeled as a nonequilibrium system cannot provide either a context or control object for EDP. The nonequilibrium is only exacerbated by a policy of dissipating resources at traffic bottlenecks.

The lessons and experiences from self-organizing system modeling show that it is impossible to get them adopted before the institutions and organizations that must use them are ready. The models developed over the years since the mid-seventies in collaboration initially with the USDOT, and later with the European Commission were not used or adopted. They clearly showed the importance of an integrated approach to social and economic development, demonstrating that land-use, transport, the geographical extension of the urban zones, retail and commercial development as well as housing and public services all interact. This leads to a complex cascade of effects as the spatial multipliers work through the system, changing patterns of demand and supply of goods and services, and the transport flows of these. Only now, some twenty-five years after the work was initiated has the first tentative use been made of these methods. This has been by the Asian Development Bank in attempting to assess the spatial social and economic impacts of new transportation infrastructure investments in West Bengal. Also, Guy Engelen, now at the RIKS institute in Maastricht, the Netherlands, working with Roger White has developed some versions of the self-organizing models of spatial interaction that the Netherlands government has successfully used. There are impressive presentation and display tools for the interaction of users with the models and this certainly contributes to their success. In the US, the idea of coupled land-use/transportation models is only now being developed by the Oregon DOT. If planning can be done at a truly strategic level, integrated models with the real freedom of geography to self-organize can be admitted.

# 4 An Allocation of Scaled Objects

The ambiguities in EDP, the political and technical challenges to procedure, and the limits of modeling in a complex geography all demand a clearer allocation of scaled objects to decisions and decision-support models: The uncertainty demonstrable in "strategic planning" must be matched to appropriately scaled strategic decisions, not "projects". The current EDP process is ill-adapted to geographical variety, especially central city versus suburb, versus rural areas. The self-organizing models explicitly include the dynamics of geography whose complexity forces the basic question of how to allocate distributed, adaptive, but coordinated decisions about geography.

A prescription refers back to the basic scale hierarchy of Figure 1. A valid conception of the MPO is as a focal *collective* of interactions among local jurisdictions over project planning. If an artificially defined "region" is reified as a non-elective jurisdiction, it is too big to deal with projects and too small to deal with "strategies"—especially inter-regional resource redistribution—in a megalopolitan world. In any case, states are the next legitimate sovereignty above local governments (and the true sovereignty of localities varies greatly among the United States). Since there are many important multi-state (and even international) urbanized areas, the interaction of sovereignties at any scale cannot be avoided. Each of these scales needs a proper context for the interaction, but synthesizing a pseudo-jurisdiction to reify the peer collective muddles both the sovereign context and the proper degree of freedom of the constituents (as we see empirically with the MPOs). The project scale properly belongs to the locality of one or more infra-regional jurisdictions. The strategic level is above that, and of indefinitely large—certainly multi-metropolitan and by the constitutional doctrine of the commerce power, national—scale.

Federal-aid transport thesis implements such a contextual strategy in several ways. However, the redistributed resources mediated by EDP are part of the muddle of the TOTC: By being divorced from "local" development and traffic-generation accountability, most of the funding is for more capacity for more traffic congestion. Looking back on the origins of the interstate freeway program that was the proximal motivation for the MPOs [12], the explicit strategy of reforming metropolitan areas for the sake of automobiles has met its limits and is obsolete. The *increasing* use of federal thesis by Congress to mandate specific "earmarked" projects is a clear violation of scale on behalf of political power concentration.

Strategic context should not decide plans or projects, but rather the *rules* of interaction relevant to geography. We should be looking in nomos for the rules regarding externalities in physical space-time. Among those are the evolved rules of rights of way and of damages. We have seen the *rights-of-way* laws (i.e., traveler rights), still on the books for pedestrians and bicycles, eradicated *de facto* by the automobile's more powerful consumption of space (i.e., speed, momentum) accompanied by the externalities of noise, pollution and physical threat. Further, the eminent domain *right-of-ways* (i.e., strips of land, pavements and structures) for freeways and arterials disrupt local link networks and the pattern of development. This exercise of EDP is contrary to the historical co-evolution of access and place, and as reflected in historical procedures for laying out of roads. This dichotomization of "important roads" from places leads to the typical "sprawl" pattern of new development on cul-de-sacs or strip-malls debouching onto clogged arterials: We haphazardly develop networks of minimal local connectivity and maximum traffic generation. In terms of externalized damages, we need to apply accountability at the individual level and at the scale of local jurisdictions: The case of an upstream polluter needs to be extended to generators of traffic. By focusing only on link LOS, the externality (congestion) is accommodated rather than called to account, and it is the damaging party, not the damaged party, that EDP now favors with more capacity for the new traffic. Congestion tolls have long been advocated, but in the U.S. federal-aid law has long prevented tolling of the major highways (excepting separate toll-authority roads not under federal aid and recent "HOT" demonstrations mostly associated with *new* freeway construction).

Resource redistribution for development goals is a decision-object at the highest level of eminent domain. The federal-aid highway program, and its transit appendages, started out with that purpose. But the surface transportation program has devolved to a supposedly "equitable" (but in fact categorically and modally biased) channeling of funds from source back to source. If there are no differential inter-regional strategies as objects at the federal level, then the imposition of top-down eminent domain has outlived its rationale. EDP instead has entered the muddle of the TOTC of excess traffic, through a cross-subsidized consumption of metropolitan space. Rather than assuming some centralized plan—or even regional programming of projects to expend top-down resources—we need explicit context for the interactions among politically enfranchised jurisdictions. Our very abhorrence of state and federal intervention in local affairs demands that their decision objects be truly contextual and allow local degrees of freedom. The self-organizing models demonstrate, rather than obscure, the predictive uncertainty in regional and inter-regional evolution. Geography is too complex for direct large-scale intervention, and must evolve as a ruly interaction among localities with well-defined accountability and local information. This prevents large-scale intervention at the project level, but exposes the dynamical parameters and feedbacks of interaction that need to be addressed contextually. That most likely includes breaking the positive feedbacks that now exist between generating congestion and getting top-down funding. The self-organization of geography is a matter of growth *and* limits.

## Bibliography


[1] Prigogine, Ilya, 1980, *From Being to Becoming—Time and Complexity*, W.H. Freeman, NY.

[2] Weiner, Edward, 1992, Urban Transportation Planning in the United States, USDOT, Washington, DC.

[3] Salthe, S.N., 1985, Evolving Hierarchical Systems, Columbia University Press, NY; Development and Evolution: Complexity and Change in Biology, MIT Press, 1994. There are several later papers either individually or joint by Nelson and Salthe that further elaborate the scale hierarchy concept.

[4] Hayek, 1978, Law Liberty and Legislation, University of Chicago Press.

[5] FHWA (Federal Highway Administration), 2000, Highway Statistics, Washington, DC.

[6] Mowbray, A.Q., 1969, Road to Ruin, Lippincott, NY.

[7] FHWA, 2002, TRANSIMS information at http://tmip.fhwa.dot.gov/transims/

[8] Barton-Aschman Associates, Inc., and Cambridge Systematics, Inc., 1997, Model Validation and Reasonableness Checking Manual, http://tmip.fhwa.dot.gov/clearinghouse/docs/mvrcm/

[9] Allen, P.M. et. al., *The Dynamics of Urban Evolution, Volume 1: Inter-Urban Evolution and Volume 2: Intra-Urban Evolution*, Final Report, October 1978. USDOT-RSPA, Cambridge, MA.

[10] ________, 1981, *Urban Evolution, Self-Organisation and Decision Making*, Environment and Planning A, pp 167-183.

[11] Allen, P.M., 1997, Cities and Regions as Self-Organizing Systems: Models of Complexity, Gordon and Breach, Environmental Problems and Social Dynamics Series.

[12] BPR (Bureau of Public Roads), 1939, Toll Roads and Free Roads, Report to Congress, Washington, DC.

# Chapter 27

# City of Slums: self-organisation across scales

**Joana Barros**
Centre for Advanced Spatial Analysis - CASA
University College London
j.barros@ucl.ac.uk

**Fabiano Sobreira**
Programa de Pós-graduação em Desenvolvimento Urbano
Universidade Federal de Pernambuco
fjasobreira@yahoo.co.uk

Paper presented at the International Conference on Complex Systems (ICCS2002), Nashua, NH, USA, June 9-14, 2002.

## 1. Introduction

The city is certainly a fine example of a complex system, where the parts can only be understood through the whole, and the whole is more than the simple sum of the parts. In the present paper we explore the idea that some of these parts are themselves complex systems and the interrelation between complex subsystems with the overall system is a necessary issue to the understanding of the urban complex system.

Spontaneous settlements are clear examples of complex subsystems within a complex urban system. Their morphological characteristics combined with their development process are traditionally understood as chaotic and unorganised. And so are Third World cities, traditionally known for their inherent chaotic and discontinuous spatial patterns and rapid and unorganised development process.

The paper consists in a brief theoretical analysis developed on the interrelationship between two urban processes across scales: the local process of formation of inner-city squatter settlements and the global process of urban growth. What is the role that spontaneous settlements play in the global dynamics of the city? We explore this issue by analysing experiments of 'City-of-slums', an agent-based model that focuses on the process of consolidation of inner-city squatter settlements within a peripherisation process.

The paper also includes two previous studies on these topics where the dynamics of these two urban processes are examined as two isolated complex systems and an analysis of the morphological fragmentation of the distribution of spontaneous settlements within the overall city and within the spontaneous settlements themselves. Based on these analyses, we conclude with a brief discussion on the role of self-organisation in the socio-spatial dynamics of Third World cities.

## 2. Latin American cities: growth and fragmentation

The urbanization process in cities of developing countries is often insufficiently planned and poorly coordinated. The morphological result is a fragmented set of patches, with different morphological patterns often disconnected from each other. This fragmented pattern has its origins in the successive superposition of different urban typologies, including planned areas, spontaneous settlements, housing tracts, slums, vacant sites, institutional areas, shopping malls, informal town centres and so on. The Third World city is the result of the combined dynamic of fragments that are in constant mutation and evolution.

Spontaneous settlements fill some of the gaps in this erratic development, at the same time creating obstacles for any attempts to rationalize the development process and introduce effective land-use control measures (UNCHS, 1982). Hence, land occupation by spontaneous settlements not only adds to this haphazard growth, but it is also partly a result of it. Spontaneous settlements can be classified according to locational and morphological characteristics in *inner city* and *peripheral settlements.*

Yet there is no generally accepted theory of spontaneous settlement location. However, there is an agreement that land availability and proximity of high-intensity mixed land use, usually jobs opportunities, have strong influence on it (Dwyer, 1975; Ulack, 1978). The most interesting characteristic of those settlements, however, is their evolution in time. At the same time that the housing stock and services are improving, or being 'upgraded', the city grows as a whole, changing the relative location of such settlements. Peripheral settlements are incorporated to the inner city by urban growth. Thus, spontaneous settlements that developed on what once was the city's periphery are often on land that has become very valuable, as the city expands. (UNCHS, 1996).

## 3. Complex systems of complex objects

The study of urban systems in the light of complexity theory is now well established. Cities are clearly complex systems and with advances and popularisation of computer tools, the possibilities to explore this viewpoint are constantly increasing.

The morphological structure of the city is built from the interplay of different dynamics, offering an extra level of complexity to these systems. As Holland (1995:1) suggests "a city's coherence is somehow imposed on a perpetual flux of *people and structures*". From Holland's words one can identify two different kinds of fluxes: the flux of *people* and the flux (or change) of *structures*. The ever-changing nature of cities, however, seems to require both interpretations for a better understanding. Not only it is necessary to understand the complex nature of each one of these fluxes, but it also seems to be necessary to understand the connections (or interactions) between these complex layers that together produce the emergent structure of urban space.

Complexity theory came to shift the approach in the use of computational models and quantitative measures, which have been traditionally used in quantitative urban morphology research. Cellular automata models replaced traditional causal models, shifting the paradigm of urban models towards a complexity approach. The idea of a structure emerging from a bottom-up process where local actions and interactions produce the global pattern has been widely developed ever since. CA models, however, explore only the spatial layer of the city and, although transition rules often were representations of human decision making, this representation is not explicit.

In order to explore the second layer of urban complexity, the flux of people, agent-based models were introduced in urban simulation. This came to meet the understanding that human decision making plays a major role. Although a number of models have been developed using agent-based techniques to simulate urban scenarios, including land use, pedestrian modelling, and so on, the application of agent-based simulation to urban spatial change is not a consensus in the research community.

Agent-based modelling can be seen as an approach in which benefits "exceed the considerable cost of the added dimensions of complexity introduced into the modelling effort" (Couclelis, 2001). Agent-based models can also be seen as models of 'mobile cells' (Batty & Torrens, 2001). This point of view suggests that these models would be suitable to simulations focusing on the human behaviour in a given spatial environment, as it is the case of pedestrian modelling, for example, rather than to urban spatial change.

In the present paper an agent-based model is seen as a cellular based model (raster) like cellular automaton, in which the transition rules of the CA are replaced by actual decision-making rules. Like in CA models, the choice of increasing the degree of complexity of the model or keeping it simple depends entirely on the researcher and the purposes of the model in hand. We argue that agent-based models, viewed as such, open up an avenue for analysis of dynamic processes that link spatial development with social issues. This kind of analysis is of fundamental importance when dealing with cases of strong social differentiation as the case of urban development in the Third World.

## 4. City of Slums

City-of-slums comes to combine the ideas behind two previous simulation exercises, Favela Project (Sobreira, 2002) and Peripherisation Project (Barros, 2002), to be detailed in the next sections. The aim of these three projects is to develop heuristic-descriptive models on the decentralised process underlying the spatial development of squatter settlements and growth of Latin American cities. Models are seen here as testable theories, in this case, built upon the assumption that the systems in hand are complex systems, and therefore, are systems in which local simple rules generate a complex global pattern. Thus, the models were elaborated in such a way that the behaviour rules were as simple as possible. They are totally based on the relationship agent-environment and do not explore either environment-environment or agent-agent relationships. All projects were developed in a STARLOGO platform that is a friendly user parallel programming tool developed by the Epistemology and Learning Group of the Massachusetts Institute of Technology (Resnick, 2000).

### 4.1. Favela project

The Favela project simulates the spatial development of spontaneous settlements in a local scale. The experiment is based on randomly walking agents over a cellular space, constrained by attractive and non-attractive boundaries. This is based on the features of most of the inner-city settlements, which grow in empty sites within urban areas resulted from the fragmented and discontinuous development of Third World cities as described previously. These settlements develop in a self-organized way, starting from "attractive boundaries" (streets of the existent city, which bound the site). Following this logic, the model's rules are based on the idea that the spatial development of spontaneous settlements is both constrained and stimulated by the boundaries. The built structure is developed prior to any network and rough foot tracks arise in between built structures and often consolidate, connecting houses to local services situated on the site's borders (Sobreira, 2002).

In the favela project the agent's rules resemble the behaviour of actual people looking for attractive urban sites to settle. The behaviour rules tell the agents to wonder around the site and, when reach an attractive boundary, to find an available place to settle. The model presents a "feedback" procedure in which the agents are "fed" with information about the environment and, based on that information, change their behaviour, which in turn drives the spatial development to a different path. In this case, according to a density threshold, the agents change their settling patterns (dwelling typology) and searching features. As it can be observed on the snapshots in the figure 1, the sequence of outputs from the Favela model (figure 1a) resembles the development process of a settlement in Acera, Ghana (figure 1b). This resemblance is not just related to the static features (spatial configuration) but to the dynamics (development process) of the settlement, as well: starting with isolated building units combined with open areas and as the density increases, the clustering and densification are inevitable. As the settlement gets dense and more agents come to the site searching for available space, agents take longer in the searching process, finally settling in more restricted spaces, occupying vacant spaces between existing dwellers and, thus, causing in this case the diversity of size of building clusters.

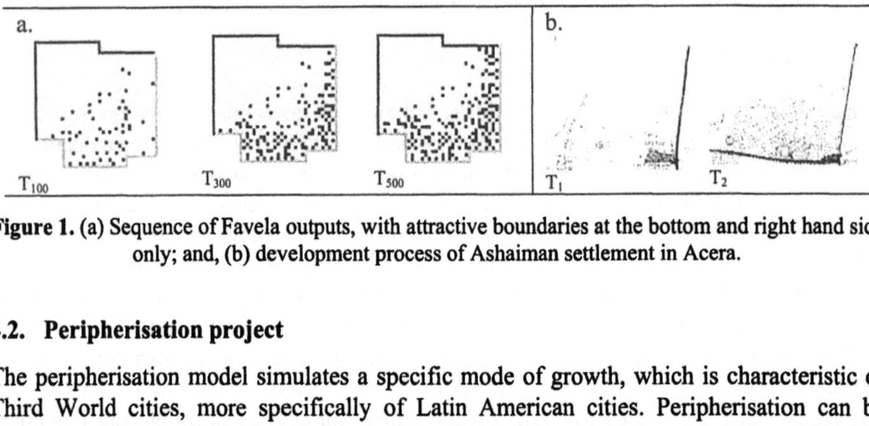


**Figure 1.** (a) Sequence of Favela outputs, with attractive boundaries at the bottom and right hand side only; and, (b) development process of Ashaiman settlement in Acera.

### 4.2. Peripherisation project

The peripherisation model simulates a specific mode of growth, which is characteristic of Third World cities, more specifically of Latin American cities. Peripherisation can be defined as a kind of growth process characterised by the expansion of borders of the city through the formation of peripheral settlements, which are, in most cases, low-income residential areas. These areas are incorporated to the city by a long-term process of expansion in which some of the low-income areas are recontextualised within the urban system and occupied by a higher economic group while new low-income settlements keep emerging on the periphery (Barros, 2002).

The model reproduces the process of expulsion and expansion by simulating the locational process of different economic groups in an attempt to reproduce the residential patterns of these cities. In the model, the population is divided in three distinct economic groups according to the pyramidal model of distribution of income in these countries. The model assumes that, despite the economic differences, all agents have the same locational preferences, that is, they all want to locate close to the areas that are served by infrastructure, with nearby commerce, job opportunities and so on. As in Third World cities these facilities are found mostly close to the high-income residential areas, the preference of location is to be close to a high-income group residential area. What differentiates the behaviour of the three income groups is the restrictions imposed by their economic power. Thus, the high-income group (represented in the model in red) is able to locate in any place of its preference. The medium-income group (in yellow) can locate everywhere except where the high-income group is already located and, in turn, the low-income group (in blue) can locate only in the vacant space.

In the model there are agents divided into three breeds (and colours) in a proportion based on the division of Latin American society by income. These societies have a triangle-like structure where the high-income group are minority on the top of the triangle, the middle-income group are the middle part of the triangle and the low-income group is on the bottom of the triangle. All the agents have the same objective that is to be as closer as possible to the red patches but they present different restrictions to the place they can locate. Since some agents can occupy another agent's patch, it means that the latter is "evicted" and must find another place to settle.

### 4.3. City of Slums: consolidation in a peripherisation context

City of Slums was built upon the peripherisation model by combining the original peripherisation logic to a *consolidation rule*. This rule refers to a process in which spontaneous settlements are gradually upgraded, and, as time passes, turn into consolidated *favelas* or, in other words, spontaneous settlements that are harder to evict. As a result of the introduction of the consolidation logic, the city of slums model generates a more fragmented landscape than the homogeneous concentric-like spatial distribution of classes in which consolidated spontaneous settlements are spread all over the city.

The consolidation process is built into the model through a *cons* variable. This *cons* variable has its value increased at each iteration of the model and, at a certain threshold, the blue patch turns into the consolidation state, represented by the brown color in the model. If a red or a yellow agent tries to settle on the blue patch in a stage previous to the consolidation threshold, the blue patch is replaced by the respective new occupant's patch color. Otherwise, brown patches are 'immune' to eviction.

Three basic parameters were tested in the development of the model: proportion of agents per breed, consolidation threshold, and steps; the latter concerning the number of steps each agent walks in its searching for a place to settle. The result of such variations can be observed on the figure 2 where step = 2 presents a clearly more fragmented pattern than step = 1. The parameter step = 2 also leads to a faster spatial development in the simulation. This is due to the fact that the larger the step, the more empty spaces are left between patches, making the search virtually easier, that is, the agents find an appropriate place to settle faster. As a consequence, the simulation grows rapidly, and a combination of empty spaces and more mixed pattern produce a fragmented spatial result. This process resembles what actually can be observed in the Third World cities, where, despite the general tendency for economic segregation, there are 'fragments' of low and middle income residential areas within high-income zones, and vice-versa, what is caused by the accelerated and discontinuous process of development.

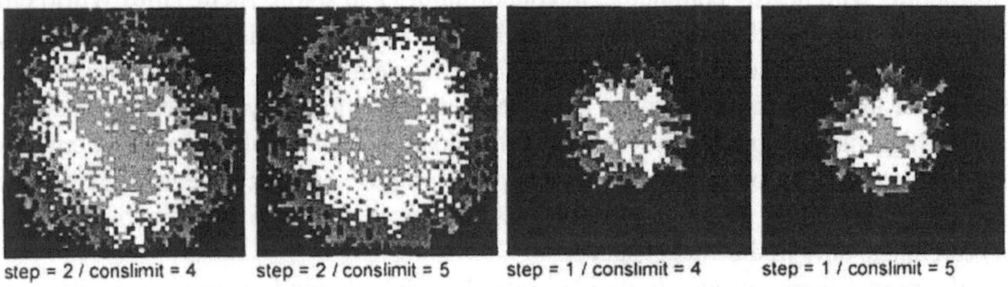


**Figure 2.** Variations of step and consolidation threshold parameters, time = 2000.

It is important to mention that in this kind of model the 'time' can only be measured through the number of iterations of the agents within the model. This condition opens up new possibilities of analysis of the model considering that, at the same number of iterations (t), the spatial development of the system will present variations depending on the parameters. This can be observed in the figure 2, where the variation of parameters at t = 2000 were tested.

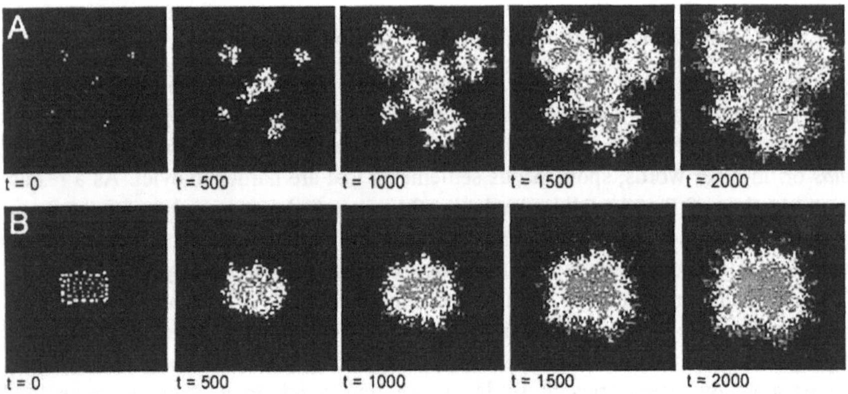


**Figure 3.** Experience with different initial conditions, polycentric (A) and colonial grid (B)

Different initial conditions were also tested, in an attempt to explore to what extend path dependence influences the model's behaviour. The figure 3 presents two sequences of the *city of slums* model with different initial conditions. As it can be observed, at the beginning of both simulations there are no brown patches (consolidated spontaneous settlements) in the virtual urban landscape. After some iterations, brown cells appear in all the three social-economic zones, resembling what occurs in actual cities. In the two last snapshot (t=1500 and t=2000) one can identify a very peculiar pattern, which seems quite similar to the typical distribution of spontaneous settlements in Third World, in special Latin American cities.

## 5. Fragmentation: statistical properties of spatial complexity

In recent years a great deal of effort in pure and applied science has been devoted to the study of nontrivial spatial and temporal scaling laws which are robust, i.e. independent of the details of particular systems (Bak, P. 1997; Batty, M. and Longley, P. 1994; Gomes, M. et all, 1999). Spontaneous settlements tend to follow these scaling laws in both scales, local and global (Sobreira & Gomes, 2001; Sobreira, 2002). This multiscaling order is analysed here by a fragmentation measure which is related to the diversity of sizes of 'fragments' (built units) in these systems. Diversity is understood here as a measure of complexity (Gomes et all, 1999) and an expression of universal dynamics.

In the settlement scale the fragmentation pattern refers to the diversity of size of islands (cluster of connected dwellings) while in the global scale it concerns the size distribution of patches of spontaneous settlements within the city.

**Figure 4.** Graphic representations of three squatter settlements situated, respectively, in Bangkok (Thailand), Nairobi (Kenya) and Recife (Brazil) and fragmentation graph .

**Figure 5.** Favela project samples and fragmentation graph.

The figure 4 presents graphic representations of three squatter settlements and a graph which describe the average scaling pattern of their islands. The discrete variable s gives a measure of the size or area of an island. The figure 5 presents three samples run through the Favela project. The snapshots are related to the time when the development reached approximately the same number of houses of the real settlements of figure 4 (around 250 dwellings), what allows us a more precise statistical comparison. The graphs in figure 5 describe the same variables and coefficients of the figure 4. So, when analysing the samples generated under such local-rule parameters, we find the same statistical pattern of fragmentation and diversity, which reinforce our conjecture, which connects boundaries, packing and diversity as the interrelated key aspects to the internal development of squatter settlements. This distribution f(s) in the graphs of figures 4 and 5 obeys a scaling relation given by $f(s) \sim s^{-\tau}$, with $\tau = 1.6 \pm 0.2$. The exponent $\tau$ is robust and refers to the degree of fragmentation of the settlement.

In figure 6 the fragmentation pattern is analysed through the size distribution of settlements in three Third World cities and compared to the size distribution of settlements in the *City of Slums* simulations in figure 7. In particular, the settlements in each city were grouped according to their area, and the relation between number of settlements ($N_{(a)}$) and respective size interval (a) were plotted in a log-log graph. As one can observe from the graph of figure 6, the scaling law which describe the settlements size distribution in the real cities falls in the same statistical fluctuation of the scaling law which describe the size distribution of the *city of slums* simulations. The graphs in figure 6 and figure 7 describe the same scaling relation $N(a) \sim a^{-\alpha}$., where $\alpha = 1{,}4 \pm 0{,}2$.

**Figure 6.** Fragmentation pattern of settlements in three Third World cities: Kuala Lumpur, in Malaysia; Manila, in Philippines; and Lima, in Peru.

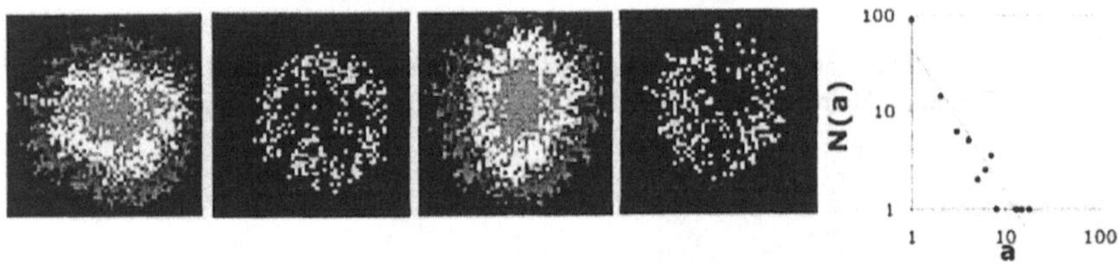


**Figure 7.** Distribution of settlements in the City of Slums model with fragmentation graph

Both global scale fragmentation patterns (real and simulated) are statistically the same, found for the local scale internal morphologies of the settlements. The negative exponents ($\alpha$ and $\tau$) indicate a non-linear scaling order, in which there is a great number of small units (islands, at the local scale, and settlements at the global scale), a small number of big units, and a consistent distribution between them. In this aspect, we argue that such similarity of patterns is clearly an empirical evidence of a multiscaling relation between local and global urban systems and it is a suggestive indication that the agent-based models generate simulations which truly describe the fragmented features of these self-organised systems.

## 6. Conclusions

Third World cities have been traditionally studied as chaotic and uncontrolled spatial structures. Furthermore, they have always called attention for their high rates of growth and intriguing spatial structures with odd elements. Spontaneous settlements, in this context, have been seen as isolated structures within this messy system, usually approached as anomaly rather than as an inherent global feature.

The present paper presents a change to this perspective, focusing on the role that spontaneous settlements – as complex subsystems – play in the global dynamics of development. We understand that spontaneous settlements are constantly shaping and being shaped by a self-organised process which drives the system to a fragmented pattern that can be verified across scales. Therefore, they are key elements to understand the spatial pattern of Third World cities.

From a socio-spatial point of view, the existence of spontaneous settlements can be understood as instability pockets which are necessary for the structural stability of the global system (Portugali, 2000). If we consider that spontaneous settlements actually absorb part of the existent social instability - translated here as housing deficit - in unstable pockets within the city, one could say that they are necessary for the structural stability of the global system. Viewed as such, spontaneous settlements are fragments that keep the system away from what otherwise would be a breakdown of the already fragile and unstable equilibrium of Third World cities socio-spatial structure. This idea comes to reinforce Turner's (1988) argument that spontaneous settlements can be seen as an alternative solution, rather than a problem for the housing deficit. In the Third World urban context, spontaneous settlements play a paramount role within a system in which the parts do explain the whole, but only when seen in the light of a self-organised process.

## References


Bak, P. (1997), "How Nature Works: The science of self-organized criticality", Oxford, University Press.

Barros, J. (2002), "Research Report" (unpublished), University College London.

Batty, M., Longley, P. (1994), "Fractal Cities: A Geometry of Form and Function", London, Academic Press.

Batty, M., Torrens, P. (2001), "Modeling Complexity: The Limits to Prediction", CASA Working Paper Series, **36**, available on-line at www.casa.ucl.ac.uk.

Couclelis, H. (2001), "Why I no longer work with Agents", in Special Workshop on Agent-Based Models of Land-Use, Land-Cover Change, CSISS, Irvine, available on-line at www.csiss.org/events/other/agent-based/papers/couclelis.pdf.

Dwyer, D. J. (1975), "People and housing in Third World cities: perspectives on the problem of spontaneous settlements", London, Longman.

Gomes, M., Garcia, J., Jyh, T., Rent, T., Sales, T. (1999), "Diversity and Complexity: Two sides of the same coin?", in The Evolution of Complexity, 8, 117-123, Dordrecht, Kluwer Academic.

Holland, J. (1995), "Hidden Order: how adaptation builds complexity", Massachusetts, Helix Books.

Portugali, J. (2000), "Self-organization and the City", London, Springer-Verlag.

Resnick, M. (2000), "Turtles, termites, and traffic jams: explorations in massively parallel microworlds", Cambridge, The MIT Press.

Sobreira, F., Gomes, M. (2001), "The Geometry of Slums: boundaries, packing and diversity", CASA Working Paper Series, **30**, available on-line at www.casa.ucl.ac.uk.

Sobreira, F. (2002), "The Logic of Diversity: complexity and dynamic in spontaneous settlements". Doctorate thesis. Federal University of Pernambuco.

Turner, J. (1988), "An introductory perspective", in Turner, B. (edit), A Third World case book, London, BCB/HFB.

Ulack, R. (1978), "The role of urban squatter settlements", Annals of the Association of American Geographers, **4**, 68, 535-550.

UNCHS – Habitat (1982), "Survey of slum and squatter settlements", Dublin, Tycooly International Publishing Limited.

# Part III:

# Applications

# Chapter 1

# Programmable Pattern-Formation and Scale-Independence

**Radhika Nagpal**
PostDoctoral Lecturer, MIT Artificial Intelligence Lab
radhi@ai.mit.edu

This paper presents a programming language for pattern-formation on a surface of locally-interacting, identically-programmed agents, by combining local organization primitives from developmental biology with combination rules from geometry. The approach is significantly different from current approaches to the design of self-organizing systems: the desired global shape is specified using an abstract geometry-based language, and the agent program is *directly compiled* from the global specification. Using this approach any 2D Euclidean construction can be formed from local-interactions of the agents. The resulting process is extremely reliable in the face of random agent distributions and varying agent numbers. In addition the process is *scale-independent*, which implies that the pattern scales as the number of agents increases, with no modification of the agent program. The pattern also scales asymmetrically to produce related patterns, such as D'Arcy Thompson's famous transformations.

## 1 Introduction

Cells cooperate to form complex structures, such as ourselves, with incredible reliability and precision in the face of constantly dying and replacing parts. Emerging technologies, such as MEMs, are making it possible to embed millions of tiny computing and sensing devices into materials and the environment. We would like to be able to build novel applications from these technologies that achieve the kind of complexity and reliability that cells achieve. These new environments pose significant challenges: a) How does one achieve a particular global goal from the purely local interactions of vast numbers of parts? b) What are the appropriate local and global paradigms for engineering such systems?

This paper presents a programming language approach to self-assembling complex structures, using techniques inspired by developmental biology. We present a programming language for instructing a surface of locally-interacting, identically-programmed agents to differentiate into a particular pattern. The language specifies the desired global pattern as a construction on a continuous sheet, using a set of axioms from paper-folding (origami) mathematics [3]. In contrast to approaches based on cellular automata or evolution, the program executed by an agent is *automatically compiled* from the global shape description. With this language, *any plane Euclidean construction* pattern can be be specified at an abstract level, compiled into agent programs, and then synthesized using purely local interactions between identically-programmed agents. The process relies on the composition of a small set of general and robust biologically-inspired primitives. The resulting process is not only reliable in the face of random agent distributions and random agent death but is also theoretically analyzable[4].

The process is also *scale-independent.* Scale-independence implies that the pattern, however complex, scales as the number of agents increases, with no modification of the agents program. The pattern also scales asymmetrically, allowing a single program to generate many related patterns, such as D'Arcy Thompson's famous coordinate transformations which he used to explain shape differences in related species[5]. Scale-independence is common in biology. The pattern-formation process provides insights into how complex morphology emerging from local behavior can exhibit global properties such as scale-independence.

This research is motivated by emerging technologies, such as MEMs[1] devices, that are making it possible to bulk-manufacture millions of tiny computing elements integrated with sensors and actuators and embed these into materials to build novel applications: smart materials, self-reconfiguring robots, self-assembling nanostructures. Approaches within the applications community have been dominated by a centralized, hierarchical mind-set. Centralized control hierarchies are not easily made scalable and fault-tolerant, and centralized/heuristic searches quickly become intractable for large numbers of agents. The tendency to depend on centralized information, such as global clocks or external beacons for triangulating position, puts severe restrictions on the applications while exposing easily attackable points of failure. Currently, however, few alternatives exist. Decentralized approaches based on cellular automata models of natural phenomena and artificial life research have been difficult to extend to engineering systems; local rules are constructed empirically without providing a framework for constructing local rules to obtain any desired goal. Evolutionary and genetic approaches are more general but the local rules are evolved without any understanding of how or why they work. This makes the correctness and robustness

[1]Micro-electronic Mechanical Devices. Integrates mechanical sensors/actuators with silicon based integrated circuits.

of the evolved system difficult to analyze.

Biological systems achieve incredible robustness in the face of constantly dying and replacing parts. The precision and reliability of embryogenesis in the face of unreliable cells, variations in cell numbers, and changes in the environment, is enough to make any engineer green with envy. We propose to use morphogenesis and developmental biology as a source of mechanisms for organizing complex behavior. Our approach is to formalize these general principles as *programming languages* — with explicit primitives, means of combination, and means of abstraction — thus providing a framework for the design and analysis of self-organizing systems. This work is part of larger vision called Amorphous Computing to explore new programming models for collective behavior[1, 2].

# 2 A Programmable Surface

Our model for a programmable surface consists of randomly distributed agents on a 2D surface. All agents have the *identical* program, but execute it autonomously based on local communication and internal state. Communication is strictly local: an agent can communicate only with a small local neighborhood of agents within a distance $r$. The surface starts out with a few simple initial conditions, but there are no external beacons for triangulating position or global clocks. Individual agents have limited resources and no unique identifiers, instead they have random number generators to break symmetry. The motivation comes from the applications — we would like to cheaply bulk manufacture billions of smart sensors and embed them in the environment. Assumptions such as globally unique identifiers, global clocks or coordinates, regular grids and perfectly reliable elements are unrealistic in this setting. Furthermore developmental biology suggests that it should be possible to construct complex structures without such assumptions.

# 3 Global Specification Language

The key to this process is the language for specifying the desired global pattern. Paper-folding (origami) provides a natural, although somewhat unusual, way for describing how to create patterns starting from a blank continuous sheet. Huzita presented a set of axioms for origami, a subset of which are equivalent to the plane Euclidean axioms (ruler and compass axioms)[3]. Each axiom generates new lines starting from an existing set of lines and points.

1. `crease-lbp`: Fold a line between two points $p1$ and $p2$.

2. `crease-p2p`: Fold $p1$ onto $p2$ to create a line (perpendicular bisector).

3. `crease-l2l`: Fold line $L1$ onto $L2$ to create a line (bisector of the angle between $L1$ and $L2$).

4. `crease-l2self`: Fold $L1$ onto itself through $p1$.

```
;; OSL PROGRAM for a CMOS INVERTER
(define v1 (crease-l2l e23 e41))
(define v2 (crease-l2l v1 e23))
(define v3 (crease-l2l v1 e41))
(define IN  (create-region e41 v3))
(define OUT (create-region e23 v1))
(define MID (create-region v1 (or v2 v3))
;; Similarly create horizontal regions...

;; Lay down Material (differentiate)
(within-region IN (color h1 "poly-red"))
(within-region MID (color (or h2 h3) "poly-red"))
(within-region OUT (color h1 "poly-red"))
(within-region CNTR (color v3 "poly-red"))
(within-region UP (color v1 "n-diff-yellow"))
(within-region DOWN (color v1 "p-diff-green"))
(define contacts (intersect v1 (or e12 h1 e34))
(color contacts "contacts-black")
```

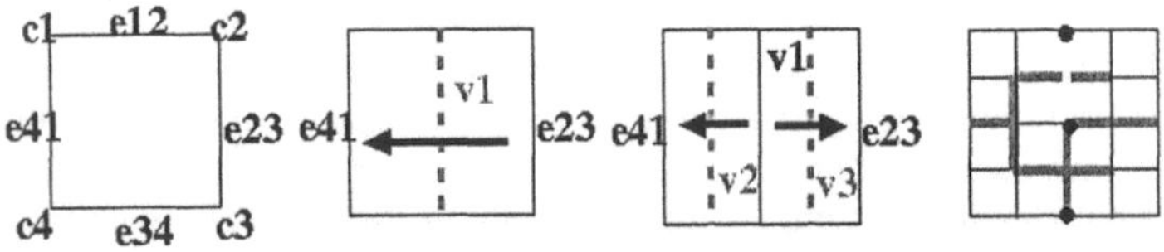


**Figure 1**: OSL code and diagram for a CMOS Inverter

In addition, there are two more operations - intersecting two lines to find a new point and defining a region. Given a line that divides the sheet into two parts, one side can named as a region using a landmark point on that side.

We have developed a programming language based on these axioms and paper-folding practice, called the Origami Shape Language (OSL). The language is described in detail in [4]; here we have illustrated it with a simple example, a caricature pattern of a CMOS inverter. Figure X shows a diagram of how the pattern is created and the corresponding OSL code. The sheet always starts out blank but with four sides and corners. Using the axioms the sheet is recursively subdivided and a grid pattern of lines is created. The lines are used to define different regions and the regions are used to set the colors (state) of the agents. In an actual inverter, the different colors refer to different materials.

Rather than specify the desired pattern directly, this language specifies a "generative program" or a process for creating the pattern. However this specification is abstract — the process is on a continuous sheet with no notion of agents or self-assembly. By specifying the pattern this way, we can take advantage of known and new results in geometry to analyze the power of our system.

# 4 Global to Local Compilation

The agent program is directly compiled from the global shape specification. This is done by using a small set of primitives for organization at the local level and combining these primitives in well-understood ways to produce robust and predictable behavior.

## 4.1 Biologically-inspired Primitives

**Gradients:** Gradients are analogous to chemical gradients secreted by biological cells; the concentration provides an estimate of distance from the source of the chemical. Gradients are believed to play an important role in providing position information in morphogenesis[6]. An agent creates a gradient by sending a message to its local neighborhood with the gradient name and a value of zero. The neighboring agents forward the message to their neighbors with the value incremented by one and so on, until the gradient has propagated over the entire sheet. Each agent stores the minimum value it has heard for a particular gradient name, thus the gradient value increases away from the source. Because agents communicate with only neighboring agents within a small radius, the gradient provides an estimate of distance from the source. The source of a gradient could be a group of agents, in which case the gradient value reflects the shortest distance to any of the sources. Thus, the shape and positions of the sources affects the spatial pattern of gradient values. For example if a single agent emits a gradient then the value increases as one moves radially away from the agent but if a line of agents emits a gradient then the gradient value increases as one moves perpendicularly away from the line.

**Neighborhood Query:** This primitive allows an agent to query its local neighborhood and collect information about their state. For example an agent may collect neighboring values of a gradient for comparison. This primitive is from cellular automata. An agent can also broadcast a message to its local neighborhood.

## 4.2 Composition into Local Rules

Each of the global OSL operations can be implemented as a simple agent program (also called a local rule) using the above primitives. OSL points and lines are represented by groups of agents; all agents the group are equal. Each agent has a boolean variable in its internal state for each distinct point/line. Initially the sheet starts out with four distinct lines and points (edges and corners). The initial conditions are very simple; agents do not know where they are within an edge and the remainder of the sheet is homogeneous, like a blank sheet of paper.

The axioms use gradients to determine which agents belong to the crease. The axioms make use of the fact that gradients provide a distance estimate as well as reflect the shape of the source. For example, axiom 1 creates a line from one point to another by having one point generate a gradient and the other point "grow" a line towards increasing values of the gradient, similar to Coore[2].

Axiom 2 creates a line such that any point on the crease is equidistant from points $p1$ and $p2$. Therefore if $p1$ and $p2$ generate two different gradients, each agent can compare if the gradient levels are approximately equal to determine if it is in the crease line. The other two axioms are similar, but take advantage of the fact that lines produces gradient values that increase away from the line.

Regions allow the user to *restrict the context* in which a local rule applies. A region is created by using bounded gradients, which implies that certain types of agents will not forward the gradient message; the intuition being that certain agents can act as barriers to particular gradients. A gradient is created from point that cannot pass a particular line, marks the region on one side of the line.

The final compilation of an OSL program involves creating local boolean state variables for each distinct point and line and then translating each OSL operation into a call to the corresponding agent procedure with the appropriate arguments. The compiler assigns different gradient names for each call. For example, `(define d1 (crease-p2p c1 c3))` becomes `(define d1 #f) (set! d1 (axiom2-rule c1 c3 gradient1 gradient2))`. Thus, the agent program mirrors the original OSL program. However note that at the OSL level there is no notion of gradients, or even agents. The compilation process from global to local is easy to understand. However the agent programs generated are no different from any other emergent systems — the eventual shape "emerges" as a result of local interactions between the agents and the initial conditions.

## 5 Examples

The examples presented were generated by specifying the pattern using OSL, compiling the OSL program to generate the agent program and then executing the agent program on the simulated programmable sheet. The initial conditions are always the same, and the number of agents is between 1000-8000 with local neighborhoods of 15 agents on average. Figure 2 shows the result of simulating the inverter program. Complex patterns can be created in a modular fashion: for example a chain if inverters can be created by segmenting the sheet into regions and invoking the inverter program within each region. Many examples of patterns are presented in [4].

## 6 Scale-Independence and Related Shapes

The formation of the same structure at many different scales is common throughout biology. Many species develop normally over large variations in egg sizes and large morphological differences can occur between species with little genetic difference. Genetic analysis is unlikely to reveal much information in such cases. However artificial systems can give us insight into how complex patterns may achieve scale-independence and what limitations exist on the scaling.

The Origami Shape Language is a scale-independent description of shape, the program specifies the shape but not the size of the sheet to use. At the cell level

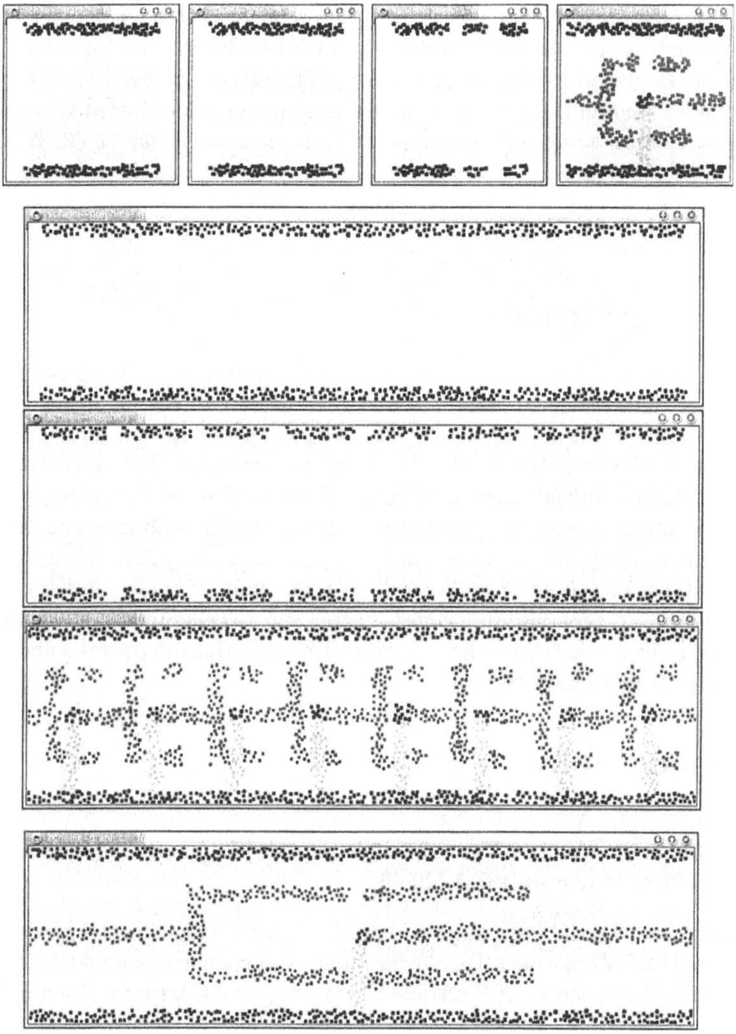

**Figure 2**: Simulation images of pattern-formation: single inverter, chain of inverters, the single inverter run on a long sheet.

scale-independence is achieved by using primitives that depend on comparisons of gradients values, never absolute values. Axiom 2 is similar to Wolpert's model of balancing gradients, and axiom 1 grows a line until the end is reached - so both primitives scale. At the global level scale-independence is achieved by recursively creating nesting structures, starting from the original boundary. Thus one can create highly complex structures without reference to size.

This leads to an interesting observation, which is that by changing the *shape* of the boundary we can also change the shape generated. The inverter program executed on a long sheet produces a stretched inverter, a kite shaped surface produces a distorted inverter. D'Arcy Thompson observed that the forms of many related animals (crabs, fish, skulls) could be transformed into one another by stretching along different axes, and that the relationships extended even to internal structures. He envisioned differential growth as the mechanism responsible. The origami shape language suggests another mechanism — changing the shape of initial boundary.

# 7 Conclusions

This work represents a different approach to engineering self-organizing systems. Rather than trying to map a desired goal directly to the behavior of individual agents, the problem is broken up into two pieces: a) how to achieve the goal globally b) how to map the construction steps to local rules. The compilation process confers many advantages: we can use theoretical results from paper-folding to reason about the kinds of shapes can and cannot be self-assembled and we can use the decomposition into primitives and means of combination to analyze the robustness of the system. We believe that many of these mechanisms will be applicable to programming smart matter and reconfigurable substrate applications, as well as provide a basis for directing experiments towards understanding biological morphogenesis.

# Bibliography

[1] ABELSON, ALLEN, COORE, HANSON, HOMSY, KNIGHT, NAGPAL, RAUCH, SUSSMAN, WEISS "Amorphous Coomputing" *Communications of the ACM* **volume 43, number 5** (2000).

[2] COORE, Daniel, "Botanical Computing: A Developmental Approach to Generating Interconnect Topologies on an Amorphous Computer", *PhD Thesis*, MIT, Department of Electrical Engineering and Computer Science, Feb 1999.

[3] HUZITA, Humiaki, "The Algebra of Paper-folding", *1st International Meeting of Origami Science and Technology* (1989).

[4] NAGPAL, Radhika, "Programmable Self-Assembly: Constructing Global Shape using Biologically-inspired Local Interactions and Origami Mathematics" *PhD Thesis*, MIT, Department of Electrical Engineering and Computer Science, June 2001.

[5] THOMPSON, D'Arcy, *On Growth and Form*, Cambridge University Press, U.K., 1961

[6] WOLPERT, Lewis, *Principles of Development*, Oxford University Press, 1998.

# Chapter 2

# Amorphous Predictive Nets

**Regina Estkowski, Michael Howard, David Payton,**
HRL Laboratories, LLC
Malibu, CA 90265
{estkowski, howard, payton}@hrl.com

## 1. Introduction

This paper describes our approaches for coordinating the actions of extremely large numbers of distributed, loosely connected, embedded computing elements. In such networks, centralized control and information processing is impractical. If control and processing can be decentralized, the communications bottleneck is removed and the system becomes more robust. Since conventional computing paradigms provide limited insight into such decentralized control, we look to biology for inspiration.

Due to progress in the miniaturization of sensors and computing elements and in the development of necessary power sources, large arrays of networked wireless sensor elements may soon be realizable. The challenge is to develop software that enables such amorphous arrays to self-organize in ways that enable the sensing capabilities of the whole to exceed that of any individual sensor.

Our goal is to devise local rules of interaction that cause useful computational structures to emerge out of an array of distributed sensor nodes. These distributed logical structures appear in the form of local differences in sensor node state. These local state differences serve to form distributed circuits among nodes, allowing groups of nodes to perform cooperative sensing and computing functions that are not possible at any single node. Further, since the local differences emerge and are not pre-programmed, there is never a need to assign specific functions to specific nodes.

In this paper we describe two methods, each using only local interactions between nodes to detect the presence and heading of some local transient property of the environment (e.g., presence of a warm body). These methods provide a purely distributed means of computing the direction and likely destination of a sensed movement, with no need for centralized data analysis or explicit data fusion. Such a prediction could activate sensors ahead of the movement of the sensed object, turning on more expensive sensing functions that are normally dormant to save power. An active minefield could use the techniques to attract mines to the most likely avenue of approach. Streetlights could be turned on ahead of cars on a road less traveled.

## 2. Related Research

Our research focus is on future applications of sensor networks wherein the sensor nodes themselves will be extremely small, cheap, and simple, and will be deployed somewhat haphazardly or randomly. Examples of such networks can be found at UC Berkley, where the goal is to create sensor nodes the size of dust particles that can be released in large quantities from the air. Such sensor networks have very limited computational capabilities, and are unlikely to have sophisticated on-board position location capabilities such as GPS.

A number of other methods for monitoring object presence and movement have been developed, but many of these methods are limited in scope and related to a narrow application, or require sophisticated sensors and centralized processing. The most relevant is the work being done at MIT on amorphous computing and the work being done at USC on directed diffusion in sensor networks, although neither of the two encompasses our system.

MIT is making progress on pattern formation in amorphous networks in the context of Paintable Computing and "shape formation" via the use of origami mathematics [Nagpal 2001]. These patterns are not used in the context of object tracking. MIT uses some of the same basic primitives we use in pattern formation, but the overall methods are different.

The USC work [Intanagonwiwat 2000] uses directed diffusion for object tracking, but it assumes that each sensor knows its location, and can inform a user of an intruder's position via directed diffusion. We make no assumptions about node location.

Our use of a virtual pattern sets up a virtual heterogeneous network in which different sensors have different functions depending upon their position in the pattern. We are aware of no previous attempt to use the sensor distribution and network structure itself to track objects and predict movement direction.

## 3. Pheromone Messaging

We use a diffusion-based messaging paradigm called *virtual pheromones* [Payton, et al 2000, 2002]. Virtual pheromones provide a simple messaging scheme that establishes a gradient among a large number of distributed, locally communicating nodes. In earlier work, we have used a custom-made IR transceiver unit, as shown in Figure 1, to transmit and receive virtual pheromones in 8 distinct directions. However, in most of the sensor node applications described in this paper, we envision using RF communications between nodes, and therefore constrain ourselves to omni-directional transmission and reception of virtual pheromone signals. A virtual pheromone is encoded as a single modulated message packet consisting of a type field, a hop-count field, and a data field. The type is an integer that identifies a unique pheromone class for the message. The data field may be used to optionally transmit a few bytes of data. The hop-count is used to establish how far a pheromone may travel from its originating source and to establish simple pheromone gradients. The originating node sets the hop-count to an integral number of times the message is to be relayed, and sends the message to its local neighbors. Upon receipt, the hop-count field is decremented and the message retransmitted without any need for acknowledgement. If a node receives the same type of pheromone from multiple sources, only the message with the highest hop-count value is selected for re-transmission. This results in a uniform gradient leading away from the source. These rules for message propagation provide a distributed version of the wavefront propagation method used in Dijkstra's shortest-path algorithm [Dijkstra 1959].

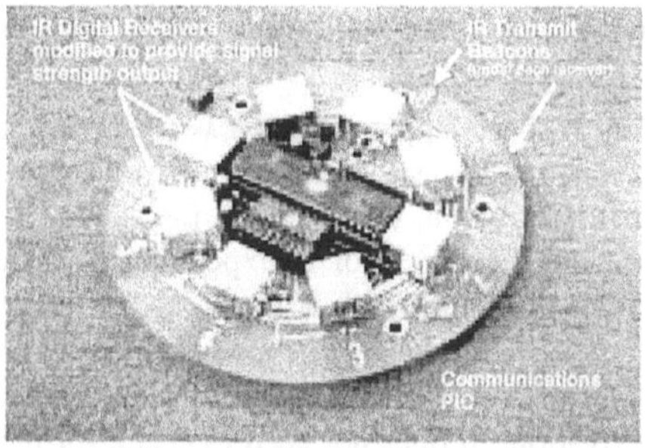


**Figure 1: Transceiver for virtual pheromones**

## 4. Motion Prediction Methods

In the following, we describe two different motion prediction methods. In the first method, a temporal differencing technique is used to obtain very coarse motion detection and prediction for objects moving across a sensor array. In the second method, patterns are formed within the sensor array to differentiate nodes. Interactions between such differentiated nodes produce more precise motion detection and prediction.

Both methods are applicable to a distributed network of locally connected sensor nodes, commonly called a sensor network. Each sensor measures some local transient property of the environment (e.g., presence of a warm body or an object), at a limited range, so the optimal distribution of sensor nodes would be at inter-node distances just less than double their maximum sensing range. In each of the methods below, sensors do not directly measure velocity - only the presence or absence of the object. The network connections are considered to be wireless, and nodes have no knowledge of their neighbors or even of their own location. All communication is by means of unreliable short range broadcast.

If there is a way to determine distance between nodes, e.g. signal strength, it is possible to select a more uniformly distributed subset of nodes in order to obtain a better gradient. One or more nodes emit a special distribution pheromone message, containing a minimum and maximum range parameter. Receiving nodes that are within the specified range will join the active subset of nodes. Those outside the specified limits become inactive. Only active nodes relay the distribution pheromone. Sometimes this type of pre-conditioning results in better pattern formation in the subsequent steps.

### 4.1. A Temporal Differencing Technique

The first technique creates a very simple pattern in the sensor array when each node compares its activation level from one moment to the next. In a two-state implementation, we will label the states ON and OFF. An ON node is either directly sensing the object or is experiencing an increasing activation level indicating it is potentially in the path of the object's motion. All other nodes are OFF.

Any node that senses the object turns ON and originates a pheromone with a high activation. It does not need to check for incoming messages. The activation message is diffused throughout the sensor node population creating a complete gradient of activation, using the algorithm described in Section 3. We assume that message diffusion is much faster than the movement of the objects the nodes sense. As an object moves, different nodes sense it and take over the job of initiating the pheromone gradient.

Nodes that do not directly sense the object base their state on the difference between the activation level of the last message they received and the level of the current message. As the object moves, different nodes sense it and become pheromone initiators, and others that no longer sense it stop sending their pheromones. This causes the gradient to slide across the ion of each node from one moment to the next is positive in front of the movement, zero to the sides, and negative to the rear. In Figure 2, nodes that sense an object at geographic position A at time 1 create a gradient field. At time 2, other nodes at point B sense the object, resulting in the second gradient. When each node receives the gradient message at time 2, it subtracts the activation level from time 1 from the new time 2 activation, and gets a value that is either positive, negative or zero.

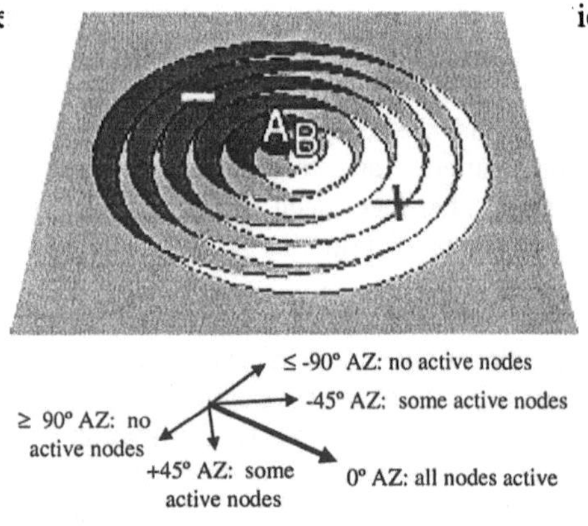


**Figure 2. Temporal difference pattern in gradient after object moves from A to B, and % of active nodes as a function of azimuth**

Therefore, the simple activation rule for nodes that do not directly sense the object is to turn ON if the temporal difference is positive, and turn OFF otherwise. The resulting activation pattern is useful for waking up nodes in a sensor net to track an object that may not continue to travel on a linear path. Imagine that the background of Figure 2 is a large number of randomly placed nodes. If the azimuth vectors are superimposed on the temporal difference pattern, it can be seen that all nodes in the center of the path of the object (at 0º AZ) will be turned on, and the percentage of activated nodes will drop as azimuth increases. In other words, the percent of activated nodes in a particular direction roughly corresponds to the likelihood that the object will move in that direction.

It is difficult to extend the object's motion vector to nodes ahead of the object in a more focused way without requiring directional messaging. One strategy using directional messaging would be for each node to keep track of the gradient vector as the object moves toward it. If any node subsequently senses the object directly, it can conclude the object came from the direction of steepest ascent, which is still in memory. The sensing node sends activation pheromone as before, but the message that goes in the direction of the object's movement is annotated to tell nodes that receive it to turn ON. Nodes that receive the annotated message forward it in the same direction, while again sending unannotated messages to all other neighbors so they can track changes in the gradient over time.

### 4.2. Focused Predictions Using Pattern Formation

In the second technique, sensor nodes are differentiated into parallel spatial bands to provide motion detection along different axes. In this method, a virtual pattern of bands emerges through specially designed local interactions between nearby nodes. This results in a pattern state within individual nodes that either sensitizes or desensitizes them to particular activation/inhibition signals from neighboring nodes. Activation/inhibition rules are designed such that messages signaling the presence of an object are inhibited along bands of the same type, but are propagated into bands of a different type as shown in Figure 3. This, in effect, leads to a form of moving edge detection for objects moving across the sensor array from one spatial band to another.

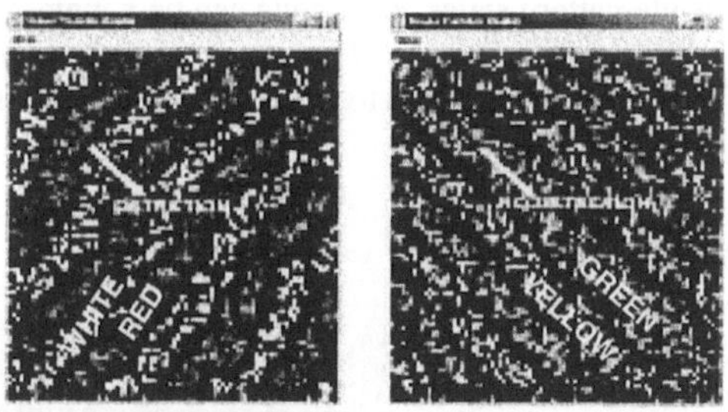


**Figure 3. Orthogonal sets of parallel bands are used to detect movement along specific directions.**

*Band Creation*

Bands are generated by first choosing two "anchor nodes" lying on opposite ends of a diameter of the sensor net. These anchor nodes determine the orientation of the initial set of bands. Starting with an identical pattern state in all nodes, one of the two anchor nodes initiates a pheromone signal that creates a gradient throughout the entire network. When this signal reaches the second anchor node, the recipient issues a second pheromone signal. This second signal propagates using rules comparable to directed diffusion [Intanagonwiwat 2000], wherein signals only advance along the axis of steepest ascent of the first gradient. We call this a "white" pheromone signal because all nodes that receive it will switch to a white pattern state, thereby forming a white band as shown in the leftmost frame in Figure 4.

After nodes have switched to a white pattern state, they transmit a limited-range "red" pheromone signal. This causes all nodes that receive it that are not already in the white pattern state to switch to the red pattern state, as shown in the middle frame of Figure 4. Likewise, nodes in the red pattern state transmit a white pheromone signal that switches all non-red nodes within range to the white pattern state. The net result is a sequence of parallel bands as shown in the rightmost frame of Figure 4.

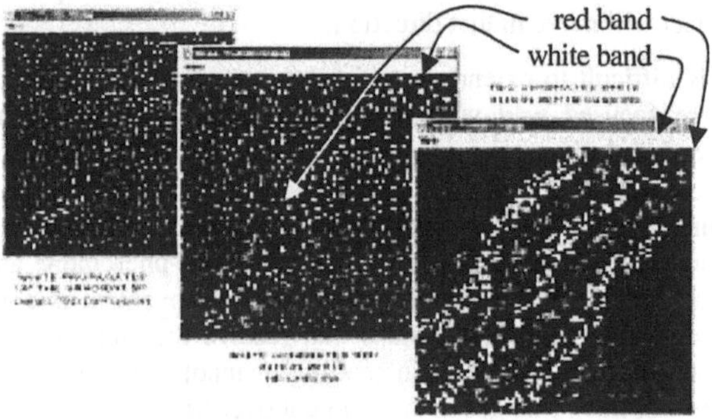


**Figure 4. Formation of initial set of parallel bands.**

A set of bands orthogonal to the first is formed using gradients initiated from both of the anchor nodes. Midway between the two anchor nodes, these gradients meet with equal hop counts, and the nodes in that region switch to a green pattern state. Just as before, nodes in the green pattern state send a pheromone that triggers neighboring nodes to switch to the yellow pattern state. Likewise, nodes in the yellow pattern state send messages to switch neighbors to the green pattern state. This results in another set of parallel yellow / green stripes that is orthogonal to the original red / white stripes. Because the green and yellow pattern states are independent of the red and white pattern states, each node can be a member of both the green/yellow and the red/white stripe patterns. The same process could be used to create a number of different band orientations to achieve any desired resolution of motion sensitivity.

*Detection and Prediction*

Once stripes have formed, the resulting pattern states can aid in detection of a moving object. When a sensor detects an object, it sends out a short-range priming pheromone labeled with its pattern state as shown in Figure 5. Nodes that receive

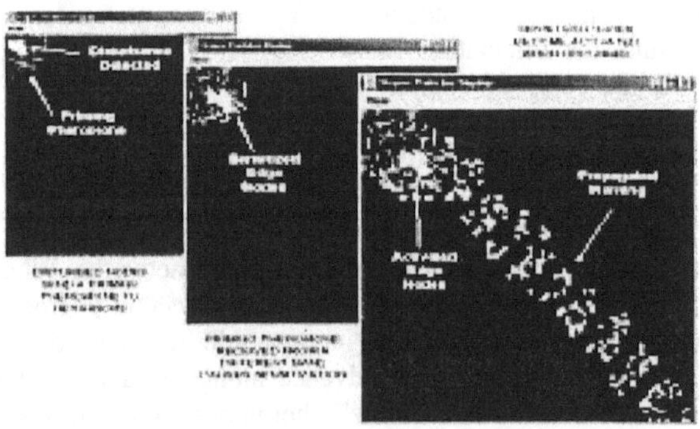

**Figure 5. A warning signal (yellow) propagates through the network after persistent motion is detected.**

this message and lie in a different stripe band become sensitized for a short time. If a sensitized node detects an object, it sends out a warning to be propagated. This warning message is accepted and propagated only by nodes that lie on the same band as both the priming sensor and the warning sensor. The initial warning message is weak and does not travel far. If the detected object continues to move in the same direction, the warning message is reinforced at receiving nodes and propagates further. If motion is no longer detected along the given direction, the warning pheromone at these nodes decays and the nodes revert to their original state. This provides a simple form of motion prediction whereby nodes far from the moving object register a warning if the object continues to move toward them.

## 5. Data Extraction

These methods provide a purely distributed means of computing the direction and likely destination of a sensed movement, with no need for centralized data analysis or explicit sensor data fusion. In the preceding discussion, we have used the results of the distributed computation only to change node state. However, it may be desirable to view the states of nodes, e.g., to follow the activation path to the sensed object. We would like the sensor array to act as a distributed display embedded in the environment. In effect, each node becomes a pixel, or an annotation, on the immediate environment. The node's position within the environment provides context to interpret the meaning of the transmitted information.

The easiest solution would be to put colored blinking lights on each node; but this limits the type of data that can be represented. Our approach, called the World-Embedded Display [Payton et al, 2002], is to visualize the distributed data in the collection of nodes directly, *in situ*. The user wears an augmented reality (AR) head-mounted display, shown in Figure 7. Each node transmits a character of data via infrared, and a camera mounted on the head mounted display with an IR filter reads the data. The system converts the character into a symbol that is drawn on the video shown in the display. Figure 7 illustrates the effect: two of our mobile robots as they

**Figure 7. AR head-mounted display has pencil-cam to image IR data transmitted from individual sensor nodes. Gradient superimposed on nodes.**

are viewed in the AR display, with the pheromone gradient arrows floating above them.

## 6. Conclusion

We have described two purely distributed methods for computing the direction and likely destination of an object of interest, with no need for centralized data analysis or explicit sensor data fusion. These algorithms apply to spatially distributed collections of simple sensing nodes with only local communication and rudimentary processing capabilities.

The temporal differencing scheme is very simple and has the nice property of waking up the most nodes in directions that are likeliest to be in the object's path. This results in a sort of heuristic search pattern for the future movements of the object, which is desirable in many applications. The prediction can be somewhat more focused using a more constrained type of messaging.

The second technique produces a tightly focused motion prediction without requiring directed messaging. It detects object movement that crosses an oriented pattern, and uses the pattern to activate nodes in the path of motion. The pattern must be oriented correctly for motion to be detected; it is possible for several different orientations to coexist in the network.

This paper is concerned with motion extrapolation, not the identification problem. However, without the ability to uniquely identify objects, these approaches can be fooled much like the eye is fooled into perceiving motion with a movie marquee. If sensors can accurately identify objects, pheromones could be labeled with a feature ID, making node state ID-specific. These issues are left for future work.

## References

Dijkstra, E.W. 1959, A Note on Two Problems in Connection with Graph Theory, *Numerische Mathematik*, 1:269-271.

Intanagonwiwat, C., Govindan, R., Estrin, D., 2000, Directed Diffusion: A Scalable and Robust Communication Paradigm for Sensor Networks, *MobiCOM 2000*

Nagpal, N., 2001, Programmable Self-Assembly: Constructing Global Shape using Biologically-inspired Local Interactions and Origami Mathematics, PhD Thesis, MIT Dept. of EE and CS

Payton, Daily, Estkowski, Howard, Lee, 2001, Pheromone Robots, *Autonomous Robots*, Kluwer Academic Publishers, Boston, MA, Vol. 11, No. 3.

Payton, Estkowski, Howard, 2002, Progress in Pheromone Robotics, *Proceedings 7th International Conference on Intelligent Autonomous Systems*, Marina del Rey, CA, 25-27.

# Chapter 3

# Multidimensional Network Monitoring for Intrusion Detection

**Vladimir Gudkov and Joseph E. Johnson**
Department of Physics and Astronomy
University of South Carolina
Columbia, SC 29208
gudkov@sc.edu; jjohnson@sc.edu

An approach for real-time network monitoring in terms of numerical time-dependant functions of protocol parameters is suggested. Applying complex systems theory for information flow analysis of networks, the information traffic is described as a trajectory in multi-dimensional parameter-time space with about 10-12 dimensions. The network traffic description is synthesized by applying methods of theoretical physics and complex systems theory, to provide a robust approach for network monitoring that detects known intrusions, and supports developing real systems for detection of unknown intrusions. The methods of data analysis and pattern recognition presented are the basis of a technology study for an automatic intrusion detection system that detects the attack in the reconnaissance stage.

## 1 Introduction

Understanding the behavior of an information network and describing its main features are very important for information exchange protection on computerized information systems. Existing approaches for the study of network attack tolerance usually include the study of the dependance of network stability on network complexity and topology (see, for example [1, 2] and references therein); signature-based analysis technique; and statistical analysis and modelling of network traffic (see, for example [3, 4, 5, 6]). Recently, methods to study spatial traffic flows[7] and correlation functions of irregular sequences of numbers occurring in the operation of computer networks [8] have been proposed.

Herein we discuss properties related to information exchange on the network rather than network structure and topology. Using general properties of information flow on a network we suggest a new approach for network monitoring and intrusion detection, an approach based on complete network monitoring. For detailed analysis of information exchange on a network we apply methods used in physics to analyze complex systems. These methods are rather powerful for general analysis and provide a guideline by which to apply the result for practical purposes such as real time network monitoring, and possibly, solutions for real-time intrusion detection[9].

## 2 Description of Information Flow

A careful analysis of information exchange on networks leads to the appropriate method to describe information flow in terms of numerical functions. It gives us a mathematical description of the information exchange processes, the basis for network simulations and analysis.

To describe the information flow on a network, we work on the level of packet exchange between computers. The structure of the packets and their sizes vary and depend on the process. In general, each packet consists of a header and attached (encapsulated) data. Since the data part does not affect packet propagation through the network, we consider only information included in headers. We recall that the header consists of encapsulated protocols related to different layers of communications, from a link layer to an application layer. The information contained in the headers controls all network traffic. To extract this information one uses tcpdump utilities developed with the standard of LBNL's Network Research Group [10]. This information is used to analyze network traffic to find a signature of abnormal network behavior and to detect possible intrusions.

The important difference of the proposed approach from traditionally used methods is the presentation of information contained in headers in terms of well-defined numerical functions. To do that we have developed software to read binary tcpdump files and to represent all protocol parameters as corresponding time-dependent functions. This gives us the opportunity to analyze complete information (or a chosen fraction of complete information that combines some parameters) for a given time and time window. The ability to vary the time window for the analysis is important since it makes possible extracting different scales in the time dependance of the system. Since different time scales have different sensitivities for particular modes of system behavior, the time scales could be sensitive to different methods of intrusion.

As was done in reference paper[11], we divide the protocol parameters for host-to-host communication into two separate groups with respect to the preserving or changing their values during packet propagation through the network (internet). We refer to these two groups of parameters as "dynamic" and "static". The dynamic parameters may be changed during packet propagation. For example, the "physical" address of a computer, which is the MAC parameter of the Ethernet protocol, is a dynamic parameter because it can be changed if the packet has been re-directed by a router. On the other hand, the source IP address is an example of a static parameter because its value does not change during packet propagation. To describe the information flow, we use only static parameters since they may carry intrinsic properties of the information flow and neglect the network (internet) structure. (It should be noted that the dynamic parameters may be important for study of network structure properties. Dynamic parameters will be considered separately.)

Using packets as a fundamental object for information exchange on a network and being able to describe packets in terms of functions of time for static parameters to analyze network traffic, we can apply methods developed in physics and applied mathematics to study dynamic complex systems. We present some results obtained in references [11, 12] to demonstrate the power of these methods and to recall important results for network monitoring applications.

It was shown [11] that to describe information flow on a network one can use a small number (10 - 12) of parameters. In other words, the dimension of the information flow space is less than or equal to 12 and the properties of information flow are practically independent of network structure, size and topology. To estimate the dimension of the information flow on the network one can apply the algorithm for analysis of observed chaotic data in physical systems, the algorithm suggested in paper [13] (see also ref. [14]and references therein). The main idea relates to the fact that any dynamic system with dimensionality of $N$ can be described by a set of $N$ differential equations of the second order in configuration space or by a set of $2N$ differential equations of first order in phase space.

Assuming that the information flow can be described in terms of ordinary differential equations (or by discrete-time evolution rules), for some unknown functions in a (parametric) phase space, one can analyze a time dependance of a given scalar parameter $s(t)$ that is related to the system dynamics. Then one can build $d$-dimensional vectors from the variable $s$ as

$$y^d(n) = [s(n), s(n+T), s(n+2T), \ldots, s(n+T(d-1))] \tag{1}$$

at equal-distant time intervals $T$: $s(t) \to s(T \cdot n) \equiv s(n)$, where $n$ is an integer number to numerate $s$ values at different times. Now, one can calculate a number of nearest neighbors in the vicinity of each point in the vector space and plot the dependance of the number of false nearest neighbors (FNN) as a function of time. The FNN for the $d$-dimensional space are neighbors that move far away when we increase dimension from $d$ to $d+1$ (see, for details ref.[11]).

The typical behavior of a scalar parameter and corresponding FNN plot are shown in Figs. (1) and (2). From the last plot one can see that the number of FNN rapidly decreases up to about 10 or 12 dimensions. After that it shows a slow dependency on the dimension, if at all. Fig. (2) shows that by increasing the dimension $d$ step-by-step, the number of FNN, which occur due to projection of

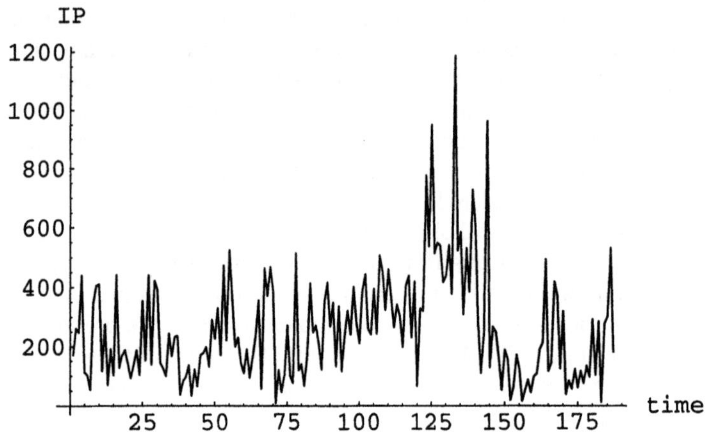


**Figure 1**: Protocol type ID in the IP protocol as a function of time (in $\tau = 5sec$ units).

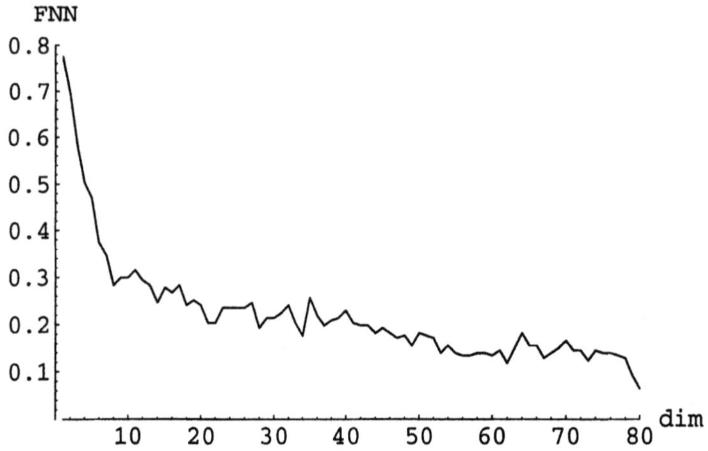


**Figure 2**: Relative number of false nearest neighbors as a function of dimension of unfolded space.

far away parts of the trajectory in higher dimensional space is decreases to a level restricted by system noise that has infinite dimension. Therefore, for a complete description of the information flow one needs not more than 12 independent parameters. The dynamics of information flow can be described as a trajectory in a phase space with the dimension of about 10 - 12. Since this dimension does not depend on the network topology, its size, and the operating systems involved in the network, this is a universal characteristic and may be applied for any network.

However, we cannot identify exactly these independent parameters. Due to the complexity of the system it is natural that these unknown parameters which are real dynamic degrees of freedom of the system would have a complicated relationship with the parameters contained in the network protocols. Fortunately, the suggested technique provides very powerful methods to extract general information about the behavior of dynamic complex systems. For example, the obtained time dependence of only one parameter, the protocol ID shown on Fig.(1), is enough to reconstruct the trajectory of the information flow in its phase space. The reconstructed projection of the trajectory on 3-dimensional space is shown on Fig. (3). Therefore, one can see that the complete description of the network information traffic in terms of a small number of parameters is possible. The important point is that this trajectory (usually called as an "attractor") is well-localized. Therefore, it can be used for detailed analysis and pattern recognition techniques. It should be noted that the attractor presented here is obtained from one parameter measurement only, for that being illustrative purposes. For real analysis we use multi-dimensional high accuracy reconstruction.

# 3 Real Time Network Monitoring and Detection of Known Intrusions

The proposed approach for network traffic description provides the possibility of real-time network monitoring and detection of all known network attacks. This is because one collects from tcpdump binary output the complete information about network traffic at any given point in the network. All header parameters are converted into time dependant numerical functions. Therefore, each packet for host-to-host exchange corresponds to a point in the multidimensional parametric phase space. The set of these points (the trajectory) completely describes information transfer on the network. It is clear that this representation provides not only the total description of the network traffic at the given point but also a powerful tool for analysis in real time. Let us consider some possible scenarios for the analysis.

Suppose we are looking for known network intrusions. The signature of an intrusion is a special set of relationships among the header parameters. For example [9], the signature for the attempt to identify live hosts by those responding to the ACK scan includes a source address, an ACK and SYN flags from TCP

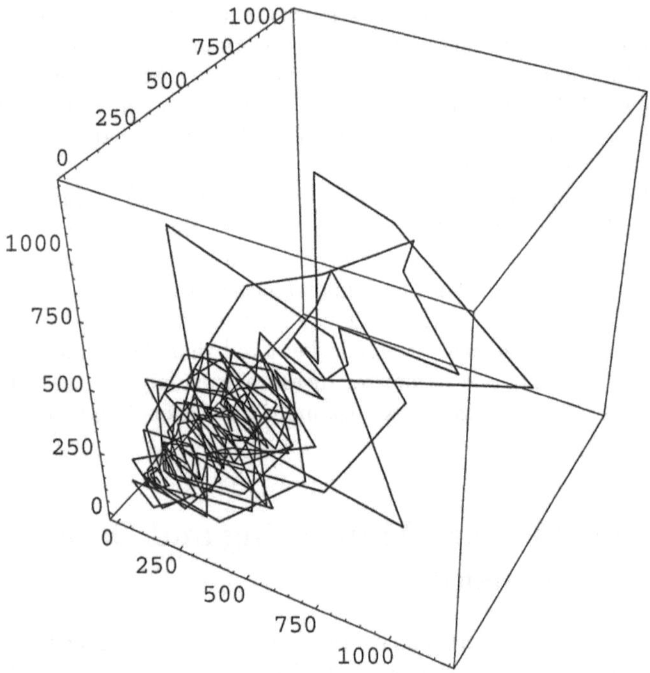


**Figure 3**: The projection of the trajectory of the information flow 3-dimensional phase space.

protocol, a target address of the internal network, sequence numbers, and source and destination port numbers. The lone ACK flag set with identical source and destination ports is the signature for the ACK scan. This is because the lone ACK flag set should be found only as the final transmission of the three-way handshake, an acknowledgement of receiving data, or data that is transmitted where the entire sending buffer has not been emptied. From this example one can see that the intrusion signature could be easily formulated in terms of logic rules and corresponding equations. Then, collecting the header parameters (this is the initial phase of network monitoring) and testing sets of them against the signatures (functions in terms of the subset of the parameters) one can filter out all known intrusions. Since we can collect any set of the parameters and easily add any signature function, it provides the way for a continuous upgrading of the intrusion detection system (IDS) built on these principles. In other words, such an IDS is universal and can be used to detect all possible network intrusions by adding new filter functions or macros in the existing testing program. It is very flexible and easily upgradable. The flexibility is important and can be achieved even in existing "traditional" IDS's. What is out of scope of traditional approaches is the mathematically optimized minimization of possible false alarms and controlled sensitivity to intrusion signals. These properties are an intrinsic feature of our approach.

The important feature of the approach is the presentation of the parameters in terms of time dependant functions. This gives the opportunity to decrease as best as possible for the particular network the false alarm probability of the IDS. This can be done using sophisticated methods already developed for noise reduction in time series. Moreover, representation of the protocol parameters as numerical functions provides the opportunity for detailed mathematical analysis and for the optimization of the signal-to-noise ratio using not only time series techniques but also numerical methods for analysis of multi-dimensional functions. The combination of these methods provides the best possible way, in terms of accuracy of the algorithms and reliability of the obtained information, to detect of known intrusions in real time.

Also, the description of the information flow in terms of numerical functions gives the opportunity to monitor network traffic for different time windows without missing information and without overflowing storage facilities. One can suggest ways to do it. One example is the use of a parallel computer environment (such as low cost powerful Linux clusters) for the simultaneous analysis of the decoded binary tcpdump output. In this case the numerical functions of the header parameters being sent to different nodes of the cluster will be analyzed by each node using similar algorithms but different scales for time averaging of signals (or functions). Thus, each node has a separate time window and, therefore, is sensitive to network behavior in the particular range of time. For example, choosing time averaging scales for the nodes from microseconds to weeks, one can trace and analyze network traffic independently and simultaneously in all these time windows. It is worthwhile to remember that the optimal signal-to-noise ratio is achieved for each time window independently thereby providing the best

**Table 1**: The parameters involved in intrusion signatures as shown on Fig.(4).

| Number | Protocol | Parameter | Frequency |
|---|---|---|---|
| 1 | IP | Destination IP Address | 3 |
| 2 | IP | Source IP Address | 1 |
| 3 | IP | Length | 1 |
| 4 | IP | More Fragment Flag | 2 |
| 5 | IP | Don't Fragment Flag | 2 |
| 6 | IP | Options | 1 |
| 7 | TCP | Source Port | 1 |
| 8 | TCP | Destination Port | 1 |
| 9 | TCP | Urgent Flag | 1 |
| 10 | TCP | RST Flag | 1 |
| 11 | TCP | ACK Flag | 2 |
| 12 | TCP | SYN Flag | 2 |
| 13 | TCP | FIN Flag | 1 |
| 14 | UDP | Destination Port | 2 |
| 15 | UDP | Source Port | 1 |
| 16 | ICMP | Type | 2 |
| 17 | ICMP | Code | 2 |

possible level of information traffic analysis for the whole network. There are three obvious advantages for this approach. The first is the possibility to detect intrusions developed on different time scales simultaneously and in real time. The second is the automatic decreasing of noise from short time fluctuations for long time windows due to time averaging. This provides detailed information analysis in each time window without loss of information. At the same time, it discards all noise related information, drastically reducing the amount of information at the storage facilities. The third advantage is the possibility to use (in real time) the output from short time scale analyzed data as additional information for long time scale analysis.

To give an idea of how many parameters are used to describe signatures of currently known intrusions we use the result of the comprehensive (but probably not complete) analysis[12] of known attacks, i.e., smurf, fraggle, pingpong, ping of death, IP Fragment overlap, BrKill , land attack , SYN flood attack, TCP session hijacking, out of band bug, IP unaligned timestamp, bonk, OOB data barf, and vulnerability scans (FIN and SYN & FIN scanning). The frequencies of the parameters involved in signatures for these intrusions are shown on Fig.(4). The numeration of the parameters is explained in Table 1. One can see that the number of parameters used for signatures of intrusions is rather small . This fact further simplifies the procedure of the analysis.

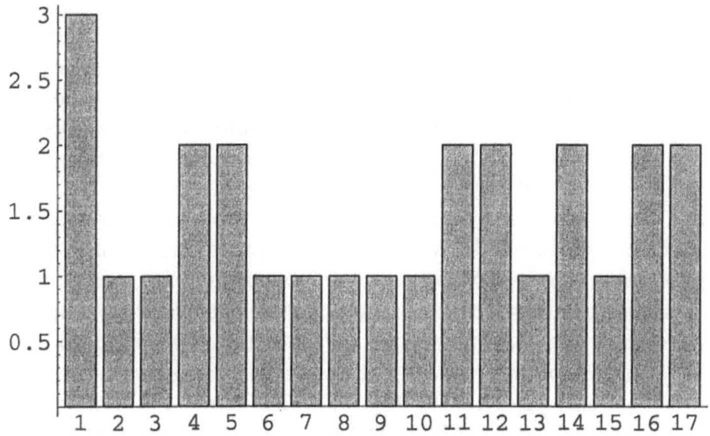


**Figure 4**: Frequencies of the parameters used in signatures of intrusions. For numbering rules see Table 1.

# 4 Detection of Unknown Intrusions

The aforementioned approach could be considered a powerful and promising method for network monitoring and detection of known network intrusions. However, the more important feature of the approach is the ability to detect previously unknown attacks on a network in a wide range of time scales. This ability is based on the method of describing information exchange on a network in terms of numerical functions of header parameters (or a trajectory in multidimensional phase space) as well as using methods of theoretical physics for the analysis of dynamics of complex systems. These methods lead to a very useful result for the small dimensionality of the information flow space. Since the number of parameters used in packet header is large (on the order of hundreds), the practical search for unknown (even very abnormal) signals would be a difficult problem, if not impossible. Therefore, the small dimension of the parametric space of the information flow is a crucial point for the practical approach for the detection of unknown intrusions.

To build a real time intrusion detection system that is capable of detecting unknown attacks, we exploit the fact that we need to analyze only a small number of parameters. Furthermore, as is known from complex systems theory, the choice of the parameters is not important unless they are sensitive to system behavior. The last statement needs to be explained in more detail. Generally, hundreds different parameters could be encapsulated in the packet headers. The question is which parameters we need to choose for the right description of the information flow. Following the discussion in the previous section, one might surmise that we need to make our choice from the known quoted 17 parameters. It may be a good guess. However, the number 17 is bigger than the dimension of the phase space which we have in mind, and it could be that hackers will invent new attacks with new signature parameters that are not included in the set presented in the previous section. The right answer to these remarks follows from complex systems theory. For a complete system description one needs only the number of parameters equal to the phase space dimension (more precisely, the smallest integer number that is larger than fractal dimension of the phase space). It could be a set of any parameters that are sensitive to the system dynamics (and the 17 discussed parameters could be good candidates). We do not know, and do not suppose to know, the real set of parameters until the theory of network information flow is developed or a reliable model for information flow description is suggested. Nevertheless, a method developed to study non-linear complex systems provides tools to extract the essential information about the system from the analysis of even a small partial set of the "sensitive" parameters. As an example, one can refer to the Fig.(3) which shows the 3-dimensional projection of the reconstructed trajectory from the time dependent behavior of only one parameter (the protocol ID shown on Fig.(1)). It means that the complete description of the network information flow could be obtained even from a small set of "sensitive" parameters.

One of the ways to implement this approach is to use the multi-window method discussed in the previous section with the proper data analysis for each time scale. This method of analysis is not within the scope of the current paper and will be reported elsewhere. We will review only the general idea and the problems related to this analysis. To detect unknown attacks (unusual network behavior) we use a deviation of signals from the normal regular network behavior. For these purposes one can use a pattern recognition technique to establish patterns for normal behavior and to measure a possible deviation from this normal behavior. However, the pattern recognition problem is quite difficult for this multidimensional analysis. According to our knowledge, it is technically impossible to achieve reliable efficiency in a pattern recognition for space with a rather large dimension, such as 10. On the other hand, the more parameters we analyze the better accuracy and reliability we can obtain. Therefore, we have to choose the optimal (compromise) solution that uses pattern recognition techniques in information flow subspaces with low dimensions. By applying appropriate constraints on some header parameters one can choose these subspaces as cross sections of the total phase space defined. In this case, we will have a reasonable ratio of signal-to-noise and will simplify the pattern recognition technique and improve its reliability. For a pattern recognition we suggest using a 2-3 dimension wavelet analysis chosen on the basis of detailed study of the information traffic on the set of networks. The wavelet approach is promising because it reduces drastically and simultaneously the computational time and memory requirements. This is important for multidimensional analysis because it can be used for an additional, effective noise reduction technique.

# 5 Conclusions

We suggest a new approach for multidimensional real time network monitoring that is based on the application of complex systems theory for information flow analysis of networks. Describing network traffic in terms of numerical time dependant functions and applying methods of theoretical physics for the study of complex systems provides a robust method for network monitoring to detect known intrusions and is promising for development of real systems to detect unknown intrusions.

To effectively apply innovative technology approaches against practical attacks it is necessary to detect and identify the attack in a reconnaissance stage. Based on new methods of data analysis and pattern recognition, we are studying a technology to build an automatic intrusion detection system. The system will be able to help maintain a high level of confidence in the protection of networks.

We thank the staff of the Advanced Solutions Group for its technical support. This work was supported by the DARPA Information Assurance and Survivability Program and is administered by the USAF Air Force Research Laboratory via grant F30602-99-2-0513, as modified.

# Bibliography

[1] RÉKA, A., J. HAWOONG and B. ALBERT-LÁSZLÓ, *Nature* **406** (2000), 378–381.

[2] STROGATZ, S. H., *Nature* **410** (2000) 268–276.

[3] DERI, L. and S. SUIN, *Computer Networks* **34** (2000), 873–880.

[4] PORRAS, P. A. and A. VALDES, "Live Traffic Analysis of TCP/IP Gateways", *Internet Society Symposium on Network and Distributed System Security*, San Diego, California (March 11-13, 1998).

[5] CABRERA, J. B. D., B. RAVICHANDRAM and R. K. MEHRA ," Statistical Traffic Modeling for Network Intrusion Detection", *Proceedings of the International Simposium on Modeling, Ananlysis and Simulation of Computer and Telecommunication Systems*, IEEE (2000).

[6] HUISINGA, T. *et al.*, *arXiv:cond-mat/0102516* (2000).

[7] DUFFIELD, N. G. and M. GROSSGLAUSER, *IEEE/ACM Transactions on Networking* **9 No 3** (2001) 280–292.

[8] AYEDEMIR, M. *et al.*, *Computer Networks* **36** (2001) 169–179.

[9] NORTHCUTT, S., J. NOVAK and D. MCLACHLAN, *Network Intrusion Detection, An Analyst's Handbook*, New Riders Publishing, Indiapolis, IN (2001).

[10] LBNL's Network Research Group , *http://ee.lbl.gov/*.

[11] GUDKOV, V. and J. E. JOHNSON, *arXiv: nlin.CD/0110008* (2001).

[12] GUDKOV, V. and J. E. JOHNSON, *arXiv: cs.CR/0110019* (2001).

[13] ABARBANEL, H. D. I., R. BROWN, J. J. SIDOROWICH and L. Sh. TSIMRING, *Rev. Mod. Phys.* **65** (1993) 1331–1392.

[14] ECMANN, J.-P. and D. RUELLE, *Rev. Mod. Phys.* **57** (1985) 617–656.

# Chapter 4

# Co-operative Agents in Network Defence

**Robert Ghanea-Hercock**
BTexact Technologies
Adastral Park, B61,pp5, Martlesham, Suffolk, UK
Robert.Ghanea-Hercock@bt.com

## 1. Introduction

The question addressed in this paper is how can complex information networks survive hostile attacks and what mechanisms can increase survivability and defence in large-scale computing networks. As recent evidence indicates, firms, governments and other organisations urgently require better defensive strategies in cyberspace, (Anderson et al 1999, and Briney 2000). In particular, the ability of an information network to maintain itself in the face of continuous perturbation also raises more complex issues related to system metabolism, and network topology.

A multi-agent simulation model has been developed which demonstrates spontaneous group formation and the self-maintenance of group integrity. These system parameters are proposed as critical aspects in the defence of network systems. Each agent is susceptible to a modelled virus infection, passed between the agents. We then introduced an artificial immune system to each agent, which allows 'antibody' solutions to be exchanged between the agents within a social group.

The interest in this behaviour stems from the concept that by linking together the sensory and intelligence capabilities of a large number of agents distributed across a network we can *amplify* the ability of the network to resist attacks or intrusion. Specifically through social co-operation, agents can benefit from the combined defensive capabilities of their particular group.

# 2. Collective Defence

## 2.1. Introduction

The process of survival in natural agents is intimately linked with collaborative behaviour. Stereotypical examples include schooling in fish, and herding behaviour in land mammals. Natural systems have also widely employed distributed sensing and defensive processes, as in the social insects (Wilson 1971) and within multi-cellular organisms in the form of immune systems (e.g. Segel 1998). This paper therefore considers how to utilise social co-operation in software agents in order to achieve a group defensive capability.

## 2.2. The Problem

There is an urgent demand within the computing industry for robust and secure systems and networks. Unfortunately the number of successful attacks is increasing via an increasing number of channels, e.g. email, worms, instant messaging, mobile devices, and wireless links. The almost religious belief in rigid corporate firewalls as a perfect defence is finally succumbing to the realisation that no static defence mechanism will ever suffice (Cohen 1997).

# 3. System Design

The agent system presented in this paper models the secure transmission of useful pre-processed intrusion data through a large-scale computing network. The system is intended to operate as part of an intrusion detection system (IDS). Within proposed IDS (Cohen et al 1998), this approach is defined as "behavioural change detection".

In order to develop some understanding of the dynamics of agent interactions and group cohesion on the integrity of complex networks we created a multi-agent model as an experimental platform. Using this tool we investigated a range of behaviours which might influence the robustness and integrity of such a society of interacting agents. Within the agent group an attack on any one agent or host machine should be visible to other agents within the local domain of the agent. Hence warnings and defence solutions can be rapidly propagated throughout the network.

## 3.1. Metabolic Rate Sensors

The key concept is to create an independent measure of the normal operational state of the network to be defended. By comparison with biological agents we need to create a measure of the *metabolic rate* of the services and data flow on the host network. This is achieved through each defence agent monitoring the flow of inter-agent traffic, in addition to monitoring any local host-specific intrusion sensors, such as port scanning alerts. By analogy with natural agents the first stage of most medical tests is to measure a few macro-scale variables, which are strong predicative indicators of the health of the organism, i.e. temperature, pulse and blood pressure.

# 4. Security Agents

Co-operative software agents have been frequently proposed as a tool for a variety of information processing tasks, i.e. e-commerce transactions (Maes et al. 1999), work-flow modelling or as personal assistants (Etzioni 1996). However, relatively little research has considered their role as an element of network security, exceptions being (Crosbie & Spafford 1995, Helmer et al 1998).

## 4.1. Social Agent Models

The first question we need to address is what properties of a software agent make them suitable for inclusion in a security system. Software agents therefore possess a number of useful properties that would be beneficial in the construction of a distributed adaptive security system. In particular the ability to sense their environment and take proactive decisions against potential threats. The particular aspect of multi-agent systems that we have focused on is their ability to form dynamic social groups. The interest in this behaviour stems from the concept that by linking together the sensory and intelligence capabilities of a large number of agents distributed across a network we can *amplify* the ability of the network to resist attacks or intrusion. Specifically through social co-operation, agents can benefit from the combined defensive capabilities of their particular group.

# 5. Agent Defence Systems

Using a collective formation of smart software agents to form an adaptive immune-response structure within a network has been discussed in existing literature. Some preliminary work in this field has already demonstrated the effectiveness of such methods (Filman & Linden 1996, and Yialelis, Lupo & Sloman 1996). In particular work by Helmer et al (1998) demonstrates a multi-agent network defence system in which software agents monitor low-level network activity and report it to higher-level software agents for analysis. In the system proposed by Crosbie and Spafford (1995) a similar distributed set of agents monitors network traffic and machine activity, including CPU utilisation. The work by Carver et al (2000) demonstrates the use of a distributed heterogeneous group of agents as an IDS solution. The focus is on dynamic and adaptive response to varying levels of security threats. Work by Balasubramaniyan et al (1998) discusses a detailed design and methodology with common features to the proposed COSMOS system and model. The work by Qi He and Sycara (1998) demonstrates the use of encrypted KQML message exchange among a networked group of agents which is used for secure PKI certificate management. Parallel work on artificial immune systems has also been considered, (Forrest et al 1994, and Kephart 1994).

## 5.1 Comparison

The COSMOS project shares the concept of using a distributed set of co-operative software agents and adds the following novel feature:

Metabolic monitoring. COSMOS uses macro-scale changes in the behaviour and processes in a network in order to detect anomalous states, corresponding to attacks.

The system can then respond to any type of attack whether intrusions or viral attacks, unlike existing IDS systems.

**5.2 Experiments**

The agent simulation was developed using the REPAST agent toolkit from the University of Chicago (http://repast.sourceforge.net/). We first constructed a two-dimensional discrete spatial world model, in which a population of artificial agents could interact and move, based on the Sugarscape model (Epstein and Axtell 1996). This model was selected as it represents a suitable test case environment for investigating complex multi-agent simulations. The model is based on a population of agents, which are initialised randomly with the following set of variables:

i) **Vision** – an agent can sense other agents and food objects within a specified radius from its own co-ordinates.

ii) **Metabolism** – agents have an integer counter which represents their rate of energy consumption. Assigned randomly in a specified range. Increased whenever an agent is infected with a pathogen.

iii) **Lifespan** – agents are initialised with a fixed lifespan, randomly assigned, typically between 20 – 200 time steps.

iv) **Sugar** – agents require sugar to survive, which is an environmental resource. Sugar re-grows once consumed by an agent at some specified rate. Agents consume sugar by decrementing the value proportional to their metabolic rate. This would translate into an agents consumption of local CPU and machine resources.

v) **Spice** – as described in the Epstein & Axtell model, a second commodity was introduced into the world which is only available from other agents, and is required for agent survival. Agents can only acquire spice when they engage in a trade interaction with another agent. The rules of trade are described in the following section.

vi) **Immune system** – agents have an array of N characters, which represents a simplified immune system.

vii) **Pathogens** – agents may be initialised with a dynamic array of viral infections, composed of short random character strings.

The simulation uses a tagging scheme on each agent in order to distinguish separate social groups of agents. The purpose of this design is to enable multiple groups of agents to coexist within the same physical Intranet environment and maintain independent operations and behaviours.

**5.3 Experimental Objectives**

The first experiments were designed to study under what conditions socially co-operative groups of agents would spontaneously develop, using the defined model.

***Immune System Development***

The second stage of the work involved adding an artificial immune model to the agents. During each trade interaction between two agents, the initiating agent is passed a vector of N disease strings. Each string is a short sequence of characters, which the receiving agent then attempts to find a match for from its internal immune system, (a large array of character strings). Each string which the agents fails to match results in an increment to its metabolic rate. This results in a gradual degradation of an agent's ability to survive and a reduction in its lifespan. The agents can also undergo random mutation of elements in their immune system, in order to generate potentially novel "anti-body" solutions to current diseases in the population.

A second process was available in the simulation to allow agents to exchange a copy of the anti-bodies they have acquired, to each agent they trade with. Figure 1 illustrates the impact of allowing this co-operative inter-agent exchange to occur. The average number of social connections in the population more than doubles, indicating a significant increase in the agents state of health. This is also reflected in greater stability and lifespan of their social groups.

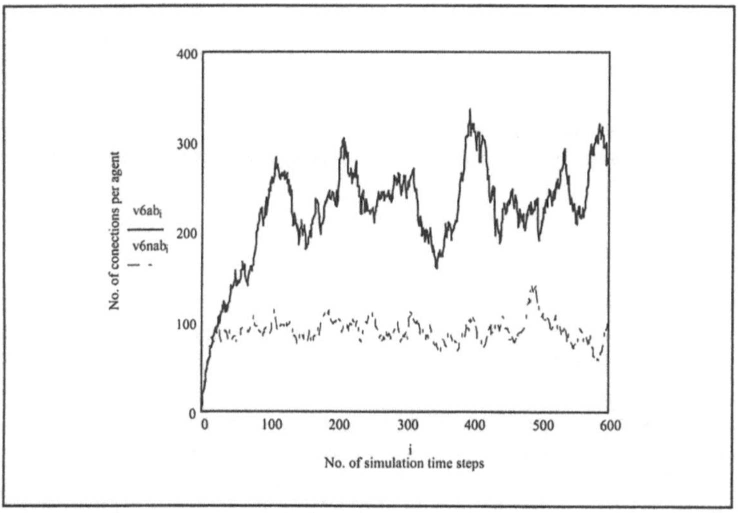


*Figure. 1. Graph of average number of connections per agent with 6 disease strings per agent. Upper trace shows the effect on the average health of the agents of allowing a co-operative exchange of anti-body vectors between agents during trading interactions.*

Figure 2 illustrates how sharing useful solutions to infections the agent population is able to eliminate the majority of infections in the case of a high degree of trust between the agents. The residual level is due to new agents joining the network and introducing new infections.

The metabolic conversions of such a cluster/group therefore contribute to defining its sense of self, (i.e. ability to recognise self-elements). Hence abnormal perturbations of the metabolic rate may be one method for agents to detect when attacks or intrusions are in progress.

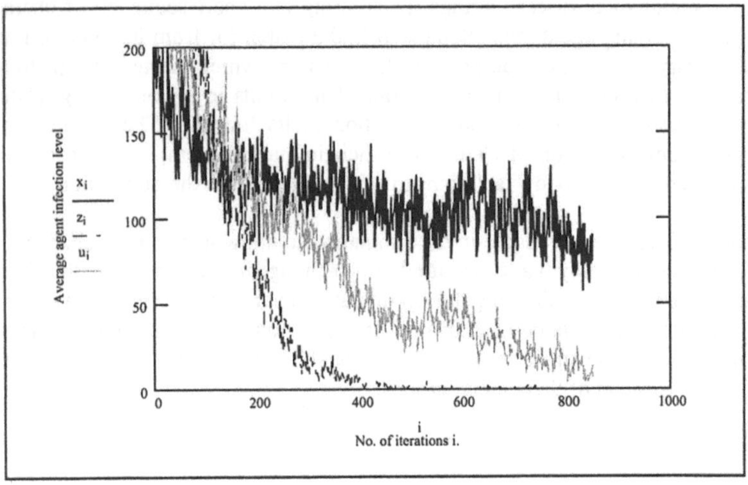


*Figure 2. Graph showing decrease in average infection level with shared antibody process enabled between agents, (curve $x_i$ trust level = 1 (low), $u_i$ trust level = 4 (medium), $z_i$ trust level = 9 (high).*

### 5.4 Future Development

We are currently implementing a multi-agent IDS system across our local Intranet environment with the aim of detecting hostile behaviour and attacks in the network. This prototype is based on the FIPA JADE agent platform, which provides core messaging and agent visualisation services. In order to transport the artificial antibody signatures between the agents we have designed an encrypted XML formated ACL message structure. A specific application domain for the project is in defending peer to peer (P2P) networks. In particular P2P class networks are currently difficult to secure using existing network security methods as they bypass traditional firewall mechanisms, and may span multiple corporate networks.

## 6 Conclusions

This work indicates that a cohesive network of socially interacting agents can create a highly robust and adaptive defence system for information networks. The agent simulation we have developed demonstrates that it is possible to create a population of autonomous agents, which form self-healing social groups with greater resistance to attacks and perturbation than isolated agents. A key parameter of such co-operation is the degree of trust which is established between agents within the same domain, as increased levels of trust can assist in the rapid diffusion of anti-body solutions. (Although at the risk of corrupted agents exploiting such trust).

The process of continuous inter-agent meme transfer enables the agents to maintain a measure of group identity, which is essential to the process of distinguishing self from non-self. In addition the cooperative exchange of recognised patterns for hostile pathogens/viruses greatly improves the immune response of such an agent community.

## REFERENCES

Anderson R., Feldman P., Gerwehr S., Houghton B., Mesic R., Pinder J, Rothenberg J., and Chiesa J. (1999). "Securing the U.S. Defense Information Infrastructure: A Proposed Approach." MR-993-OSD/NSA/DARPA. www.rand.org/publications/electronic/info.html

Balasubramaniyan J., Jose Omar Garcia-Fernandez, Spafford E., and Zamboni D. "An Architecture for Intrusion Detection using Autonomous Agents". Department of Computer Sciences, Purdue University; Coast TR 98-05; 1998.

Briney A., "Security Focused", Online report on Information system security, from http://www.infosecuritymag.com 2000.

Carver C.A., Hill J.M, Surdu J.R., and Pooch U.W., "A Methodology for using Intelligent Agents to provide Automated Intrusion Response," IEEE Systems, Man, and Cybernetics Information Assurance and Security Workshop, West Point, NY, June 6-7 2000, pp. 110-116.

Cohen F., " 50 Ways to Defeat your Intrusion Detection System", onlien paper at http://all.net/journal/netsec/1997-12.html.

Crosbie M. and Spafford E. "Defending a Computer System using Autonomous Agents", In 18th National Information Systems Security Conference, oct 1995. http://www.cs.purdue.edu/homes/mcrosbie/research/NISSC95.ps.

Dittrich D., " The "Tribe Flood Network" distributed denial of service attack tool ", online report at http://staff.washington.edu/dittrich/misc/tfn.analysis.txt.

Epstein J., Axtell R., "Growing Artificial Societies: Social Science from the Bottom Up", MIT Press, 1996.

Etzioni O., "Moving up the information food chain: Deploying softbots on the world-wide web". In Proceedings of the Thirteenth National Conference on Artificial Intelligence (AAAI96) , Portland, OR, 1996.

Filman R., and Linden T., "Communicating Security Agents", Proc. WET ICE 1996.

Forrest S., Perelson S., Allen L., and Cherukuri R., "Self-Nonself Discrimination in a Computer". In Proceedings of IEEE Symposium on Research in Security and Privacy, pages 202-- 212, Oakland, CA, 16-18 May 1994.

Helmer G.G., Wong J.S., Honavar V., and Miller L. "Intelligent agents for intrusion detection". In Proceedings, IEEE Information Technology Conference, pages 121--124, Syracuse, NY, September 1998.

Kephart J.O., " A Biologically Inspired Immune System for Computers", Artificial Life IV, Proceedings of the Fourth International Workshop on Synthesis and Simulatoin of Living Systems, Rodney A. Brooks and Pattie Maes, eds., MIT Press, Cambridge, Massachusetts, 1994, pp. 130-139

Maes P., Guttman R. and Moukas A., "Agents that Buy and Sell: Transforming Commerce as we Know It." Communications of the ACM, Mar 1999 Issue, available online from http://ecommerce.media.mit.edu /Kasbah/.

MAFTIA "Malicious and Accidental Fault Tolerance for Internet Applications", IST Programme RTD Research project, 2001, http://www.newcastle.research.ec.org/maftia/summary.html

Moody J.& White R.D.,"Social Cohesion and Embeddedness: A Hierarchical Conception of Social Groups", Submitted to American Journal of Sociology 2000.

Segel, A., and R. Lev Bar-Or. "Immunology Viewed as the Study of an Autonomous Decentralized System." In Articial Immune Systems and Their Applications, edited by D. Dasgupta, 65-88. Berlin: SpringerVerlag, 1998.

Watts, D. & Strogatz S. "Collective Dynamics of 'small-world' networks", Nature 393, 440-442 (1998).

Wilikens M., Jackson T. "Survivability of Networked Information Systems and Infrastructures", EU DG III/F, Deliverable report on Survivability, Joint research Centre Italy, 1997.

Wilson, E.O. (1971) The insect societies. Harvard University Press.

# Chapter 5

# An Algorithm for Bootstrapping Communications

**Jacob Beal**
MIT AI Lab
jakebeal@mit.edu

In a distributed model of intelligence, peer components need to communicate with one another. I present a system which enables two agents connected by a thick twisted bundle of wires to bootstrap a simple communication system from observations of a shared environment. The agents learn a large vocabulary of symbols, as well as inflections on those symbols which allow thematic role-frames to be transmitted. Language acquisition time is rapid and linear in the number of symbols and inflections. The final communication system is robust and performance degrades gradually in the face of problems.

## 1 Introduction

Neuroscience has postulated that the brain has many "organs" — internal subdivisions which specialize in one area. If we accept this view, then we need some sort of mechanism to interface these components. The design of this mechanism is limited by the hardware which the brain is constructed out of, as well as the size of the blueprints specifying how it is built. Neurons, as hardware, are relatively slow and imprecise devices, but they are very cheap, and it's easy to throw a lot of them at a problem in parallel. Our DNA is only about 1 gigabyte, too small to encode the full complexity of interfaces between all of the different components.

I approached this design problem from a hardware hacking point of view, with the question, "If I were designing the human brain, how would I build

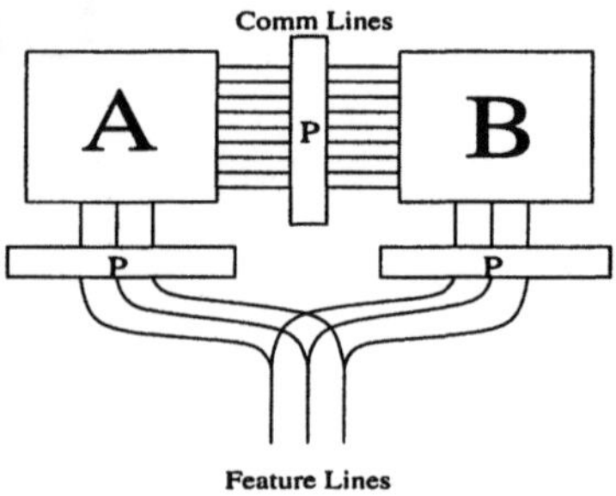


**Figure 1.1**: The agents labelled A and B are interconnected by *comm lines* — a bundle of wires with an arbitrary and unknown permutation. The agents also share some *feature lines* with the outside world, again with unknown permutations.

this interface?" It needs to be self-configuring, to beat the limited blueprints problem, and it needs to learn quickly. On the other hand, hardware is very cheap, and I can design in a domain with a huge number of interface wires between two components.

I have developed an algorithm which bootstraps communications solely from shared experience and I present it here as an existence proof and a tool for thinking about how a brain might be composed out of independent parts that learn to communicate with each other: it is possible for two agents to rapidly construct a language which enables them to communicate robustly.

# 2 System Model

There are two agents in the system, connected by a very large bundle of wires called *comm lines*. Each agent has another set of wires called *feature lines*, over which external information can arrive. Metaphorically, the comm lines are a nerve bundle connecting two regions of the brain and the feature lines are nerve bundles that carry part of the brain's observation of the outside world. The actual wires in the bundles might be arbitrarily twisted and rearranged between the two ends of the system, so we add an unknown permutation to each bundle to model this effect and prevent any implicit sharing of ordering information between the two agents.

The comm lines have four states: 1,-1,0, and X. When undriven, the line reads as 0. Information is sent over the line by driving it to 1 or -1, and if the line is being driven to both 1 and -1, it reads as a conflict — X. In the experiments conducted here, I used a set of 10,000 comm lines.

Feature lines are named with a symbol which they represent and read as undriven, driven, or driven with another symbol. In the experiments I conducted, names of feature lines are things or actions and the symbols driven on them are roles. So typical feature lines might be **bob**, **mary**, or **push**, and typical roles might be **subject**, **object**, or **verb**. The set of feature lines is of undefined size

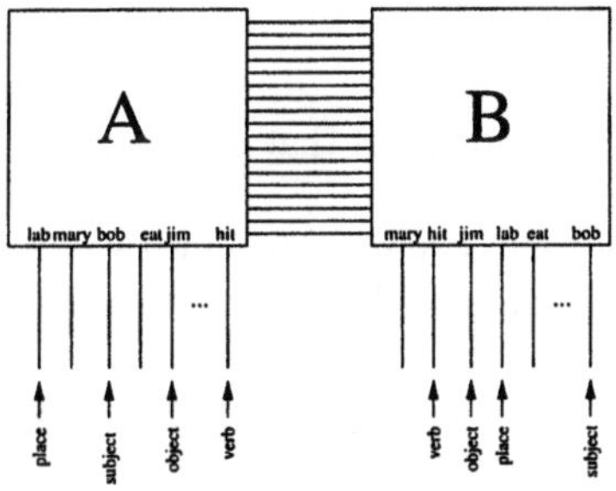


**Figure 1.2**: During a training cycle, the feature lines of both units are driven. Each agent attempts to learn from the comm lines driven by the other agent.

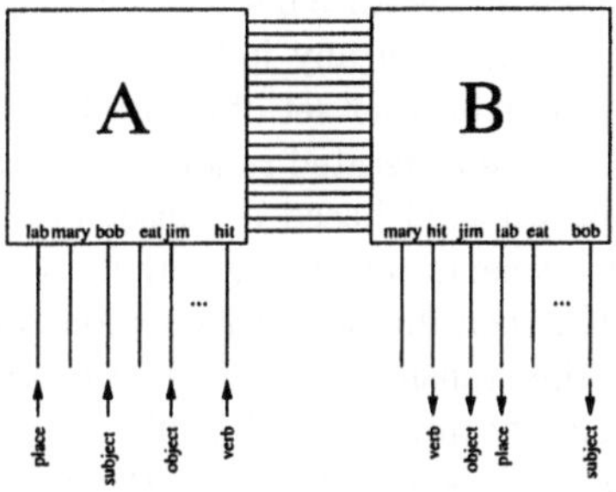


**Figure 1.3**: During a test cycle, the feature lines of one unit, say A, are driven and the feature lines of the other unit are observed. The test is scored by number of mistakes in B's reproduction of A's feature lines.

— the agents have no knowledge of what feature names or roles exist until they encounter them in practice.

An agent can read and drive both comm and feature lines, but are constrained to synchronous schedule. Each cycle of the system has a "talk" phase and a "listen" phase. Agents can read lines at any time, but can only drive comm lines during the talk phase and feature lines during the listen phase. At the end of the talk phase, for symmetry breaking purposes one agent is randomly selected to have spoken first. The agent which spoke first can read the results of both agents speaking in its listen phase, while the one which spoke second reads only what the first agent spoke.

There are two types of cycles — training cycles and test cycles. In a training cycle, the data on the feature lines is sent to both agents. In a test cycle, one agent is randomly selected to receive input from the feature lines, while the other receives no input. Performance may then be evaluated on the basis of how well the output of the agent receiving no input matches the values on the feature lines.

# 3 Algorithm

The key idea driving this algorithm is that sparseness makes it easy to separate the stimuli.

Knowledge in the system is represented by two sets of mappings: symbol mappings and inflection mappings. An inflection mapping links a symbol carried on a feature line to a real value between 0 and 1. A symbol mapping links a feature line with two sets of comm lines, designated as *certain* and *uncertain*, and includes an integer designated *certainty*.

These mappings are used symmetrically for production and interpretation of messages. In the "talk" phase, each driven feature line selects the *certain* comm lines associated via the symbol mapping and drives them with the unary fraction associated with the symbol on the feature line via the inflection mapping. In the "listen" phase, if enough of a feature line's associated comm lines are driven, then the feature line is driven with any inflection mapping symbols within a fixed radius of the unary code on that set of comm lines.

Both types of mappings are generated randomly when a feature or inflection is first encountered, then adjusted based on observations of the other agent's transmissions. These adjustments take place only if an agent spoke second; if it was the first one to speak, then its own transmissions are on the lines as well, inextricably mixed with the transmissions of the second agent, and this would make accurate learning significantly more difficult.

Inflection mappings are adjusted with a very simple agreement algorithm: if the received unary code is significantly different from expected, the code in the mapping is set to the received value. If a unary code matches which should not have, then it is purged and generated anew.

Symbol mappings are slightly more complicated. The first time an agent hears a given symbol spoken by the other agent, it adds every driven comm line to the *uncertain* lines for that symbol. Each time thereafter that it hears the symbol again, it intersects the driven lines with its *uncertain* lines, thereby eliminating lines associated with other symbols. After several iterations of this, it assumes that there is nothing left but lines which should be associated with the symbol, adds the *uncertain* lines to the *certain* lines, and begins to use them for communication. A few more iterations after that, it begins paring down the *certain* lines the same way, so that the two agents can be assured that they have identical mappings for the symbol.

A more detailed description of the algorithm, including code implementing it, may be found in [1] and [2].

# 4 Results

To test the algorithm, I used a system with an $n_w$ of 10000 comm-lines and a $n_{wps}$ of 100 random wires selected to generate a new symbol mapping.

I trained the system for 1000 cycles, then evaluated its performance over an additional 200 cycles. Each cycle, an example is generated and presented to the system. In the training phase, there is an 80% chance it will be presented to both agents and a 20% chance it will be presented to only one (That is, 80% training cycles, 20% test cycles). During the evaluation phase, the first 100 are presented to the first agent only, and the second 100 are presented to the second agent only. A test is considered successful if the input feature set is exactly reproduced by the listening agent.

The examples input to the feature lines are thematic role frames generated from a set of 50 nouns, 20 verbs, and 4 noun-roles. Each example is randomly generated with 0-2 verbs assigned the "verb" role and 2-4 nouns assigned noun-roles. No noun, verb, or noun-role can appear more than once in an example. A typical scene, then, might be `'((approach verb) (jim subject) (shovel instrument) (lab object))`, which corresponds loosely to "Jim approached the lab with the shovel." All told, there are more than 1.2 billion examples which can be generated by the system, so in general an agent will never see a given scene twice.

In a typical run of this system, after about 200 cycles most symbols will have entered the shared vocabulary and can be successfully communicated between the two agents. After about 500 cycles, the set of inflections will have stabilized as well. In the final round of 200 tests, the success rate is usually 100%, although occasionally due to the stochastic nature of the algorithm, the inflections will not yet have converged by the end of 1000 tests and consequently one or more will not be transmitted correctly.

## 4.1 Convergence Time

The time needed to develop a shared vocabulary is proportional to the number of symbols in the vocabulary. A symbol is learned when both agents have *certainty* for that symbol greater than $t_c$. An agent increases *certainty* when it speaks second, which is determined randomly, so we may estimate this as a Poisson process. Thus, we may calculate the expected number of cycles, $c$, as follows:

$$E(c) = 2t_c\frac{\binom{2t_c}{t_c}}{2^{2t_c}} + \sum_{n=2t_c+1}^{\infty} n\frac{\binom{n-1}{t_c-1}}{2^{n-1}}$$

Evaluating this for $t_c = 4$, we find an expectation of 10.187 uses of a symbol before both *certainty* thresholds are reached.

For these experiments then, with an average of 3 nouns and 1 verb per training cycle, then, we can calculate the expected number of shared symbols $S$ as a function of elapsed cycles $t$:

$$S(t) = n_{nouns} * (1 - P(10.187, t\frac{3}{n_{nouns}}0.8)) + n_{verbs} * (1 - P(10.187, t\frac{1}{n_{verbs}}0.8))$$

where $P$ is the incomplete gamma function. Since this function is linear in the number of symbols, we see that the time to build a shared vocabulary is

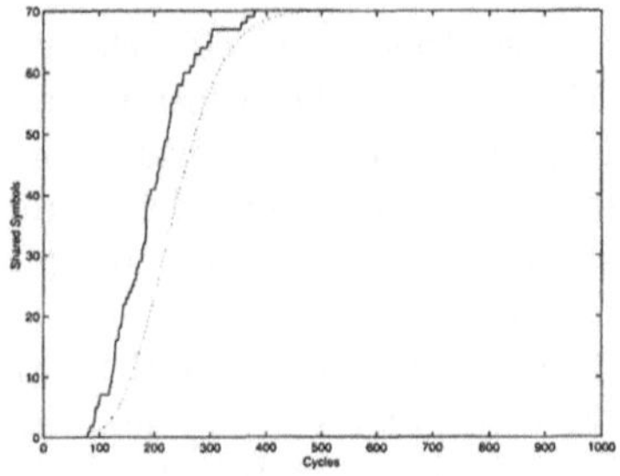


**Figure 1.4**: Number of shared symbols versus elapsed time for 50 nouns and 20 verbs. Dotted line is theoretical estimate S(t), solid line is experimental data.

linear in the number of symbols. Figure 1.4 shows experimental data confirming this estimate.

Once a shared vocabulary of symbols exists, the algorithm can begin learning inflections. If $n_i$ is the number of inflections to be learned, and $r_i$ is chosen such that $r_i * n_i \leq 0.5$, then we can show that the time to develop a shared set of inflections is $O(n_i)$.

An inflection may be learned any time a symbol is successfully transmitted in a training cycle. This occurs if the new inflection does not conflict with any of the previously learned inflections - that is, if $n$ symbols have already been learned, then it must be the case that for all $v_i$ s.t. $1 \leq i \leq n$, $|v_{n+1} - v_i| < 2r_i$. Since the value of the new symbol, $v_{n+1}$, is chosen by a uniform random process on the interval $[0, 1]$, the probability $p_{n+1}$ of choosing an acceptable inflection value is no less than $1 - (2r_i * n)$. The $n_i$th inflection, then, has the least probability of success, $p_{n_i} = 1 - (2r_i * (n_i - 1)) \geq 2r_i$, and $p_n$ is generally bounded below by $2r_i$.

For these experiments then, we can calculate the expected number of inflections, assuming a shared vocabulary, as a function $I(t)$ of elapsed cycles $t$. There are expected to be 3 noun inflections and 1 verb inflection per training cycle, so the least frequent inflection is expected to appear at with frequency at least $1/n_i$. Thus, we obtain

$$I(t) = n_i * (1 - P(1, 2r_i t \frac{1}{n_i} 0.8))$$

where $P$ is the incomplete gamma function. Since this function is linear in the number of inflections, we see that the time to build a shared set of inflections is linear in the number of inflections. Figure 1.5 shows experimental data confirming this estimate.

Thus, the algorithm is expected to converge in $O(s + n_i)$ time, where $s$ is the size of the vocabulary and $n_i$ is the number of inflections.

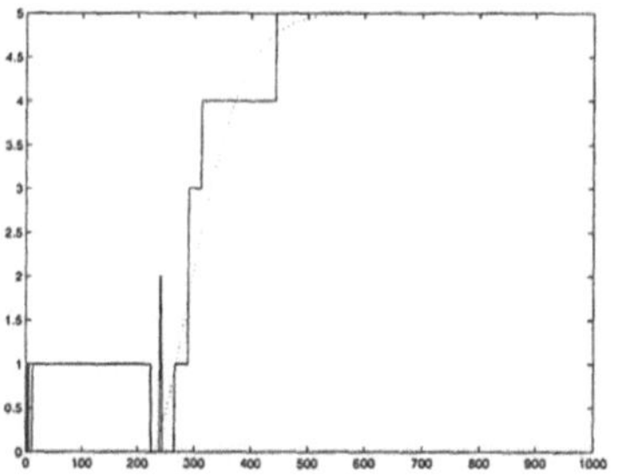


**Figure 1.5**: Number of shared inflections versus elapsed time for 4 noun inflections and 1 verb inflection, in a system with 50 nouns and 20 verbs. The dotted line is theoretical estimate I(t), beginning with cycle 230, where S(t) predicts half the vocabulary to be learned. The solid line is experimental data.

## 4.2 Channel Capacity

The number of symbols and roles which can be learned without false symbol detection and inflection misinterpretation is dependent on the number of wires $n_w$, the number of wires per symbol $n_{wps}$, and the percent stimulus necessary to recognize a symbol $p_s$.

If we want no combination of symbols to be able to generate a spurious recognition, then each symbol must have at least $n_{wps}(1-p_s)$ wires not used by any other symbol. This means that a vocabulary would have a maximum size of only $\frac{n_w}{n_{wps}(1-p_s)}$. In practice, however, we can assume that only a few symbols are being transmitted simultaneously. If we assume that no more than $m$ symbols will be transmitted at once, then we can conservatively estimate capacity by allowing any two symbols to overlap by no more than $n_{wps} * p_s/m$ wires. Thus any given symbol covers a portion of symbol space with volume:

$$\sum_{i=0}^{n_{wps}(1-\frac{p_s}{m})} \binom{n_{wps}}{i}\binom{n_w - n_{wps}}{i}$$

The whole symbol space has volume $\binom{n_w}{n_{wps}}$, so a conservative estimate of the maximum number of symbols that can exist is:

$$\frac{\binom{n_w}{n_{wps}}}{\sum_{i=0}^{n_{wps}(1-\frac{p_s}{m})} \binom{n_{wps}}{i}\binom{n_w-n_{wps}}{i}}$$

This yields a satisfactorily large capacity for symbols. For the experiments described above, with $n_w = 10000$, $n_{wps} = 100$, $p_s = 0.8$ and a maximum of 6 concurrent symbols, we find that the capacity is $1.167 \times 10^{12}$ distinct symbols.

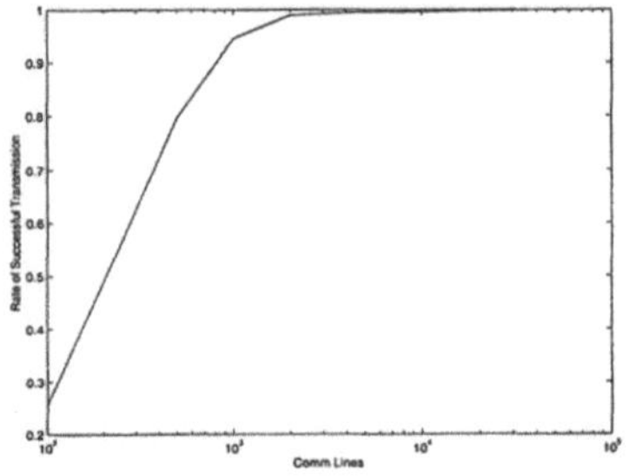


**Figure 1.6**: Number of comm lines versus transmission robustness. Horizontal axis is $n_w$ from 100 to 100,000. Vertical axis shows the symbols and inflections correctly received per symbol and inflection transmitted (spurious receptions count against this as well) over the course of 200 test cycles on a system trained with 50 nouns, 20 verbs and 4 noun-roles, $n_{wps} = 20$, $p_s = 0.8$. Accuracy degrades smoothly with decreased channel capacity.

## 4.3 Performance Degradation

We expect that the performance of the algorithm will degrade gracefully as the channel capacity is reduced. As the average Hamming distance between symbols drops, the chance that a combination of other symbols will overlap to produce a spurious recognition or interfere with the inflection being transmitted rises. Since symbols receiving too much interference are discarded, the algorithm will tend to break up clusters of symbols and move toward an efficient filling of symbol space. Thus, reducing the ratio $n_w/n_{wps}$ ought to cause the transmission errors to rise gradually and smoothly. In practice we find that this is in fact the case, as shown in Figure 1.6.

## 4.4 Parameter Variation

The values of the parameters used in the experiments above were not carefully chosen. Rather, I made a guess at a reasonable value for each parameter, expecting that the algorithm should not be very sensitive to the parameter values. (If it were, then I could hardly claim it was a robust algorithm!)

To test this, I ran a series of experiments in which I trained and tested the system with one of the parameters set to a different value. For each value for each parameter I ran 10 experiments: Figure 1.7 shows the performance of the algorithm as a function of parameter value for each of the six parameters $p_s$, $r_i$, $t_c$, $t_p$, $w_m$, and $n_{wps}$. ($n_w$ is excluded because its variation is evaluated in the preceding section) As predicted, the performance of the algorithm is good over a wide range of values for each variable.

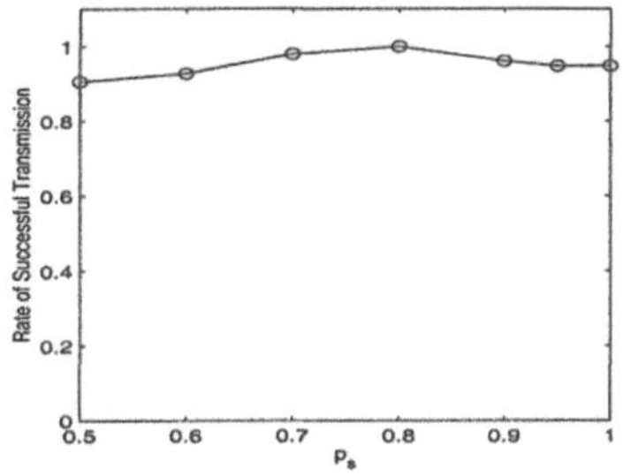


(a) Variation of $p_s$

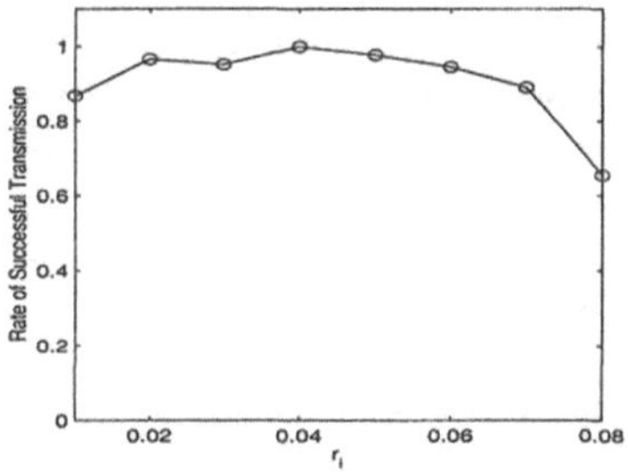


(b) Variation of $r_i$

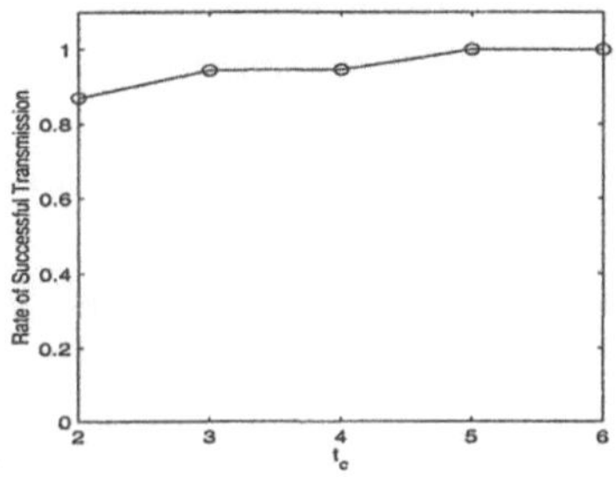


(c) Variation of $t_c$

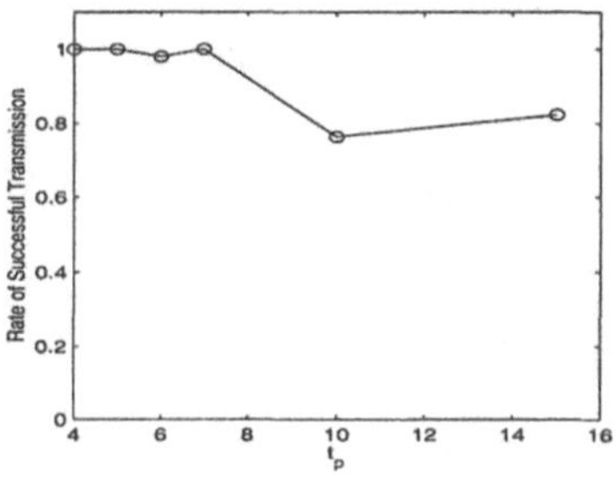


(d) Variation of $t_p$

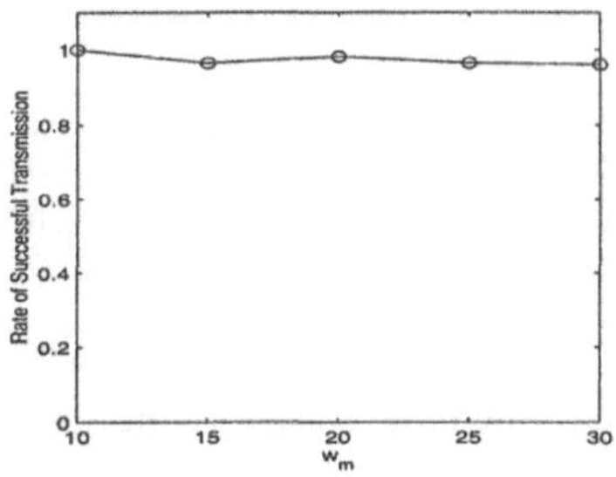


(e) Variation of $w_m$

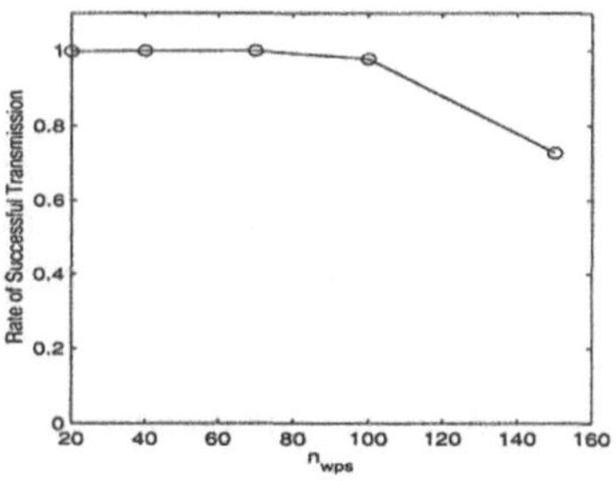


(f) Variation of $n_{wps}$

**Figure 1.7**: Variation in performance as each parameter is varied. For each graph, the horizontal axis shows the value of the parameter being varied and the vertical axis shows the fraction of symbols and inflections correctly received per symbol and inflection transmitted. Measurements are the average values over the course of 10 runs of 200 test cycles, as in 1.6. For each run of test cycles, the systems were trained with 50 nouns, 20 verbs and 4 noun-roles, with base parameter values $p_s = 0.8$, $r_i = 0.05$, $t_c = 4$, $t_p = 6$, $w_m = 20$, $n_{wps} = 100$, and $n_w = 10000$. All parameters in the system can tolerate small variations without serious degradation in performance.

## 5 Contributions

I have built an algorithm which allows two agents to generate a shared language on the basis of shared experiences only. The behavior of this algorithm can be analyzed and performs as predicted by theoretical analysis.

## 6 Acknowledgements

Particular note should be given to the help from several people. Gerry Sussman started me thinking about the problem and pointed me in this direction. My research is part of a continuing effort started by Yip and Sussman to build a "TTL databook for the mind" — a compatible system of modules that capture aspects of mental activity and can be combined to build ever more powerful systems. Thanks also to Catherine Havasi for hacking some of the early code with me and being a needling presence to keep me from slacking off.

## Bibliography

[1] BEAL, Jacob. "An Algorithm for Bootstrapping Communications" MIT AI Memo 2001-016, August, 2001.

[2] BEAL, Jacob. "Generating Communications Systems Through Shared Context" MIT AI Technical Report 2002-002, January, 2002.

[3] KIRBY, Simon. "Language evolution without natural selection: From vocabulary to syntax in a population of learners." Edinburgh Occasional Paper in Linguistics EOPL-98-1, 1998. University of Edinburgh Department of Linguistics.

[4] KIRBY, Simon. "Learning, Bottlenecks and the Evolution of Recursive Syntax." Linguistic Evolution through Language Acquisition: Formal and Computational Models edited by Ted Briscoe. Cambridge University Press, in Press.

[5] MINSKY, Marvin. *The Society of Mind.* Simon & Schuster, Inc, New York, 1985.

[6] MOOERS, Calvin. "Putting Probability to Work in Coding Punched Cards: Zatocoding (Zator Technical Bulletin No. 10), 1947. Reprinted as Zator Technical Bulletin No. 65 (1951).

[7] YIP, Kenneth and SUSSMAN, Gerald Jay. "Sparse Representations for Fast, One-Shot Learning." MIT AI Lab Memo 1633, May 1998.

# Chapter 6

# Intelligent Broadcast For Large-Scale Sensor Networks

**Rajkumar Arumugam, Vinod Subramanian and Ali A Minai**
Complex Adaptive System Laboratory
ECECS Department
University of Cincinnati
Cincinnati, OH 45221
ali.minai@uc.edu

With advances in miniaturization, wireless communication, and the theory of self-organizing systems, it has become possible to consider scenarios where a very large number of networkable sensors are deployed randomly over an extended environment and organize themselves into a network. Such networks — which we term *large-scale sensor networks (LSSN's)* — can be useful in many situations, including military surveillance, environmental monitoring, disaster relief, etc. The idea is that, by deploying an LSSN, an extended environment can be rendered observable for an external user (e.g., a monitoring station) or for users within the system (e.g., persons walking around with palm-sized devices). Unlike custom-designed networks, these randomly deployed networks need no pre-design and configure themselves through a process of self-organization. The sensor nodes themselves are typically anonymous, and information is addressed by location or attribute rather than by node ID. This approach provides several advantages, including: 1) Scalability; 2) Robustness; 3) Flexibility; 4) Expandability; and 5) Versatility. Indeed, this abstraction is implicit in such ideas as smart paint, smart dust, and smart matter.

The purpose of our research is to explore how a system comprising a very large number of randomly distributed nodes can organize itself to communicate information between designated geographical locations. To keep the system realistic, we assume that each node has only limited reliability, energy resources, wireless communication capabilities, and computational capacity. Thus, direct long-range communication between nodes is not possible, and most messaging involves a large number of "hops" between neighboring nodes. In particular, we are interested in obtaining reliable communication at the system level from simple, unreliable nodes.

# 1 Introduction

Wireless networks that operate without fixed infrastructure are called ad-hoc networks, and are a very active focus of research by the wireless community. However, most of the work focuses on networks with tens or hundreds of nodes, where most message paths are only a few hops long. All data messages in such a system are unicast, i.e., they are between specific pairs of nodes. There are two major formulations for this. In some message routing algorithms, a path discovery process is used to first find a route between the source and destination nodes (or locations), and the message is then sent along this path [7, 8, 4]. This is clearly a top-down approach with limited scalability. Other routing protocols use next-hop routing, where each node, knowing the destination of an incoming message, only determines the next node to forward the message to [3]. These protocols scale much better, but at the cost of maintaining and updating extensive amounts of information about network topology. This can be expensive in terms of energy, and can often lead to problems if the individual nodes are unreliable, causing broken links and lost messages. From a complex systems viewpoint, the problem with unicast-based next-hop methods is that they do not exploit the inherent parallelism of the system to achieve robustness. This is the issue we consider in our research.

Rather than using directed unicast between nodes, we study the possibilities of broadcast. In the simplest case, this corresponds to flooding, where every message received by a non-destination node is "flooded" to all the node's neighbors. While this is a simple approach, it is extremely wasteful of bandwidth and creates a lot of collisions — the simultaneous use of the wireless channel by multiple messages, all of which are lost as a consequence. To overcome the problems of flooding while retaining its inherent parallelism, we explore the method of intelligent broadcast. In this approach, each node receiving a message decides whether to re-broadcast it to all its neighbors or to ignore it. Note that the decision does not involve selecting *which* neighbor the message is forwarded to, but only *whether* to forward the message. The latter is a much simpler decision, and can be made on the basis of the information carried by the message in combination with that available within the potential forwarding node. This approach leads to a self-organized communication process where local decisions by the nodes produce global availability of information.

In this paper, we present a well-developed paradigm for random LSSN's, including a model for the nodes and viable broadcast-based protocols for channel access and network organization. We evaluate the performance of the network in the case of simple flooding, and then study the effect of a simple decision heuristic that allows nodes to limit message re-broadcast based on how many hops the message has already traveled. We show that this heuristic leads to a dramatic improvement in performance, making the broadcast-based system a viable — and more robust — alternative to more complicated systems under some conditions.

# 2 Medium Access Protocols

Traditional networks make use of point-to-point channels, in which interference-free communication is established between an ordered pair of users. But, under scenarios where such channels are not economical or are not available, broadcast channels are used. When nodes communicate through a broadcast channel, a single transmission by a node is heard by all nodes within the transmitting node's radius of communication. In essence, the success of a transmission between a pair of nodes is no longer independent of other transmissions: The channel becomes a shared resource whose allocation is critical for network operation. Schemes for channel allocation are known as *medium access protocols.* Medium access protocols can be broadly classified as *conflict-free* and *contention* protocols. Conflict-free protocols ensure successful transmission at all times. This can be achieved by allocating the channel to the users either statically or dynamically. In contention based schemes, a transmitting user is not guaranteed to be successful. The protocol must prescribe a way to resolve conflicts once they occur. An example is the *Aloha* protocol, which was the first random access protocol implemented. In aloha, a newly generated packet is transmitted immediately hoping for no interference from other users. Another contention-based protocol, CSMA, is described below.

## 2.1 Carrier Sense Multiple Access (CSMA)

In CSMA [6], nodes in the network sense the channel prior to transmitting packets. If the channel is sensed to be busy, the sensing node refrains from transmitting, avoiding a collision in the process. The node reschedules the packet for a later time. This delay is termed the *back-off.* If the channel is sensed to be idle, the node transmits its packet. In *slotted CSMA*, the channel is divided into mini-slots, the width of a mini-slot being equal to the maximum propagation delay in the system. Nodes using slotted CSMA are restricted to start transmissions only at mini-slot boundaries.

Even though slotted CSMA improves over the unslotted version by synchronizing the nodes and thus reducing the length of the unsuccessful transmission periods, it does little to improve the quality of channel access. Nodes with packets ready for transmission in a neighborhood, upon sensing a busy channel, back-off for a random duration (depending upon the back-off strategy). Our argument is that by backing-off to a later time, the nodes are not preventing but just deferring the collisions to a later time. Since nodes back-off to integral multiples of the slot, each slot is subject to collision among messages, generated at various times in the past, that happen to choose the same back-off point. We call this the problem of *colliding backoffs*, and it is an inherent disadvantage of slotting.

Consider another case in a multi-hop network when the channel in a neighborhood has been idle for a length of time. A node, upon generating a packet or receiving one for forwarding, would sense the channel and would transmit the packet if the channel was idle. In a broadcast CSMA channel, all the neighbors of this node would receive the packet and hence sense the carrier at the next mini-slot boundary. In all probability, many of the nodes would sense the channel free and hence transmit at the same instant. Since some of these nodes are also neighbors of each other, this would result in collision of the packets. Thus, in this protocol, a successful transmission would trigger a spate of collisions in a relatively calm neighborhood. In the next section, we propose a novel channel access scheme that solves this problem by allowing each node a fair chance to earn a slot.

## 2.2 CSMA with Mini-Backoff (CSMA-mb)

To overcome the problems with slotted CSMA in the context of multi-hop networks, we propose a novel persistent scheme termed *CSMA with mini-backoff (CSMA-mb)*. CSMA-mb is built over the slotted CSMA protocol and shares the same philosophy with regard to the maximum propagation delay in the system, i.e., in a multi-hop network, since the communication is localized (between one-hop neighbors), the maximum propagation delay, $a$, between the nodes is assumed to be not too large. In pure slotted CSMA, a collision could occur either between messages contending in a slot for the first time or among messages generated at various times in the past that happen to choose the same back-off point. In CSMA-mb, we give these messages a further opportunity to resolve their contentions by allowing them to backoff yet another time, albeit for a relatively smaller duration called mini-backoff, and hence avoid a collision. The idea behind CSMA-mb is to provide a method whereby contention can be resolved *within* a slot — thus greatly reducing the chances for collision due to colliding backoffs. The CSMA-mb protocol is described below.

- In CSMA-mb, we divide the channel into major slots and further divide each major slot into mini-slots. The structure of a major slot in CSMA-mb is shown in Fig. 1.

- The size of a mini-slot is equal to the maximum propagation delay $a$ in the system. The length of a major slot is an integral multiple of the mini-slot.

- The first $m$ mini-slots in a major slot are reserved for contention while the rest are set aside for data.

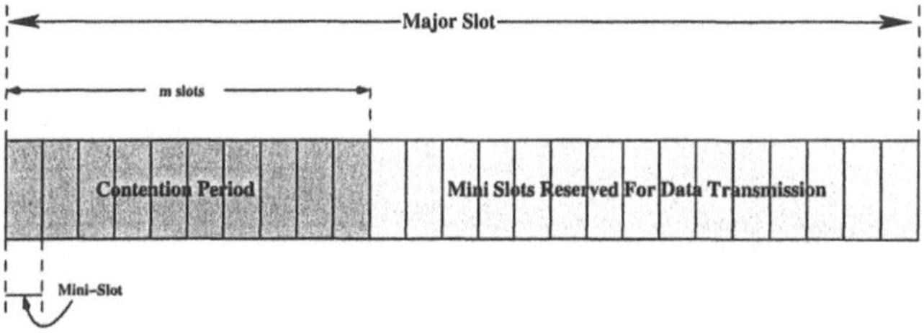


**Figure 1**: Data structure of a slot in CSMA-mb

- Nodes are allowed to contend only at major slot boundaries. Nodes sense the channel in the first mini-slot of every major slot.
- If a node senses a busy channel at the beginning of a major slot, it backs-off to the next major slot as a consequence of the persistent nature of the protocol.
- If a node senses the channel free, it does not transmit right away. Rather, it sets a random mini-backoff, the value of which lies in the range [1, m], where $m$ is the maximum number of mini-slots reserved for contention. Thus, each one of the nodes that contend at the beginning of a major slot would set a mini-backoff.
- A node then senses the carrier again when the mini-backoff expires and starts transmitting in the event of an idle channel. Thus, the node that sets the smallest mini-backoff in a neighborhood wins the major slot.
- All nodes that set a higher back-off value would then sense the channel busy when their respective mini-backoff expires and hence back-off to the next major slot.
- Collision in CSMA-mb is a possibility only when two or more nodes with messages to send in the same slot set the same mini-backoff value. Thus, the performance of the system can be improved further by increasing the length of the *contention period.*

CSMA-mb is explained with an example scenario in Fig. 2. Assume that neighbors $p, q, r$ and $s$ have packets to send at time $t$ and hence sense the carrier. Assuming that the nodes sensed a free carrier, each of the nodes would set a mini-backoff. Suppose, $p, q, r$ and $s$, backed-off to times $t_1, t_2, t_3$ and $t_4$ respectively such that $t_2 < t_1 < t_4 < t_3$. Thus, $q$ would sense the carrier free when its mini-backoff expires and would start transmission immediately afterwards. Nodes $p, r$ and $s$ would sense a busy channel when their respective mini-backoffs expire and hence would back-off to the next major slot. If pure slotted CSMA had been used for channel access, the packets from $p, q, r$ and $s$ would have collided.

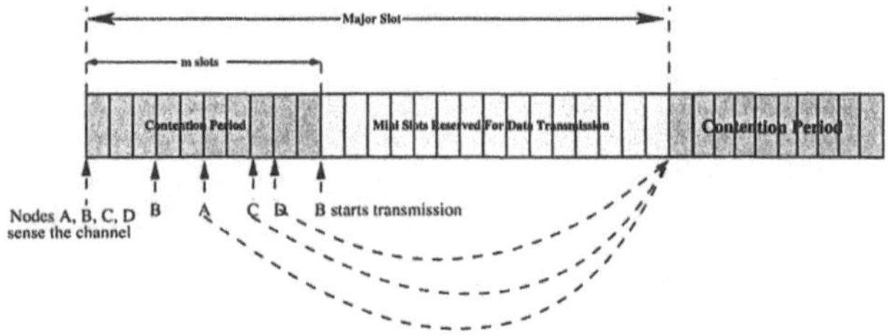


**Figure 2**: Channel Access in CSMA-mb

In conclusion, if Aloha is *impolite*, pure CSMA is *impatient*. In a similar vein, CSMA-mb can be characterized as a *cautious* protocol: it is focussed on avoiding collisions even among messages that have equal "rights" to a slot. From a complex systems viewpoint, CSMA-mb is attractive because, the efficiency in the messaging process is achieved by resolving the contentions and consequent collisions locally.

Next, we describe a simple heuristic that tries to improve network performance further by reducing collisions. It does so by controlling undesirable redundancy of message paths in the system.

# 3 Hop limit for messages

Broadcast-based communication is wasteful of resources, but it also obviates the need for gathering, storing and frequently updating massive amounts of information about a vast network. As these networks grow very large, keeping track of routes or link states will not be possible, but broadcast will always work. Also, very large-scale networks will need to use extremely cheap and, therefore, unreliable nodes. Frequent random failures of nodes in the network would make precisely directed communication inefficient unless nodes monitored each other very frequently, which would be too expensive. Using intelligent broadcast is a way to use the inherent parallelism of broadcast and the massive distributedness of the system to ensure communication.

Intelligence in broadcast can be derived either from the local information available at a node or from the information carried by the data messages. In addition, nodes can also exchange information by dedicated messages, called control or *hello* messages. These messages are usually exchanged by the nodes at periodic intervals for gathering and updating information about their local neighborhoods.

In this work, we study the effect of nodes exchanging information about the state of a message on the performance of flooding. A node, upon generating a message, assigns an upper limit to the number of hops the message can traverse in the system. This maximum limit on the hops is generally know as *hop limit*. The notion of hop limit has been explored by researchers in the past. In dynamic source routing [4], the initiator of the route request specifies the maximum number of hops over which the packet may be propagated. The packets are initially assigned a hop limit of one. If no route reply is received for this request, a new route request with the hop limit set to a predefined "maximum" is sent out. In the store and flood protocol [2], a pre-defined higher limit on the number of hops over which a message may propagate is assigned.

Our contention is that, in an LSSN, some messages might be directed to a node only a few hops away while others might have to traverse the entire length of the system. Thus, assigning a predetermined maximum to the hop limit (based on the maximum number of hops possible in the system) would benefit only messages involved in end-to-end transmission. Besides, arriving at an optimum value for hop limit off-hand is a non-trivial problem. We believe, the problem is best left to the source of the message. In our system, the messages are assigned a hop limit proportional to the expected number of hops to the destination by the source node at generation. Hence, each message has a different hop limit depending upon the geographical distance it needs to traverse.

The hop limit assigned to a message is decremented at every hop it makes. A non-destination node, upon receiving a message for forwarding, rebroadcasts the message only if the hop limit has a non-zero value. The results of this simple heuristic on the performance of flooding are discussed in §5. The performance obtained in this way has proved to be as good as, if not better than, assigning a hop limit equal to the average hop length of successful messages in the system — a statistic that is difficult to calculate at the node level. The next section gives a brief description of the system.

# 4 System Description

The key attributes of the system we consider are:

1. The user has minimal control over the distribution of the sensors beyond selecting the general region of their deployment. The sensors are then randomly distributed over this geographical space.

2. The direct communication range of individual sensors is very small compared to the size of the deployment region. In general, any pair of sensors is separated by a large number of hops. The implication of this constraint is that all algorithms must be able to function primarily with local data, and should not have to rely on global information.

3. The sensors are anonymous and all addressing in the system is purely geographical, i.e., sensors are identified only by their location.

4. The sensors have minimal computational capability and limited energy. This means that there is no room for complex signal processing or optimization algorithms at the sensor level.

# 5 Simulation and Results

Nodes are randomly distributed with a uniform density in a non-discretized environment of linear size $l$. Each node can communicate only over a limited radius, $R$. We choose an $R$, which just about guarantees full connectivity for a given network size and density. Messages are generated according to a Poisson process with mean rate $\lambda$, with each message being assigned a unique identifier. Each node receiving a packet caches this identifier. A message whose identifier is found in the cache will not be broadcast on the assumption that it was either re-broadcast earlier or rejected for re-broadcast earlier. This loop control in the system is based on the principle that no message will be re-broadcast more than once. Moreover, the message generation rate is chosen such that the network is not overloaded.

The network was simulated for $l = 50$ units, $\lambda = 0.05$ to $0.24$, $R = 8$ and 3 units for 100 and 500 nodes respectively. The width of the major slot in CSMA-mb is 85, with $m = 30$ dedicated for contention. We chose two metrics to analyze the performance of flooding:

1. *Average Message Delivery:* The ratio of total messages generated to the number successfully received by the intended final destination.

2. *Average Wait Time:* The average time a message waits in a node's queue before being forwarded.

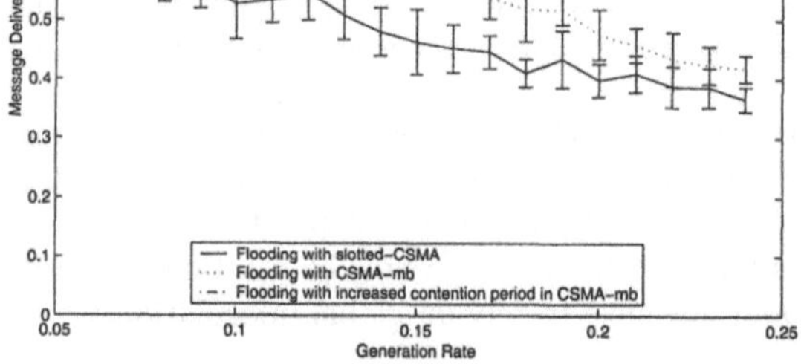


**Figure 3**: Slotted CSMA versus CSMA-mb:100 nodes

It is apparent from Fig. 3 that CSMA-mb is much superior to pure slotted CSMA in terms of delivery. In addition, Fig. 3 illustrates that by increasing the contention period for the nodes, a much higher delivery rate can be achieved.

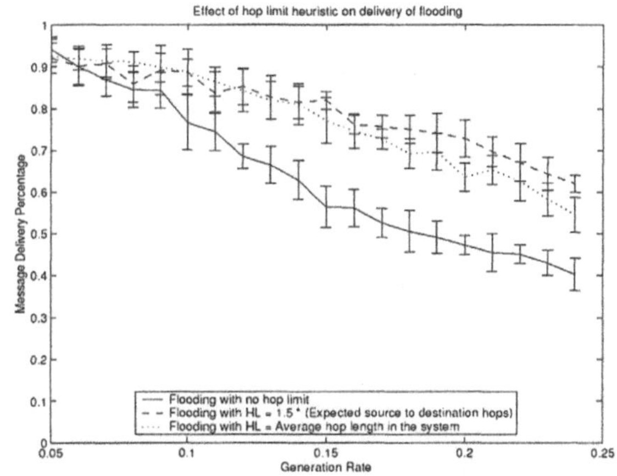


**Figure 4**: Effect of hop limit on delivery of flooding for 100 nodes

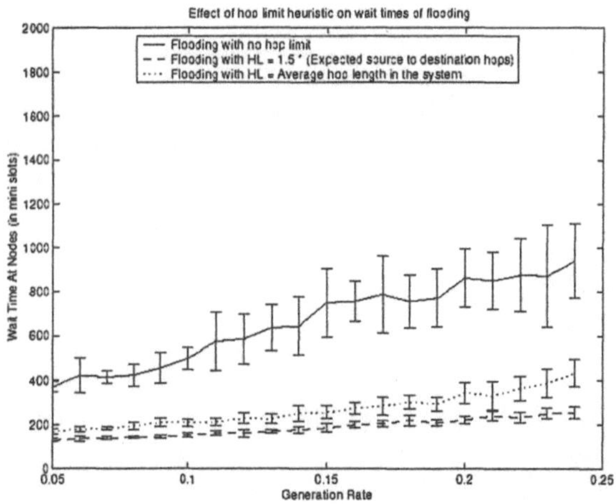


**Figure 5**: Effect of hop limit on wait times of flooding for 100 nodes

Fig. 4 and Fig. 7 show that a simple heuristic like hop limit can improve the performance of flooding dramatically. Moreover, our contention that assigning a hop limit based on the geographical separation of the source and destination of a message is a more intelligent approach than using a predefined maximum value appears to be correct. Very importantly, we have been ale to demonstrate that a *locally computable* hop limit heuristic can match — or even surpass — the performance obtained from the approach of using the average hop length of

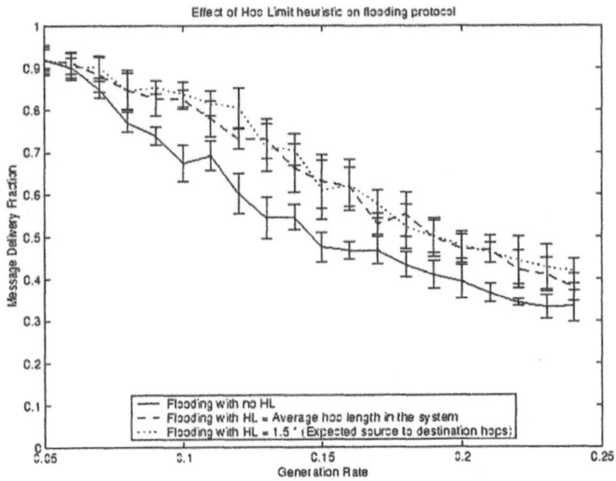


**Figure 6**: Effect of hop limit on delivery of flooding for 500 nodes

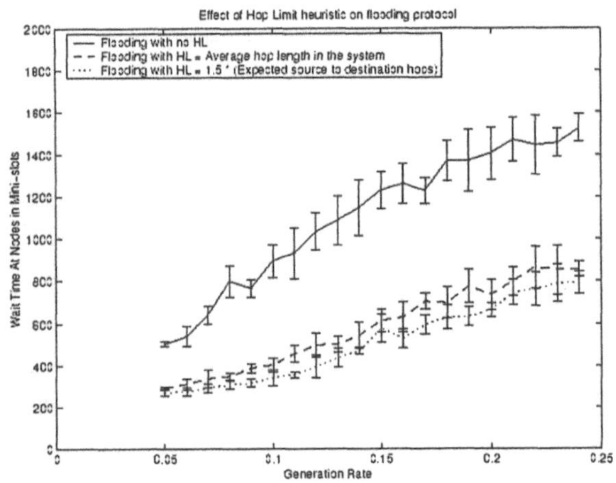


**Figure 7**: Effect of hop limit on wait times of flooding for 500 nodes

successful messages, which can only be calculated globally over the network.

# 6 Conclusion and Discussion

In this paper, we introduced CSMA-mb (Carrier Sense Multiple Access with mini-backoff), a persistent medium access protocol for broadcast-based networks.

CSMA-mb gives the nodes accessing the channel a fair chance to earn a slot by providing dedicated slots for contention. CSMA-mb trades-off wait time at a node to achieve superior delivery. Thus, messages in CSMA-mb reach the destination with increased latencies. We also presented a new heuristic for hop limit in the system and analyzed the effect of this heuristic on performance of flooding. Our method represents a practical approach to communication in large-scale sensor networks. The local nature of the method makes it much more scalable than traditional global methods, and, therefore, more suitable for self-organized complex adaptive networks as discussed in[1].

# Bibliography


[1] ARUMUGAM, Rajkumar "SCRIBE: Self-Organized Contention and Routing in Broadcast Environments", *MS Thesis*, University of Cincinnati, May, 2002

[2] HAGINO Hiroaki, Takahiro HARA, Masahiko TSUKAMOTO, Shojiro NISHIO. "Store and Flood: A Packet Routing Protocol for Frequently Changing Topology with AD-Hoc Networks" *IPSJ JOURNAL* Abstract Vol.41 No.09 - 010.

[3] JAIN, Rahul, Anuj PURI, and Raja SENGUPTA, "Geographical Routing Using Partial Information for Wireless Ad Hoc Networks", *Technical Report M99/69*, University of California, Berkeley Dec. 1999

[4] JOHNSON David B., David A. MALTZ, "Dynamic Source Routing in Ad Hoc Wireless Networks", *Mobile Computing*, Kluwer (1996), 153-181.

[5] KLEINROCK, Leonard, J. SILVESTER, "Optimum Transmission Radii for Packet Radio Networks or Why Six is a Magic Number", *Proceedings of the IEEE National Telecommunications Conference*, (Dec. 1978), 4.3.1–4.3.5.

[6] KLEINROCK, Leonard, F.A. TOBAGI, "Packet switching in radio channels: Part–I - carrier sense multiple access modes and their throughput-delay characteristics", *IEEE Transactions in Communications*, COM 23 (12)(1975), 1400–1416.

[7] PERKINS, Charles E., and Elizabeth M. ROYER, "Ad-hoc On-Demand Distance Vector Routing", *Proc. 2nd IEEE Wkshp. Mobile Comp. Sys. and Apps.*, (Feb. 1999), 90–100.

[8] PERKINS, Charles E., and Pravin BHAGWAT, "Highly Dynamic Destination-Sequenced Distance-Vector Routing (DSDV) for Mobile Computers", *Comp. Commun. Rev.*, (Oct. 1994), 234–244.

# Chapter 7

# Self-Organization of Connectivity and Geographical Routing in Large-Scale Sensor Networks

**Vinod Subramanian, Rajkumar Arumugam, and Ali A. Minai**
Complex Adaptive Systems Laboratory
ECECS Department
University of Cincinnati
Cincinnati, OH 45221
ali.minai@uc.edu

A large-scale sensor network (LSSN) is formed when a very large number of sensor nodes with short-range communication capabilities are deployed randomly over an extended region. The random distribution of nodes in an LSSN leads to regions of varying density, which means that if all nodes have an identical transmission radius, the effective connectivity would vary over the system. This leads to inefficiency in energy usage (in regions of unnecessarily high connectivity) and the danger of partitioning (in regions of low node density). In this paper, we propose a technique for adapting a node's transmission radius based on a node's local information. Through localized coordination and self-organization, nodes try to attain fairly uniform connectivity in the system to aid in efficient data messaging in the system. We study the benefits of network adaptation by incorporating it into an adaptive geographical routing algorithm called *corridor routing*. We present simulation results showing significant improvement in performance over routing algorithms that do not use network adaptation. We also propose and study several scenarios for network adaptation in the presence of node failures, and explore the effect of parameter variation.

# Introduction

A large-scale sensor network (LSSN) is formed when a very large number of sensor nodes with short-range communication capabilities are deployed randomly over an extended region. Unlike custom-designed networks, these randomly deployed sensor networks need no pre-design, need very little or no human supervision and configure themselves through a process of self-organization. These nodes are usually battery-operated and have limited computational capabilities. The vision motivating LSSN's is that using a large number of randomly deployed, locally communicating, cheap, disposable, simple and, therefore, individually unreliable nodes can produce more robust, flexible and scalable performance than a system of fewer, more powerful but expensive nodes without an appreciable loss of reliability. The focus, therefore, is in making the individual nodes as simple as possible while keeping the system "smart" [4, 6].

The problem of determining the optimal transmission radius for nodes in a wireless network is a key one for LSSNs. Researchers have argued that transmission power for each node must be sufficient to reach a "magic number", $n^*$, of other nodes in order to maximize forward progress of messages, minimize congestion, and minimize the possibility of partitioning. Kleinrock and Silvester [8] arrived at the "magic number" 6 as the optimal number of terminals to be covered by one transmission. In [13], Takagi and Kleinrock re-considered the problem and arrived at a new magic number nearly equal to 8. In LSSNs, a further reason to use no more transmission power than necessary is to conserve energy, since nodes have limited battery life.

In this paper, we propose a technique for adapting a node's transmission radius based on a node's local information. Through localized coordination and self-organization, nodes try to attain fairly uniform connectivity in the system. The proposed adaptation technique is a network level adaptation independent of the overlying routing algorithm. Our approach is based on the principle — derived from complex systems such as cellular automata and neural networks — that decisions made by nodes must be based on simple, *locally available* information rather than awareness of the wider network. In the approach proposed in this paper, we utilize the location information of nodes, which might be obtained using the global position system (GPS) or some other localization scheme [9, 3].

# System Model and Problem Statement

The network we consider comprises *n* nodes distributed uniformly in a two-dimensional region. Each node, limited by its energy capacity, can only communicate over a limited *transmission radius*. It is, thus, connected only to a small subset of the nodes, called its *neighbors*. In this paper, we assume that the nodes have variable transmitter power and receiving sensitivity, so a node's neighbors can change. The wireless network is, therefore, a random graph $G = (N, E)$, where the nodes are $N = 1, .., n$ and there is an edge $(i, j) \in E$ if node $i$ is a neighbor of node $j$ in the wireless network. Initially, we assume that all nodes have identical transmission radius, the graph is undirected.

Messages in our system are sent to geographical locations (coordinates) rather than to specific nodes. We assume that every node knows its own position. Using an initial setup process that is repeated periodically, each node in the system pre-determines the coordinates for which it is the closest node. Thereafter, it considers any message directed at any of those coordinates as intended for itself. Henceforth, we refer to this node as the *destination node*, even though the source node does not explicitly address the message to it. All data messages have a header with source and destination coordinates.

## Medium Access Control

The Medium Access Control (MAC) protocol we use, is a modification of the Slotted Carrier Sense Multiple Access (CSMA) protocol [12] and is called Carrier Sense Multiple Access with Mini-Back off (CSMA-mb) [1]. In CSMA-mb, the channel is divided into major time slots. Each major time slot is subdivided into a number of mini-slots. The size of each mini-slot is the maximum propagation delay among any neighboring nodes in the system. Nodes contend for channel access at the first mini-slot of every major time slot. The first $m$ mini-slots are reserved for contention in the system. At the beginning of every major slot, each node wishing to transmit backs off randomly to a value in the range of [1, m]. This process is termed as the mini-backoff scheme. At the end of its mini-backoff, the node senses the carrier again. If the channel is busy, it backs off to the next major slot; otherwise, it transmits. CSMA-mb thus follows a cautious persistent back off procedure. CSMA-mb has been shown to produce better results than slotted CSMA [1].

## Network Adaptation Technique

The random distribution of nodes in an LSSN leads to regions of varying density. The network adaptation technique discussed in this section intends to adapt a sensor node's transmission radius based on its 1-hop neighborhood in order to attain a pre-determined connectivity in the system, thereby making the network more efficient and robust. This algorithm is explained in detail below.

### The Algorithm

The nodes start off with a uniform high radius of communication so as to ensure good connectivity in the system. There exists a setup period in the system, known as the *system setup time*, during which the network adaptation takes place. During this period, nodes broadcast only *hello messages*. These control messages are of a smaller packet size compared to data messages. The system set-up time can be divided into two stages:

1. **Neighborhood Identification Stage:** During this stage of the system set-up time, nodes identify their neighborhood configuration using the hello messages. Each hello message carries information about its node's geographical location and its current transmission radius. When a node $A$ receives a hello message

from a new neighbor $B$, $A$ responds by sending an *acknowledgement (ack)* message as reply. This type of control messaging is termed *event-driven control messaging.* In addition to this, the system also has *periodic control messaging.* The rate of this messaging is denoted by $\lambda_{control}$. With the help of both the event-driven and periodic control messaging, nodes develop and update lists of their 1-hop neighbors (locations). The entire process helps a node determine its local connectivity.

2. **Neighbor Pruning Stage:** The second stage of the system set-up time involves the actual adaptation process.Using the information obained in the identification stage, nodes adapt their connectivity to attain a pre-determined number of neighbors denoted by $n^*$. By making a simple computation, nodes either decide to increase their transmission power by a pre-determined value or decide to decrease their transmission power so as to reach $n^*$ neighbors. Through a series of updated "hello" and "acknowledgement" messages, nodes self-organize their neighborhood and try to attain the $n^*$ neighbors.

The neighborhood identification stage and the neighbor pruning stage together constitutes the system set-up time in our algorithm. We choose the pre-determined number of neighbors ($n^*$) in our simulations following the analysis in [8, 13]. We also propose and study several scenarios for network adaptation in the presence of node failures, and explore the effect of parameter variation.

# Routing Algorithm

A primary requirement for any sensor network to function is the ability to transfer information between arbitrary points in the system, and in this paper, we study the benefits of network adaptation by incorporating it into an adaptive routing algorithm developed by us. This algorithm, called *corridor flooding*, is an intelligent broadcast algorithm that seeks to balance the redundancy and robustness of broadcast with the efficiency of directed routing. For LSSNs with a large number of very limited, unreliable nodes, we consider broadcast more appropriate than unicast because: 1) The computational resources needed for path discovery or creating and maintaining routing tables are not available to the nodes; 2) Most paths are too long (in terms of hops) to be discovered or maintained by a source node, and too transient to be worth discovering or maintaining. In this situation, a *broadcast-based* approach with its inherent redundancy of paths is the natural answer. However, it must be tempered by concerns of efficiency so that it is less wasteful than simple flooding.

Our approach is based on the principle that decisions made by nodes must be based on simple, *locally available* information rather than awareness of the wider network. Thus, each node in our network, upon receiving a message not intended for it makes a simple decision: Should it re-broadcast the message? This is simpler, for example, than determining which node to forward the message to. By intelligently controlling the basis for each local re-broadcast decision, we arrived at the *corridor flooding algorithm,* which scales effectively and is robust against node failure. The corridor routing algorithm utilizes the geographical location information of the nodes to route messages.

We define a corridor as an *imaginary* two dimensional region of a pre-determined width extending from the message source to the message destination. The pre-determined width of the corridor is termed the *corridor width.* As the imaginary corridor depends on the locations of the source and the destination of a message, the length and orientation of the corridor differs for different source-destination pairs. The corridor for a particular message is consistent at all nodes involved in routing the message because the nodes infer the same corridor based on the information received in the message itself. The source and destination location for every message is encoded in the message header.

When a node receives a message, it evaluates the position of the imaginary corridor and computes whether it lies within the region. If it lies within the corridor, the node re-broadcasts the message; otherwise, the message is discarded. This way the nodes *contain* the flood of a message to the message's imaginary corridor. The primary parameter in this algorithm is the corridor width, $w$. The corridor width determines the number of redundant paths to the destination, with the consequent robustness against node failure. A wider corridor increases the available redundancy of paths between source and destination, but increases congestion and wastes energy. A narrower corridor has the converse effect. Thus, determining an optimal corridor width is crucial to obtaining a robust and efficient system in the face of node failures.

The corridor routing algorithm exploits the *best available redundancy* in the system to achieve superior performance over broadcast-based flooding. In this paper, we show that the corridor routing algorithm outperforms even unicast-based routing protocols in networks with high rates of node failure.

## Comparison Protocols

In order to evaluate the performance of the corridor routing algorithm with the network-level adaptation, we have implemented three other routing protocols in this work. These are described below:

**Simple Flooding:** The classic flooding algorithm is the baseline case for comparison in our performance study. In this protocol, a non-destination node re-broadcasts any message it receives exactly once.

**Pseudo Unicast:** In *pseudo-unicast*, each non-destination node, upon receiving a message, unicasts the message to its *most forward neighbor* — the neighbor that provides the greatest progress towards the destination [13]. Nodes in pseudo-unicast do not employ any channel-reservation scheme [7, 2], and thus collisions are prevalent. If the most forward neighbor happens to be the immediate source node (evaluated with the information carried by the message), then the transmitting node chooses the *next* most forward node / the least backward neighbor to unicast the message. In essence, pseudo-unicast can be seen as *worst-case unicast* — the least sophisticated kind of unicast. It can also be seen as broadcast with forwarding only by a single node. By comparing this protocol with the corridor routing algorithm, one can quantify the value of path redundancy for the performance of the system: It shows whether a system with a single point of failure is sustainable under highly unreliable conditions or if it is necessary to preserve a certain degree of redundancy to achieve robustness under failures.

**Super Unicast:** *Super-unicast* works like pseudo-unicast, but assumes idealized conditions for communication. Hence, a transmission from a node $A$ to a neighbor $B$ is always successful, provided $B$ is in commission. The motivation for investigating super-unicast is to compare the corridor routing algorithm with the *best-case unicast*. Although pseudo-unicast eliminates some of the problems of broadcast, it still suffers from the classical *hidden* and *exposed* terminal problems. Researchers over the years have proposed many sophisticated unicast protocols to address these problems, e.g. MACA [7] or MACAW [2]. Instead of implementing any of these sophisticated protocols, we simply consider the case obtained when such protocols work perfectly, obtaining super-unicast.

We believe that by comparing our scheme with both pseudo-unicast and super-unicast, we cover the entire spectrum of unicast protocols.

# Simulation Framework

In this section, we describe the simulation framework and systematically present the results of performance of the corridor flooding algorithm in comparison with the simple flooding algorithm and unicast-based algorithms.

## Performance Metrics

The following performance metrics are used to evaluate the various protocols:

1. **Message Delivery** The ratio of the number of messages that reached the destination node to the number of messages generated in the system.

2. **Transmission Ratio** The number of messages transmitted in the system for every message generated. This measures the efficiency of energy usage.

3. **Message Wait Time** The average wait time of messages at every hop, expressed in terms of mini-slots.

4. **Hop Length** The average number of hops taken by messages that have reached the destination.

## Simulation Model

We performed a series of simulations with networks of 100, 500 and 1000 nodes. The nodes had an initial uniform transmission radius of 10 units, 4 units and 3 units for the 100-node, 500-node and 1000-node cases, respectively. They were distributed uniformly on a $50 \times 50$ square geographic region. The pre-determined corridor widths were chosen as 40, 16 and 12 for the 100-node, 500-node and 1000-node cases, respectively.

Messages were generated according to a Poisson process with message generation rate $\lambda$. The $\lambda$ values used in simulation ranged from 0.05 to 0.25 (per slot). Every message generated had a randomly generated source-destination coordinate pair. The simulation madc sure that the source and the destination were not identical. The $\lambda_{control}$

was fixed at 2.5 (per slot). We implemented a discrete-event simulator that kept track of message collisions in the system. The width of the major slot in CSMA-mb was 70 mini-slots, with $m = 30$ mini-slots dedicated for contention.

### Test Scenarios

The performance of the protocols was evaluated under two different scenarios:

1. *Infinite Energy Assumption:* Nodes in the system are assumed to have infinite energy for communication. The nodes are thus in operation throughout the run. We explore this scenario to analyze the best-case performance of the protocols.

2. *Random failures:* This scenario considers the possibility where certain nodes in the system fail. To model this, we divide the nodes in the system into *stable* nodes and *normal* nodes. Stable nodes are ideal, infallible nodes. These nodes are in commission throughout the lifetime of the system. In contrast, a normal node fails in each major slot with some probability $P_f$. This failure could be attributed to wear and tear or shortage of energy or even a timeout period. The node could become active and participate in routing in the next major slot. The concept of stable nodes was introduced to study the robustness of the system in relation to the degree of unreliability prevailing in the system.

## Simulation Results

Based on preliminary simulations, we determined that a value of $n^* = 8$ was appropriate for our networks. This was consistent with the recommendation in [8]. The choice of $n^*$ and the density of nodes then determined the default radius of transmission, and thus the energy usage of nodes.

### Adaptive Network: Infinite Energy Condition

Figures 1 and 2 show the performance curves of the 500 and 1000-node adaptive network under infinite energy conditions. In both cases, the broadcast-based algorithms out-perform the unicast-based algorithms in terms of message delivery. Even in larger networks, simple flooding performs exceptionally well in terms of message delivery. This is attributable to the better and fairly uniform connectivity in the system. Corridor routing performs either as well as or better than simple flooding. The savings in energy is more substantiable as the network becomes larger, emphasizing scalability. Also, there is a slight improvement in message latency in corridor routing as compared to simple flooding.

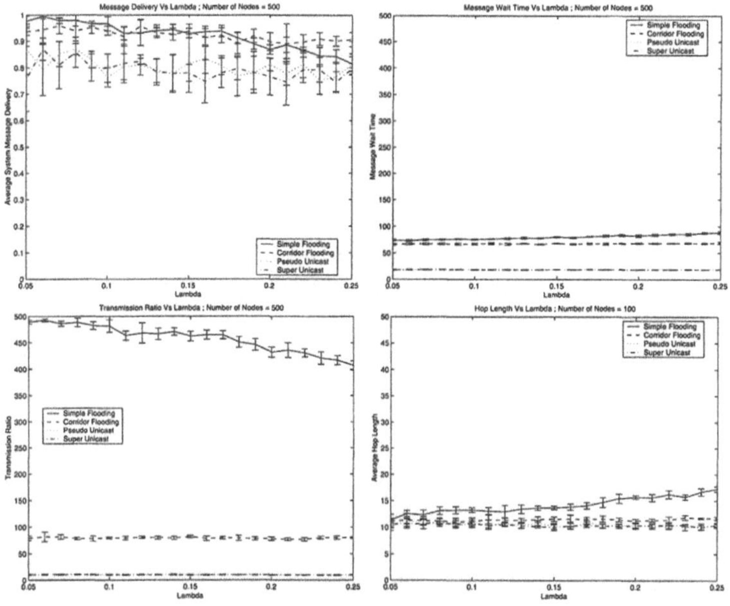


**Figure 1**: Adaptive network, infinite energy scenario: all protocols comparison with number of nodes = 500

## Adaptive Network: Random Failures

The first part of the random failure study involved keeping the message generation rate ($\lambda$) constant and varying the percentage of stable nodes. Normal nodes in the system fail with a probability of 0.5 (i.e., $P_f = 0.5$) in every major time slot.

Figure 3 shows the performance curves for a 100-node network with the message generation rate fixed at $\lambda = 0.2$. Corridor routing shows the best message delivery and wait times not much worse than the unicast protocols. Energy usage is much better than simple flooding. Overall, corridor routing clearly provides the best combination of reliability, efficiency and robustness. It can be argued that using handshaking mechanisms similar to MACA [7] or MACAW [2], one could achieve higher message delivery even under the effect of node failures. However, these methods are not suitable for the simple nodes and the failure-prone scenarios being studied here.

The second study of the random failure mode involved fixing the percentage of stable nodes in the system while varying the message generation rate ($\lambda$). Again, the normal nodes fail with a probability of 0.5 in each major time slot (i.e., $P_f = 0.5$). The percentage of stable nodes is fixed at 50% for the simulations and $\lambda$ is varied from 0.05 to 0.25.

Figure 4 shows the performance for a 500-node network. The adaptive unicast-based algorithms performs poorly in the face of such a high node failure rate (Only 50

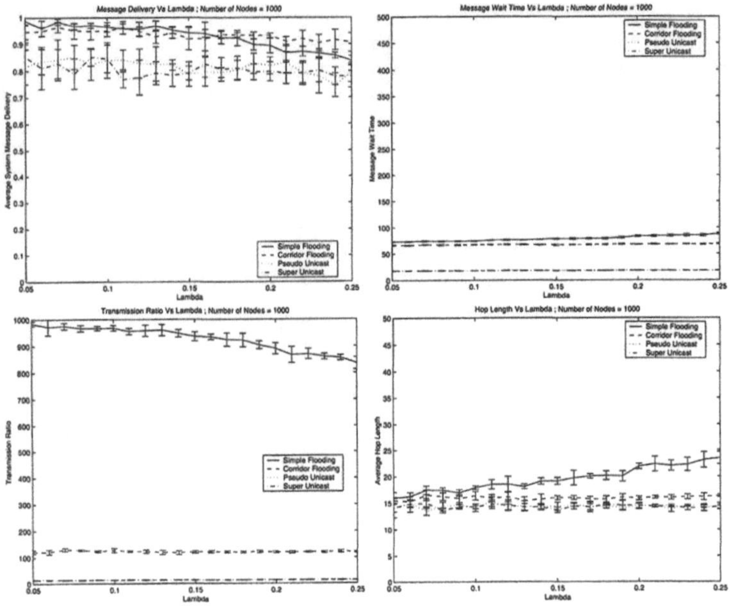


**Figure 2**: Adaptive network, infinite energy scenario: all protocols comparison with number of nodes = 1000

% of the nodes are stable). On the other hand, the broadcast-based protocols perform much better than their unicast counterparts under similar test conditions in terms of message delivery. While corridor routing performs almost as well as simple flooding in terms of message delivery, there is appreciable gain in the transmission ratio and a significant improvement in the message latency (product of message wait time and hop count). The unicast-based systems performs poorly because of the presence of a single point of failure of these algorithms coupled with the need for the messages to traverse a large number hops to reach their destination.

So far, in the analysis of the node failure modes, we used $n^* = 8$ as the pre-determined neighbor count. What happens if $n^*$ were to be increased? A higher $n^*$ would mean a greater radius of transmission for the nodes.

Next, we consider what happens if $n^*$ is raised beyond 8. One might assume that there would be a corresponding increase in the amount of traffic received at each node. However, in the presence of node failures, such an increase in a node's neighborhood actually helps performance. We chose $n^* = 10$ as the new pre-determined neighbor count for this analysis. The value of $n^*$ can be chosen based on the anticipated risk of failures of the nodes. In a scenario in which nodes have a high risk of failing, a slightly higher $n^*$ than the magic number of 6 or 8 [8, 13] would aid in the messaging.

Figure 5 shows the performance comparison of the simple flooding and corridor flooding protocols with $n^* = 8$ and $n^* = 10$ cases under node failures. In this scenario,

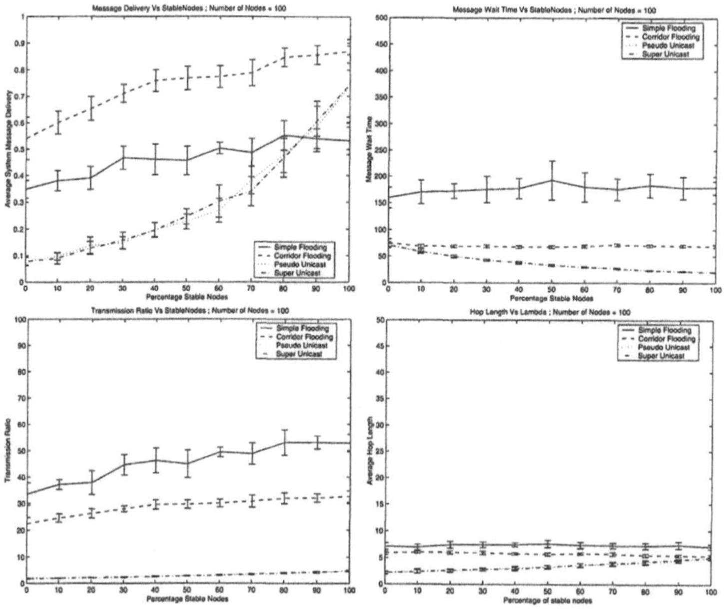


**Figure 3**: Adaptive network with node failures: all protocols comparison with number of nodes = 100 and $\lambda = 0.2$ (vs percentage of stable nodes. $P_f = 0.5$)

$\lambda$ was fixed at 0.1, the percentage of stable nodes was varied from 0 % to 100 % and the probability of failure of the nodes $P_f = 0.5$. The message delivery of simple flooding with $n^* = 10$ shows up to 35 % improvement over the message delivery of simple flooding with $n^* = 8$. Also, there is a slight improvement in the performance of the corridor routing protocol in terms of message delivery. The width of the corridor is fixed as 40 in both cases; the transmission ratio of both the corridor routing curves almost overlap due to this reason. There is a slight increase in the wait times for the $n^* = 10$ case because of the increase in contention at each hop.

Essentially, the better results for $n^* = 10$ reflect an increased redundancy of paths that compensates for node failures. Of course, there is a trade-off here — increasing the $n^*$ above a certain value will lead to performance degradation. We have not studied that phenomenon in this work.

# Summary and Future Work

The work reported here shows that adaptive broadcast-based algorithms perform significantly better than unicast-based algorithms in networks of simple nodes with high failure rates. The primary drawback of the network adaptation is the need for the system set up time. The system set up time uses up both bandwidth and energy in the system.

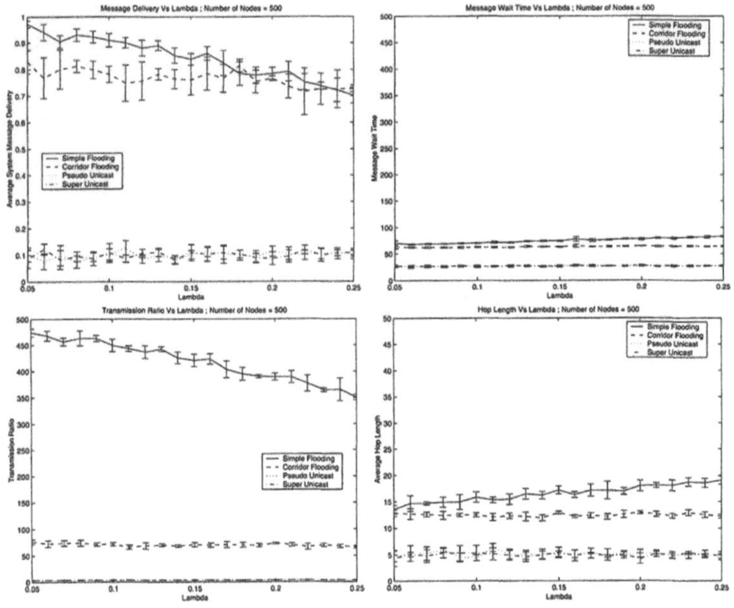


**Figure 4**: Adaptive network with node failures: all protocols comparison with number of nodes = 500, percentage of stable nodes = 50% and probability of failure $P_f = 0.5$

However, this would typically be only a small fraction of system lifetime.

Node mobility is an issue left un-explored. Actually, nodes in an LSSN are not expected to be highly mobile because of their limitations on energy, cost and complexity. Also, nodes in an embedded sensor network will generally be stationary. Nevertheless, as we rely on the geographical locations of the source and the destination nodes, node mobility would require that nodes continuously estimate their own coordinates in an absolute or relative system, e.g., using beacons or landmarks [9]. Also, the network adaptation technique would be drastically affected by node mobility because of the dynamics of node links.

# Bibliography


[1] Arumugam R. , Subramanian V. and Minai A. A., "Intelligent Broadcast for Large-Scale Sensor Networks", *Proceedings of the 4th International Conference on Complex Systems* (this volume), June 2002.

[2] Bharghavan V., Demers A. J., Shenker S. and Zhang L., "MACAW: A Media Access Protocol for Wireless LAN's", *Proceedings of the ACM SIGCOMM*, 212–225, August 1994.

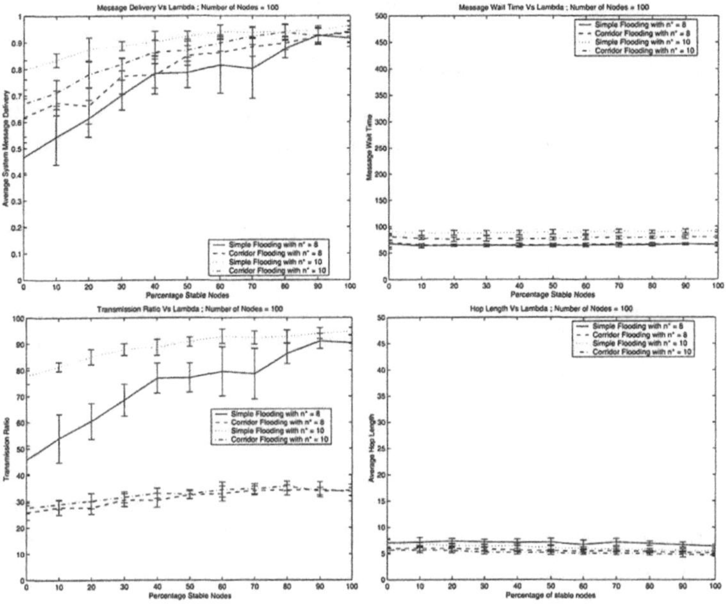


**Figure 5**: Adaptive network with node failures: comparison with higher $n^*$ - simple flooding and corridor routing with $n^* = 8$ and $n^* = 10$ and $\lambda = 0.1$ (vs percentage of stable nodes. $P_f = 0.5$)


[3] Bulusu N., Estrin D., Girod L. and Heidemann J., "Scalable Coordination for Wireless Sensor Networks: Self-Configuring Localization Systems", *Proceedings of the 6th IEEE International Symposium on Communication Theory and Application*, July 2001.

[4] Estrin D., Govindan R., Heidemann J. S. and Kumar S., "Next Century Challenges: Scalable Coordination in Sensor Networks", *Proceedings of the fifth annual ACM/IEEE International Conference on Mobile Computing and Networking*, 263–270, 1999.

[5] Johnson D. B. and Maltz D. A., "Dynamic Source Routing in Ad Hoc Wireless Networks", *Mobile Computing*. Kluwer Academic Publishers, 153–181, 1996.

[6] Kahn J. M., Katz R. H. and Pister K. S. J., "Next Century Challenges: Mobile Networking for Smart Dust", *Proceedings of the 5th Annual ACM/IEEE International Conference on Mobile Computing and Networking*", 271–278, 1999.

[7] Karn P., "MACA - A New Channel Access Method for Packet Radio", *ARRL/CRRL Amateur Radio 9th Computer Networking Conference*, September 1990.

[8] Kleinrock L. and Silvester J., "Optimum Transmission Radii for Packet Radio Networks or Why Six is a Magic Number", *IEEE National Telecommunications Conference*, 4.3.1–4.3.5, December 1978.

[9] Leonard J. L., "Large-Scale Concurrent Mapping and Localization", *Proceedings of SPIE Sensor Fusion and Decentralized Systems III*, 4196, 370–376, 2001

[10] Perkins C. E. and Bhagwat P., "Highly Dynamic Destination-Sequenced Distance-Vector Routing (DSDV) for Mobile Computers", *Computer Communication Review*, 24(4): 234–244, October 1994.

[11] Perkins C. E. and Royer E. M., "Ad-hoc On-Demand Distance Vector Routing", *Proceedings of the 2nd IEEE Workshop on Mobile Computing Systems and Applications*, 90–100, February 1999.

[12] Rom R., and Sidi M., "Multiple Access Protocols Performance and Analysis", *Springer-Verlag*, 1990.

[13] Takagi H. and Kleinrock L., "Optimal Transmission Ranges for Randomly Distributed Packet Radio Terminals", *IEEE Transactions on Communications*, 32(3), 246–257, March 1984.

# Chapter 8

# Managed Complexity in An Agent-based Vent Fan Control System Based on Dynamic Re-configuration

**Fred M. Discenzo**
Rockwell Automation, USA
Fmdiscenzo@ra.rockwell.com

**Francisco P. Maturana**
Rockwell Automation, USA
Fpmaturana@ra.rockwell.com

**Dukki Chung**
Rockwell Automation, USA
Dchung@ra.rockwell.com

## 1. Introduction

Developments in advanced control techniques are occurring in parallel with advances in sensors, algorithms, and architectures that support next-generation condition-based maintenance systems. The emergence of Multi-agent Systems in the Distributed Artificial Intelligence arena is providing important new capabilities that can significantly improve automation system performance, survivability, adaptability, and scalability. The new capabilities provided by multi-agent systems has shifted control system research into a very challenging and complex domain. A multi-agent system approach enables us to encapsulate the fundamental behavior of intelligent devices as autonomous components. These components exhibit primitive attitudes to act on behalf of equipment or complex processes. Using this approach, we have implemented a set of cooperating systems that manage the operation of an axial vent fan.

We have implemented a laboratory vent fan system (Figure 1) that operates autonomously as a fan agent in the context of a chilled water system comprised of other agents such valve, load, and pump agents.

**Figure 1.** Vent Fan Demonstration System

This paper presents the foundation technologies that are essential to realizing an adaptive, re-configurable automation system. The vent fan system serves to validate the agent methodology to manage the inherent complexity of highly distributed systems while responding dynamically to changes in operating requirements, degraded or failed components through prognostics, control alteration, and dynamic re-configuration.

The concepts above and new engineering developments have helped achieve new and important capabilities for integrating CBM technologies including diagnostics and prognostics with predictive and compensating control techniques. Integrated prognostics and control systems provide unique opportunities for optimizing system operation such as maximizing revenue generated for capital assets, maximizing component lifetime, insuring machinery survival or mission success, or minimizing total life-cycle costs.

## 2. Intelligent Machines

### 2.1. Machine Intelligence

Beginning with the famous Turing Paper about 50 years ago there has been an ongoing effort to understand cognition and intelligence in order to make machines more useful [Charniak 1985] [Nilsson 1980]. With a goal of enhancing the capability of machines, Artificial Intelligence (AI) includes techniques such as expert systems, fuzzy logic, artificial neural networks, and related model-based and model-free techniques [Charniak 1985]. Many of the automation successes reported apply biologically inspired adaptive and knowledge-based techniques to solve well-targeted, specific automation problems such as adaptive control, defect classification, and job scheduling to name a few [Zurada 1994].

The capabilities which may be provided by intelligent machines may be categorized based on the degree of embedded knowledge with the most capable systems employing real-time goal adjustment, cooperation, pre-emption, and dynamic re-configuration [Discenzo(a) 2000][Discenzo(b) 2000]. These capabilities may be effectively integrated in an agent-based system employing intelligent machines in a distributed automation system. This architecture built on a foundation of a society of locally intelligent cooperating machines provides a effective framework for the efficient and robust automation of complex systems.

### 2.2. Multi-Agent Systems

Our approach is to encapsulate the fundamental behavior of intelligent devices as autonomous components. The components employ models of primitive device behavior and enable agents to act on behalf of physical devices or complex processes.
The approach of establishing application-specific agent behavior in a reusable and scalable manner finds counterparts in other research activities such as Multi-agent Systems (MAS), Autonomous Agents, Flexible Manufacturing, and Virtual Enterprise [Shen 2001][Vasko 2000][Zhang 1999]. The evolution of web-based systems, which pushes Internet communication to accommodate agent-like services, also coincides with industrial automation requirements. Language and protocols can also serve as the basis to explore information exchange and resource discovery for agents (e.g., XML, SOAP, and UDDI [XML&DTD Specification][Xerces XML Parser][Universal Description UDDI White Paper, 2000][UDDI 2000][Simple Object Access Protocol, 2000]). In this paper, we focus on the requirements for the controller level.

The Foundation for Intelligent Physical Agents (FIPA) [FIPA][FIPA-OS] provides well-defined and widely accepted standards for multi-agent systems development that coincide with several of the premises established in this work. FIPA implementations accelerate the development of multi-agent systems.

Previous results demonstrated for autonomous control have been extended to incorporate agents that do planning, communication, diagnostics, and control [Vasko 2000][Maturana 2000][Maturana 2002]. Our focus is on the aggregation of autonomous behaviors and coordinated coalitions of smart resources. The approach we have taken is to establish a general architecture to deploy information agents for resource discovery in a highly distributed system. Resource capability may be established using an open system architecture that permits utilizing existing or future prognostic algorithms (Figure 2).

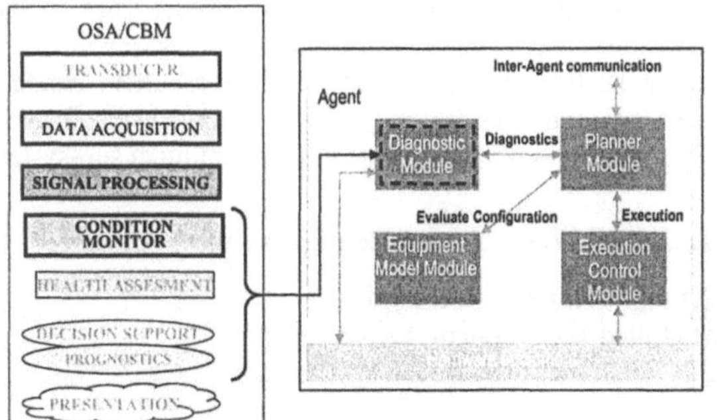


**Figure 2.** OSA-CBM in an Agent Framework

## 3. Chilled Water Demonstration System

### 3.1. Vent Fan Hardware Configuration

Fans and related air handling applications are often critical applications that occur in a wide range of automation, commercial, and military systems.

We have defined and implemented a vent fan system employing an integrated diagnostic / prognostic / controller system in an agent-based representation. This system has been implemented as one component in a hardware-in-the-loop simulation of a chilled water system. The vent fan system shown in Figure 3 is operated with a variable frequency drive (VFD) and programmable logic controller (PLC). During operation the system dynamically adjusts air flow based on state changes or new requirements and in collaboration with other agents operating in related parts of a chiller system.

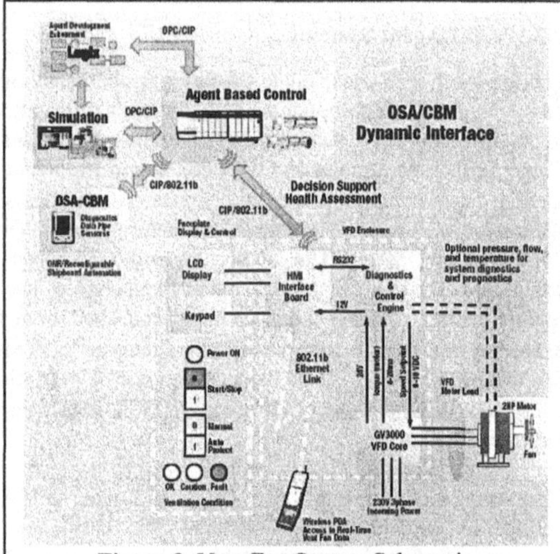

Figure 3. Vent Fan System Schematic

The system in Figure 3 includes a 3-phase, 2 hp, 230 volt, a-c induction motor coupled to a fan. The motor is an "Intelligent Motor" that includes embedded current, voltage, temperature, and vibration sensors and a processor to enable the motor to continually assess its own health. The motor and fan are mounted inside a 20" diameter steel tube as an integrated axial vent fan and the VFD motor controller is mounted on top of the structure. This hardware system is operated as an intelligent sub-system in a chilled water system.

### 3.2. Chilled Water System Configuration

#### 3.2.1 Agent types and capabilities

The chilled water system is simulated on a PLC and other than the fan system and a water pump system, all other components such as valves, pipes, and loads are simulated in a programmable logic controller. A model of the chilled water system is shown in Figure 4.

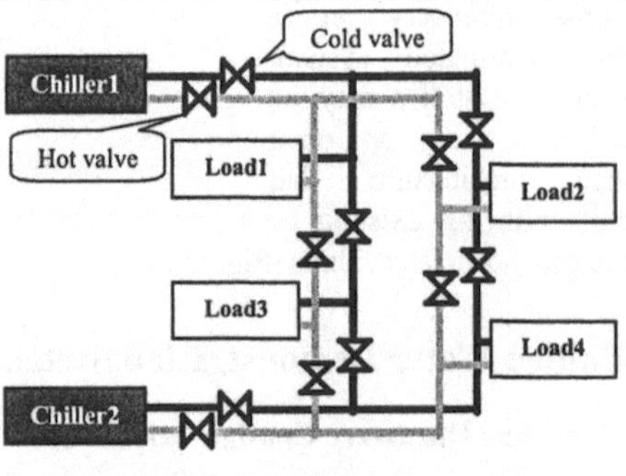

Figure 4. Chilled Water System Model

The CWS is comprised of a community of agents that control the physical equipment. In this system, the Chillers (Chiller 1 and 2) are chilled water plants, each containing a chiller unit, one or two pumps, an expansion tank, regulation valves, and flow and pressure sensors. The Loads (1, 2, 3, and 4) are associated with a heat generator and correspond to some vital or non-vital operation.

In addition, the Loads have an internal water circulation system that also provides heat-exchanging services. In this manner, the heat is evacuated from the Load area using convection heat exchange. Cold water is transmitted from the chillers into the loads through the main supply water pipes. Hot water is transmitted from the loads back to the chillers through the main return water pipes. The main water circulation system has valves to control water flow and to isolate pipe ruptures.

The function of Chiller1 is to provide water cooling. Inside the Chiller1 module there is a data repository and a connection to the physical vent fan. The vent fan increases the heat exchange capability of the chiller. The intelligent component contained in Chiller1 is a hybrid system comprised of simulation modules and the physical device. The vent fan system was connected to the simulation PLC using a wireless (IEEE 802.11b) network. Chiller2 provides water cooling capacity in simulation form only.

The main purpose of the agent system is to react to the physical system (simulation) changes to regulate the temperature of the load components. There are several considerations in this simple model:

- The agents need to be **goal-oriented** in specifying cold water requirements. This permits agents to avoid equipment damage while optimizing cold water usage.
- The agents need to **cooperate** to resolve priorities since cold water may not be 100% available to all units.
- The agents need to **adapt** their set points locally to respond to limited availability of cold water.
- The agents should be **aware** of dangerous situations occurring in the heating process to enable actions based on urgency and context.
- The agents should have a set of **self-regulating** algorithms to do low-level automatic control.

### 3.3. Chilled Water System Operation

The vent fan system with VFD may be operated in several modes one of which is under the control of an autonomous agent. In this mode, system operation changes dynamically and without regard to the set point speed or flow specified on the VFD.
The communication layer of the vent fan system handles the cooperation with the agent-based system and enables the vent fan system to be interrogated or controlled by agents. When the agent-based system starts up, the axial vent fan system is registered as a resource that can provide cooling capability. If an agent decides to use the vent fan system, it can start the fan and command to provide airflow. The low-level controller implemented in the DSP maintains air flow at a prescribed level based on system need or operating objectives. Fan speed is automatically increased or decreased to maintain flow and protect the system (e.g. self-diagnostics / prognostics) based on input from the vent fan agent. Later, the agent can check whether the fan system is providing the airflow it is asked to provide. The agent can also use the additional diagnostic information the fan system provides, such as system blockage. This

diagnostic information is published using OSA/CBM compliant XML pages. These XML pages are served by the web server which is running on the vent fan system.

### 3.4. Dynamic Virtual Clusters

We propose the study and implementation of a self-emerging organization based on the dynamic gathering of system capabilities. We conceive the system capabilities as fine granular and distributed throughout the system. The dynamic gathering of capabilities is a mechanism to achieve organizational reconfiguration. The mechanism uses dynamic clustering of agents representative of system capabilities for short periods. User and system tasks trigger the dynamic clustering. The task complexity determines the size and configuration of the clusters. Gathering multiple clusters forms a cooperation domain. An important characteristic of this mechanism is the capability of the system to aggregate resources as needed into the emergent organizations. This organizational feature provides an architecture that is robust and survivable. Several different mechanisms may be used to prescribe and manage the formation of the clusters and to coordinate communication among the components.

## 4. Dynamic Re-configuration

### 4.1. Open Systems Architecture for Condition-Based Maintenance

The utilization of open, industry standards for asset registry provides important capabilities for integrating operating information across a manufacturing plant and even across facilities. Recent developments have resulted in an Open Systems Architecture for Condition-Based Maintenance that provides a framework for the real-time integration of machinery health and prognostic information with decision support activities (Figure 2). This framework spans the range from sensor input to decision support (www.osacbm.org). This architecture specification is open to the public and may be implemented in a DCOM, CORBA, or HTTP/XML environment.

### 4.2. Integrated State Assessment and Control

There are significant operational and econimic benefits possible by closely coupling machinery health (e.g. diagnostics) and anticipated health (e.g. prognostics) information with real-time automatic control. Given a current operating state for both the machinery and the process we can drive the system to achieve a prescribed operating state at a particular time in the future. This future operating state may an improved state than would occur if we did not alter the control based on machinery health information. Furthermore, the future state we achieve is chosen to be optimal in some sense such as machinery operating cost, machinery lifetime, or mean time before failure. The prescribed operating state of a particular machine may be sub-optimal however, as part of an overall system, the system-wide operating state may be optimal with regard to energy cost, revenue generation, or asset utilization. Dynamic re-configurable agents are an effective framework for realizing the benefits of integrated prognostics and control.

### 4.3 Agent Organization

The chilled water simulation system includes fluid mechanics equations to represent the water circulation in the pipe network for both the cold and hot transmission lines. Each simulated component generates input and output (I/O) data for the equipment.

The simulation data is stored in the controllers inside data-tag symbols. The agents use the data tags as their interface with the physical equipment, The agents run in the controllers and utilize data-tags to manage the operating system and agent behavior using middleware [Maturana 2002]. Ladder logic algorithms synchronize the control events initiated by the simulation such as cooling load requirements, for response by the suitable suite of agents. Given a specific cooling demand on the chillers, the Chiller1 Agent may require an increase in the speed of the vent fan to increase the water cooling rate. This action corresponds to a first level reconfiguration. There are other levels of reconfiguration possible in this system. For example, the agents have been programmed to react to broken pipes events.

**4.4 Dynamically Created Clusters**

In order to form the emergent organization, the agents register their respective capabilities in the organizational knowledge agents known as "Directory Facilitators".

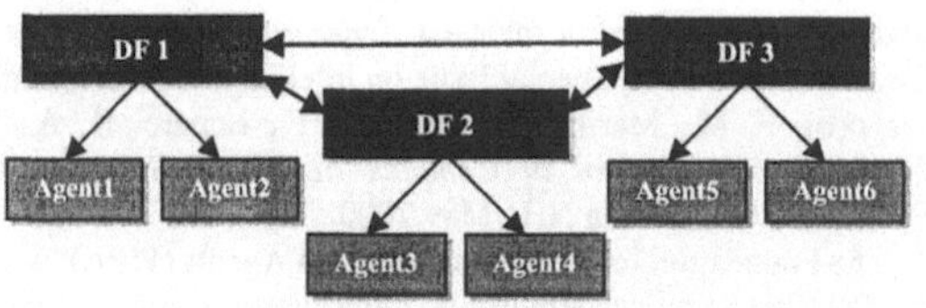

**Figure 5.** Distributed Structure of Directory Facilitators

To create decision-making organizations, agents need the ability to find suitable agents for particular tasks. There are several architectures to accomplish this search. Examples of such approaches can be found in the Blackboard, Matchmaker, Brokering System, Mediator, Federated, and Acquaintance Models Architectures [Shen 2001]. Consistent with the FIPA standard, we use a Matchmaker-based architecture which is based on Directory Facilitators (DF). On each capability request, the DF agent provides a list of agents that coincide with the requested capability. For this model, it is required that each capability provider register its capabilities with its local DF agent. Later, when a request for services is initiated, the DF agent acts as a passive agent because it only provides information services. The DF agent organization may hierarchial and agent capability registration may be made in a breadth-first or depth-first manner. Alternatively, to avoid propogation, information may only be propagated locally and new capabilitity requests will be processed by local DF agents who carry out meta-level communications to discover needed capabilities. DF agents then provide location and service capability information of remote agents to the initial requesting agent. Relevant information may be stored locally to enhance the mount of organization learning (Figure 5).

## 5. Opportunities & Challenges

Automation systems are being applied to more complex and safety-critical systems and there is a commensurate need for increased safety and reliable operation. The software content and complexity of automation systems continues to increase rapidly while software problems represent a leading cause of production breakdowns [Salimen 1992].

The new paradigm of Multi-agent / Holonic systems can provide significant benefits such as scalability, reliability, and survivability for complex critical systems. The broadscale deployment of intelligent devices, such as an intelligent vent fan system, utilizing this paradigm serves to reinforce these benefits and demonstrate an extremely flexible, adaptive, and robust system. Some of the challenges that remain include the need for a consistent framework and information model that will encompass integrating disparate agents, operating constraints, mission planning, dynamic optimization criteria, adaptive learning, and self-organizing behavior.

The technological developments cited above combined with intelligent devices implemented in an agent-based / Holonic framework promise to provide unprecedented capabilities for the automation of a broad class of complex systems. A society of autonomous, cooperating agents are well-suited to address dynamic, complex systems of the type that were previously only the domain of biological systems.

## References


Charniak, E., & McDermott, D, 1985, *Introduction to Artificial Intelligence*, Addison-Wesley

Discenzo(a), F. M., Unsworth, P. J., Loparo, K. A., Marcy, H. O., *Intelligent Motor Provides Enhanced Diagnostics and Control for Next Generation Manufacturing Systems*, IEE Computing and Control Engineering Journal, Special Issue on Intelligent Sensors, Summer 2000

Discenzo(b), F. M., Marik, V., Maturana, F., Loparo, K. A., *Intelligent Devices Enable Enhanced Modeling and Control of Complex Real-Time Systems*, International Conference on Complex Systems ICCS, Nashua NH., May 2000

FIPA, The Foundation for Intelligent Physical Agents (FIPA): www.fipa.org

FIPA - OS (Open Source), Emorphia: http://fipa-os.sourceforge.net

Jeff Bradshaw (Ed.), "Software Agents", MIT Press, Cambridge, 1997

Maturana F., Balasubramanian S., and Vasko D.,: An Autonomous Cooperative Systems for Material handling Applications. ECAI 2000, Berlin, Germany, 2000

Maturana F., Staron R., Tichy P., and Slechta P.: *Autonomous Agent Architecture for Industrial Distributed Control.* 56th Meeting of the Society for Machinery Failure Prevention Technology, Section 1A, Virginia Beach, April 15-19, 2002

Nilsson, N., 1980, *Principles of Artificial Intelligence*, Morgan Kaufmann (Los Altos)

Salimen, V, Verho, A, 1992, Multidisciplinary Problems in Mechatronics and Some Solutions, Computers Electr. Engng., Volume 18, Number 1, pp. 1-9

Shen W., Norrie D., and Barthès J.P.: Multi-Agent Systems for Concurrent Intelligent Design and Manufacturing. Taylor & Francis, London, 2001

Simple Object Access Protocol (SOAP) 1.1, W3C Note 08 May 2000: http://www.w3.org/TR/SOAP/

UDDI Data Structure Reference v1.0, UDDI Open Draft Specification 30 September 2000: http://www.uddi.org/pubs/DataStructure

Universal Description, Discovery and Integration, UDDI Technical White Paper, September 6, 2000: http://www.uddi.org

Vasko D., Maturana F., Bowles A., and Vandenberg S.: Autonomous Cooperative Systems Factory Control. PRIMA 2000, Australia, 2000

Xerces XML Parser, part of Apache XML Project: http://xml.apache.org

XML & DTD specification, World Wide Web Consortium: http://www.w3.org/XML

Zhang, X., Norrie, D.: Agentic Control at the Production and Controller Levels. IMS 99, Sept. 22-24, 1999, pp. 215-224, Leuven, Belgium 1999

Zurada, J. M., & Marks II, R. J., & Robinson, C. J., 1994, Computational Intelligence: Imitating Life, IEEE Press, ISBN 0-7803-1104-3, 1994, pp. 5-12

# Chapter 9

# Neural Net Model for Featured Word Extraction

**A. Das**
Department of Mathematics, Jadavpur University,
Calcutta 700 032India; email: atin_das@yahoo.com
**M. Marko**
Faculty of Management, Comenius University, Slovakia
**A. Probst**
Faculty of Mathematics, Physics & Informatics,
Comenius University, Slovakia
**M. A. Porter**
Center for Applied Mathematics, Cornell University, USA
**C. Gershenson**
School of Cognitive and Computer Sciences,
University of Sussex, U. K.

*Abstract:*
Search engines perform the task of retrieving information related to the user-supplied query words. This task has two parts; one is finding 'featured words' which describe an article best and the other is finding a match among these words to user-defined search terms. There are two main independent approaches to achieve this task. The first one, using the concepts of semantics, has been implemented partially. For more details see another paper of Marko et al., 2002. The second approach is reported in this paper. It is a theoretical model based on using Neural Network (NN). Instead of using keywords or reading from the first few lines from papers/articles, the present model gives emphasis on extracting 'featured words' from an article. Obviously we propose to exclude prepositions, articles and so on, that is , English words like "of, the, are, so, therefore, " etc. from such a list. A neural model is taken with its nodes pre-assigned energies. Whenever a match is found with featured words and user-defined search words, the node is fired and jumps to a higher energy. This firing continues until the model attains a steady energy level and total energy is now calculated. Clearly, higher match will generate higher energy; so on the basis of total energy, a ranking is done to the article indicating degree of relevance to the user's interest. Another important feature of the proposed model is incorporating a semantic module to refine the search words; like finding association among search words, etc. In this manner, information retrieval can be improved markedly.

## 1. Introduction

Huge collection of data in the form of article, paper, webpages, etc. on various topics are available on the internet. Search engines help us to retrieve information according to our own interest. Algorithms that govern this search operation are getting increasingly complex as complexity in searching terms as well as total volume of available data both are increasing very rapidly. Most evaluation in information retrieval is based on precision and recall using manual relevance judgements by experts. However, especially for large and dynamic document collections, it becomes intractable to get accurate recall estimates, since they require relevance judgements [May, 1999]. We study the relevance point particularly in anther paper by Gershenson at al. (2002).

Search operations are based on extraction of summary from available documents and finding those with good match with user-supplied query or search terms. Lycos, Alta Vista, and similar Web search engines have become essential tools for locating information on the ever-expanding World Wide Web (WWW). Underlying these systems are statistical methods for indexing plain text documents [Mauldin et al., 1994]

Achieving the task of finding degree of relevance of an article according to user's interest has two parts. The first one is to extract features of an article and the other is to match this feature against the user defined search terms. We discuss both the facets in detail along with our understanding of the problem in Section 2. Some existing methods will be discussed therein. Neural or connectionist computation and modeling is an emerging technology with a variety of potential applications such as classification, identification, estimation, etc [Pathak, 1995].
Here we propose a Neural Network (NN) model which will rank an article according to the degree of relevance of it against the user-defined search words. The proposed NN model does both parts of the task associated with such ranking as discussed earlier. Details of the NN model in given Section 3. Refinement of search word in terms of semantics is another important aspect to avoid unrelated retrievals and is discussed in Section 4. A few examples of searching and text summarization will be given at the end of this paper as appendix and are discussed in Section 5. It may be noted that these examples are being reproduced in relation to the paper; any sort of relative comparison is not particularly intended. Finally, we conclude with several pertinent remarks concerning this work

## 2. Extraction of Featured words

Featured words are those words that best describe the paper. Instead of using 'keywords' or extracting first few lines from an article cannot give a good representation of on what the article deals. In fact proper choice of features are most crucial as on these words- the searching is done. Research attempted to this direction carry various names; like 'Text Mining' defined as 'The knowledge discovery process which looks for identifying and analyzing useful information on data which is interesting to users from big amounts of textual data' [Atkinson, 2000] or Information Retrieval which includes tasks like automated text characterization, information extraction, text summarization etc.

Typically, the importance of a word depends both on its frequency in the document under consideration and its frequency in the entire collection of documents. If a word occurs frequently in a particular document, then it is considered salient for that document and is given a high score. In order to select such words, different approaches are in practice. For example, Justin et al. (1997) argues, that as HTML documents are very much 'structured' with tags indicating headings etc. compared to plain text, to weight parameters in the following way. Words in HTML fields should have weights (in the order of maximum to minimum) as follows

i) TITLE
ii) H1, H2, H3 (headlines)
iii) B (bold), I (italics), and BLINK
iv) underlined anchor text

Marcu (1997) represented 7 possible ways to determine the most important parts of a text, for example important sentences in a text i) contain words that are used frequently,(ii) contains words that are used in the title and section headings(iii) are located at the beginning or end of paragraphs iv) are located at positions in a text that are genre dependent—these positions can be determined automatically, through training techniques (v) use bonus words such as "greatest" and "significant" or indicator phrases such as "the main aim of this paper" and "the purpose of this article", while non-important sentences use stigma words such as "hardly" and "impossible" etc.

Fuka et al. (2001) showed that for important-term selection, many different techniques and heuristics that have developed are just a sub-set of more advanced methods originating in the field of pattern recognition.

In this work, we follow the approach of Jennings et al. (1999) to take into account the place of occurrence of a word while considering its weight. For example, a word in the title of an article carries more relevance than one used in the body text. Therefore, different weights are assigned to words according to their place of occurrence. But before that, we have to exclude preposition, article, etc., for example, exclude English words like "of, the, are, so, that, " etc. from the text which are frequently used in any article but are poor candidate to be selected as featured words. The filter shown in Fig. 1 excludes such words. See also the example of summarization, given in appendices, which shows that such exclusion does not affect the 'concept part' of summary of an article.

To form such a featured word list, the article is read first. To read a full paper as input to any processing module would be a heavy load of computation, so a choice of the first few hundred words can serve the purpose -also because beyond this limit, generally technical or scientific notations appear which are not relevant for the present purpose. Determining place of occurrence is a complex task in itself because of different file formats in use; so selection of about first 200 words is sufficient to include the title, author, affiliation and abstract etc. which will be given higher weights than those occurring in the body text.

## 3. Neural Network models

Application of neural network for data compression, feature extraction, and statistical clustering are decade old [Hammerstorm, 1994]. Neural networks are valuable on several counts than traditional programming approach because of the former's learning capacity and its capacity in producing different dynamical states regarding a system which in the present case concerns with total activation energy of excited NN model [Das et al., 2002]. Anderson et al. compared the advantages of using NN over other conventional programmatic approaches in details.

Joachimes (1999) experimentally studied new method leading to conclusive results in a WWW retrieval study and finds which search engine provides better results: Google or MSN (Microsoft Network) Search. Bruza et al. (2000) compared the search effectiveness when using query-based Internet search via Google and Yahoo search engines with the focus to evaluate aspects of the search process.

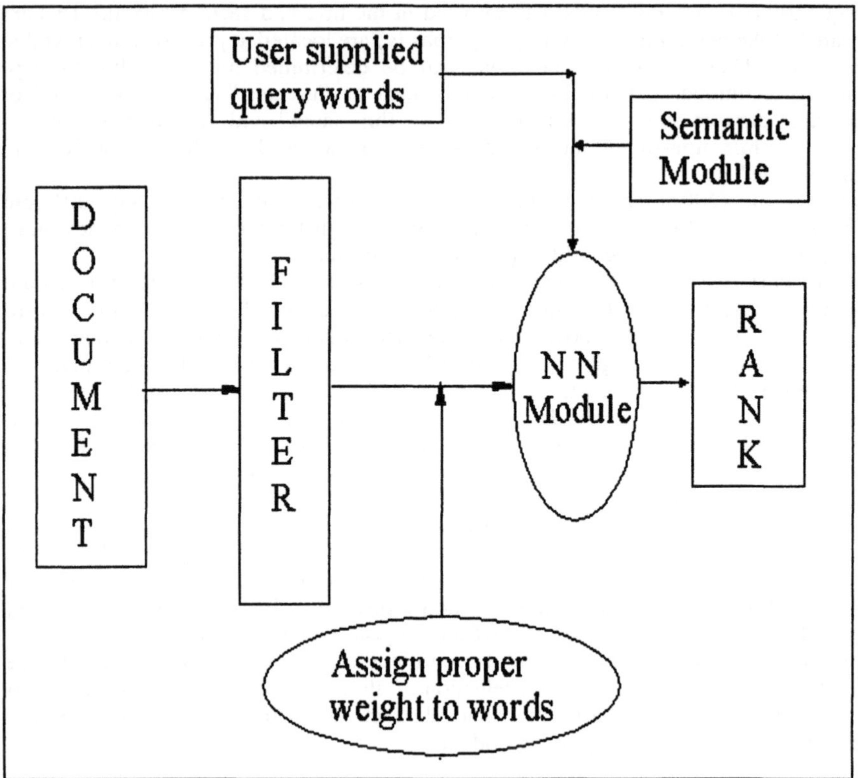


Fig 1: Showing the proposed scheme schematically.

We present here a theoretical neural network learning model with several nodes. The NN model consists of several nodes. Each node is assigned with a word (from the user defined search terms) and pre-assigned equal energy. The model reads an article and if a match between a node and a word is found, that node is fired and gets a higher energy. Here the strength of energy change depends on place of occurrence of the word in the article also. The process of firing will continue until the network settles down to an equilibrium state in accordance to its nodes. So finally we have a set of active nodes and we take the article ranking as the sum of the energy of all the active nodes. An article with higher energy clearly contains more of the search words in its featured word list. So this ranking will indicate the degree of association of the article and user's interest.

So the proposed NN model does both parts of the task associated with ranking an article according to user's interest as discussed earlier.

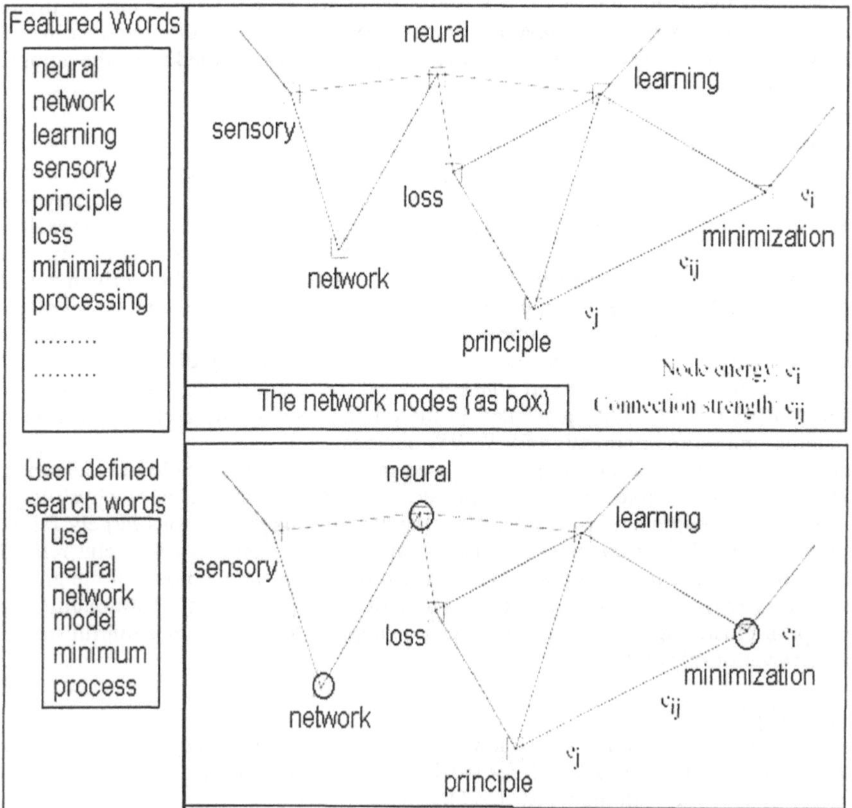


Fig2 : The network with fired nodes (as circles)

In our network model the strong linking of nodes represents the close relationship of words and their meanings (the issue of refining search words to clarify meaning is described in the section), and this restricts the connectivity of the network. Thus the number of nodes is restricted in that each node represents part of a user interest, and the connectivity is restricted in the sense that a connection is only established among the featured words.

## 3.1 Specific example of Complexity centered web

We like to give a specific example of how the present model can be used for developing a complexity based specialized media. In another paper of our group [Gerhenson et al., 2002] being presented in this conference, concepts and advantages of using specialized media to search WWW for information on specific scientific points (e.g. complexity) have been discussed, in comparison to using general (e.g. Goggle) or semi-specialized (e.g. NEC Research Index) media.

From the database consisting of issues published in last one year of Complexity Digest- a weekly science web-journal focussed on the study of complexity, most commonly occurring seventy words have been found (Gerhenson, ibid.). We can take most commonly used 25 words form this list and substitute them with user defined search words in the proposed model (refer to Fig. 1), keeping in mind that larger number will make the NN model too robust. Now the model has the task of finding match of these 25 words with featured word list of any article as discussed earlier. By this way, we can achieve the task of developing a specialized media for the complex community.

## 4. Refinement of search words

The search engines, at its core of functioning, in order to fetch increasing number of results refine search words. For example, searching for 'physics' includes 'physical chemistry' 'biophysics' 'physical strength', etc. Such refinements can sometimes drift far from the user's interest; sometimes producing meaningless results. In our proposed scheme we have included a semantic module for possible refinement of search words. The purpose is to include words for searching which are not present by the user supplied words. For example, search for word like 'disease' to be refined to include 'disease, illness, ailment' etc. Although it is a completely different domain of research to find the contexts and concept (called the 'meta-data'); some part can be offered off-hand and being presented in the other paper of our group dealing with semantics and ontology creation for WWW sources.

Another challenge is associating words used in such searching, called association". For example, words like 'neural' and 'physiology' can associate to 'neurophysiology' to give more comprehensive results. Existing search implementations largely fail at such situations. Using concept of clusters can solve this problem whereby related words are collected in a group and are activated by any of the members of the group. Agarwal (1995) gave such examples of semantic class generation. Additionally, the concept of clustering in semantic terms can be applied for a more meaningful search and retrieval of relevant results.

## 5. Examples

One can find the limitation of existing algorithms for search operation easily. For example of Google/Yahoo search for the word "ATIN" (name of one of the authors) has returned results with the word "LATIN" also.

In appendix A, we have reproduced an arbitrarily selected article, summarized the text two times independently with Microsoft Office summarizer and Copernic software. Comparison shows that MS summarization stresses on meaningful sentences where Copernic makes a concept list as well as a summary. In both cases, frequently used English words that we propose to filter are retained. This is an unnecessary computational overhead.

Referring to Appendix B where the original text is (manually) filtered and same two summarization tools are applied. Comparing concept list produced by Copernic as given in Appendix A and B, it is clear that excluding such words does not hamper process of concept building. This is an important justification of using filter in our model (refer to Fig. 1). It is also seen from Appendix B that summarization of filtered text results in something meaningless. There lies the need of NN to make a ranking in retrieving information from the article.

## Conclusion:

Extracting relevant features from text is an important challenge. In the present work, we showed that neural models can be used to preprocess an input article and match it with user-defined search terms. Though theoretical, this method can play an important role towards addressing the indicated challenge, as it is expected to lead to marked improvements in the search algorithms employed on the Internet.

**Acknowledgements**: We are grateful to Prof. Gottfried Mayer Kress, Penn. State Univ.,U.S. for giving important suggestions at various stages of this work. C. G. was partly supported by the CONACYT of México and by the School of Cognitive and Computer Sciences of the University of Sussex.

## References

Agarwal, R. 1995, *Semantic features extraction from technical texts with limited human intervention*, Ph.D. Thesis, Mississippi State Univ., (U.S.).

Anderson, D., & McNeill, G., 1995, *Artificial neural networks technology*, Kaman Sciences Corporation, U.S.

Becker, S., & Plumbley, M., 1996, *Unsupervised neural network learning procedures for feature extraction and classification,* Journal of Applied Intelligence, **6**, pp. 185-203.

Baeza-Yates, R. & Ribeiro-Neto, B., 1999, *Modern Information Retrieval.* Addison-Wesley-Longman,(Harlow, UK).

Bruza, P., McArthur, R., & Dennis, S., 2000, *Interactive Internet search: Keyword, directory and query reformulation mechanisms compared,* Proc. SIGIR Conference, Australia.

Das, A., Das, P., & A. B. Roy, 2002, *Chaos in a three dimensional neural network,* Accepted, Intl. J. Bifur. and Chaos.

K. Fuka K., & Hanka, R., 2001, *Feature Set Reduction for Document Classification Problems,* Proc. IJCAI-01 Workshop, Seattle, U.S.

Gershenson, C., Marko, M., Probst, A., Porter, M. A., & Das, A., 2002, *A Study on the Relevance of Information in Discriminative and Non-Discriminative Media,* Submitted, Inter Journal Complex System.

Hammerstrom, A., 1993, *Neural networks at work,* IEEE Spectrum.

Jennings, A., & Higuchi, H., 1992, *A User Model Neural Network for a Personal News Server,* IEICE Transactions on Information Interaction, VOL. 2, No. 4, 367-388.

Joachims, T., 2002, *Evaluating Search Engines using Clickthrough Data,* , Department of Computer Science, Cornell University, (US), Submitted.

Justin, E., Dayne, B., & Joachims, T., 1996, *A Machine Learning Architecture for Optimizing Web Search* , AAAI Workshop on Internet-Based Information Systems, (Portland, US).

Marcu, D., 1997, *From Discourse Structures to Text Summaries, The Proceedings of the ACL'97/EACL'97,* pages 82-88, Madrid, Spain.

Marko, M., Probst, A., Porter, M. A., Gershenson, C., & Das, A., 2002, *Transforming the World Wide Web into a Complexity-Based Semantic Network,* Submitted, Inter Journal Complex System.

Mauldin, M., & Leavitt, J., 1994, *Web agent related research at the Center for Machine Translation,* in Proceedings of the ACM Special Interest Group on Networked Information Discovery and Retrieval.

Phatak, D. S., 1995, *Fault Tolerant Artificial Neural Networks,* Proc. of the 5th Dual Use Technologies and Applications Conference, (N.Y., US).

## Appendix A

--------------(Original Article, First 200 hundred word chosen)-----------

Minimization of Information Loss through Neural Network Learning

M. D. Plumbley

Centre for Neural Networks, Department of Mathematics, King's College London, Strand, London, WC2R 2LS, UK

May 18, 1999

Abstract

Information-transmitting capability of such a neural network is limited both by constraints, such as the number of available units in a particular layer, and by costs, such as the average power used to transmit the information. In this article, we explore the concept of minimization of information loss (MIL) [2] as a target for neural network learning in this context. MIL is closely related to Linsker's Infomax principle [1]. By relating MIL to more familiar supervised and unsupervised learning procedures such as Error Back Propagation (`BackProp') [7] and principal component analysis (PCA) [1], we show how it can be used as a lingua franca for learning in all stages of a neural network sensory system.

1 Introduction
In recent years, a number of authors have used concepts from Information Theory to develop or explain neural network learning algorithms, particu-larly in sensory systems [1, 2, 3, 4, 5]. A neural network in a sensory system
is thought of as part of a communication system, transmitting Shannon
Information [6] about the outside world to further processing stages. The

------ Above summarized by MS Office97 (49 words; Original document 196 words).
Minimization of Information Loss through Neural
Network Learning
Strand, London, WC2R 2LS, UK
Information-transmitting capability of such a neural network is limited both all stages of a neural network sensory system.
Theory to develop or explain neural network learning algorithms, particu- A neural network in a sensory system
--------------same text summarized by Copernic Summarizer------------
**Concepts:**

PCA, lingua franca, network learning algorithms, communication system, transmitting Shannon, processing stages, minimization, loss, London, transmit, principle, MIL, sensory system, network learning, neural network.

**Summary:**

and by costs, such as the average power used to transmit the information loss (MIL) [2] as a target for neural network learning in this context.is closely related to Linsker's Infomax principle [1].
all stages of a neural network sensory system.
Information [6] about the outside world to further processing stages.

**Appendix B**

------------Filtered Original Text (see Fig. 1) --------------------------
Minimization of Information Loss through Neural Network Learning
M. D. Plumbley
Centre for Neural Networks, Department of Mathematics, King's College London,
Strand, London, WC2R 2LS, UK
May 18, 1999

Abstract
Information-transmitting capability neural network limited both constraints number available units particular layer, costs, such average power used transmit information. article explore concept minimization information loss (MIL) [2] target neural network learning context. MIL closely related Linsker's Infomax principle [1] relating MIL more familiar supervised unsupervised learning procedures Er- ror Back Propagation (`BackProp') [7] principal component analysis (PCA) [1], show how used lingua franca learning all stages neural network sensory system.

1 Introduction
recent years number authors used concepts Information Theory develop explain neural network learning algorithms, particu- larly in sensory systems [1, 2, 3, 4, 5].

neural network sensory system thought part communication system transmitting Shannon Information [6] about outside world further processing stages

-------Above summarized with MS Office97 (32 words; Original 172 Words)
Network Learning
Information-transmitting capability neural network limited both
article explore concept minimization information
loss (MIL) [2] target neural network learning context. MIL
all stages neural network sensory system.

neural network sensory system
---------Same text summarized by Copernic Summarizer

**Concepts:**

neural network, network learning, sensory system, MIL, principle, transmit, London, loss, minimization, processing stages, system transmitting Shannon, communication system transmitting, network learning algorithms, lingua franca learning, PCA.

**Summary:**

costs, such average power used transmit information.

# Chapter 10

# The distribution of agents to resources in a networked multi-player environment

**Robert L. Goldstone**
**Benjamin C. Ashpole**
Cognitive Science Program
Indiana University
{rgoldsto, bashpole}@indiana.edu

## 1. Introduction

A problem faced by all mobile organisms is how to search their environment for resources. Animals forage their environment for food, web-users surf the internet for desired data, and businesses mine the land for valuable minerals. When an individual organism forages their environment for resources, they typically employ a form of reinforcement learning to allocate their foraging time to the regions with the highest utility in terms of providing resources [Ballard 1997]. As the organism travels in its environment, it gathers information about resource distributions and uses this information to inform its subsequent movements. When an organism forages in an environment that consists, in part, of other organisms that are also foraging, then additional complexities arise. The resources available to an organism are affected not just by the foraging behavior the organism itself, but also by the simultaneous foraging behavior of all of the other organisms. The optimal resource foraging strategy for an organism is no longer a simple function of the distribution of resources and movement costs, but it is also a function of the strategies adopted by other organisms. Even if the resources are replenished with a constant rate, the optimal foraging strategy for an organism may fluctuate depending on the other organisms' behavior.

This research will collect a large volume of time-evolving data from a system composed of human agents vying for resources in a common environment with the aim of guiding the development of computational models of human resource allocation. We have developed an experimental platform that allows many human participants to interact in real-time within a common virtual world. Inside this world, two resource pools were created, and we recorded the moment-by-moment exploitation of these resources by each human agent. The research questions motivating the current study are: "How do resource foraging strategies unfold with time?," "Are there systematic suboptimalities in resource foraging?," and "How are foraging strategies affected by the distribution of resources and the agents' knowledge of the environment?"

### 1.1. Allocation of Energy to Resources

Determining how to allocate time and energy to resources is a deep issue that also has practical importance in biology, economics, psychology, and computer science. Biologists have long entertained the hypothesis that animals forage for resources in a near-optimal manner, given the distribution and replenishment rate of the resources, the animals' resource demands, and the energy expenditures required to harvest the resources [Krebs 1978]. Biologists studying individual foraging behavior have found many situations where resource patches are visited by animals with efficiency [Pleasants 1989]. For example, hummingbirds have been shown to sample the rate of return of flowers in a region, then forage one flower until the rate of return is below the average for all flowers, and then forage another flower with greater-than-average return [Pyke 1978].

Reinforcement learning is the theoretical study of systems that adapt their behavior to the contingencies of a situation. In reinforcement learning situations, a learner is not told which actions to take, but discovers these by observing the consequences of their actions over time [Sutton 1998]. A classic paradigm for studying reinforcement learning is the n-armed bandit problem, in which an agent is repeatedly offered a choice among n options [Berry 1985]. After each choice, the agent receives a quantifiable reward chosen from a stationary probability distribution based on their selected choice. The objective is to maximize the total amount of reward over a fixed time period.

In n-armed bandit problems, there is an "exploration versus exploitation" trade-off between learning about the true values of each choice and taking advantage of the knowledge gained. Selecting the option that has the largest expected reward given one's current estimates of the options will maximize the expected payoff for that one choice, but is not optimal in the long run. In the long run, it is optimal to explore the other options in order to improve the quality of one's estimate of their value. Holland's [Holland 1975] formal analysis of n-armed bandit problems led him to formulate a policy for allocating actions to options. He argued that resources should be allocated among the n choices such that the choices with highest expected payoffs are chosen with exponentially increasing probabilities. The exponential increase should be proportional to their estimated advantage compared to the other choices.

Psychological experiments often reveal human behavioral responses that contrast to the Holland's formal analysis. This analysis recommends very few choices of options with less-than-maximal expected payoffs after many trials have passed, particularly if there is a large discrepancy in payoffs among the different options. Instead, human participants often show probability matching, the tendency to distribute responses in proportion to their payoffs [Estes 1954; Grant 1951]. If Option 1 results in a payoff on 25% of trials and Option 2 results in a payoff on 75% trials, and if these payoffs are known to a participant, then the optimal strategy is to always select Option 2. The expected payoff for this strategy is .75. However, a participant who engages in probability matching will select Options 1 and 2 with probabilities of 25% and 75%, respectively, yielding an expected payoff of .25*.25+.75*.75=.625. In terms of the tradeoff between exploration and exploitation, people frequently engage in too much exploration, assuming that they know the payoff probabilities and know that these probabilities are fixed.

## 1.2. Foraging in Groups

The above analyses from biology, computer science, and psychology are based on a single individual harvesting resources without competition from other agents. This assumption is unrealistic in a world where agents typically congregate in populations. In fact, biologists have also explored the allocation of a population of agents to resources. One model in biology for studying the foraging behavior of populations rather than individuals is Ideal Free Distribution [Fretwell 1972]. According to this model, animals are free to move between resource patches and have correct ("ideal") knowledge of the rate of food occurrence at each patch. The animals distribute themselves to patches where the gained resources will be maximized. The patch that maximizes resources will depend upon the utilization of resources by all agents. Consistent with this model, groups of animals often distribute themselves in a nearly optimal manner, with their distribution matching the distribution of resources. For example, Godin and Keeleyside [Godin 1984] distributed edible larvae to two ends of a tank filled with cichlid fish. The food was distributed in ratios of 1:1, 2:1, or 5:1. The cichlids quickly distributed themselves in rough accord with the relative rates of the food distribution before many of the fish had even acquired a single larva and before most fish had acquired larvae from both ends. Similarly, mallard ducks distribute themselves in accord with the rate or amount of food thrown at two pond locations [Harper 1982]. In these situations, the population of animals behaves in the same way as a single, probability matching agent. If one patch produces two times the amount of resource as another patch, there will be two times as many animals at the larger relative to smaller resource.

Unlike probability matching in a single individual, matching distributions of resources with a population of animals is optimal. This may shed light on the discrepancy between formal analyses of n-armed bandit problems and empirical observations of probability matching. If an animal does not persistently, ideally exploit a set of resources without any competition from other animals, then that animal may not have evolved to make optimal choices in a laboratory situation where this is the task. Instead, this animal may be using the evolutionarily stable strategy of allocating its responses proportionally to the distribution of resources [Gallistel 1990]. If all animals do this, then as a collective they will optimally exploit the entire set of resources. Thus, probability matching as observed in the laboratory may reflect an adaptation that, when possessed by all animals in a group, allows the group to optimally forage.

The current research explores the foraging behavior of groups of humans. In general accord with an Ideal Free Distribution model, groups of fish, insects, and birds have been shown to efficiently distribute themselves. Are groups of people as rational as these animals? A computer-based platform for the foraging experiment allows us to manipulate experimental variables that would be difficult to manipulate in a more naturalistic environment. In the present case, we were interested in manipulating the relative outputs of the different resource patches and the knowledge possessed by the agents. With respect to the first manipulation, the central issue is whether systematic inefficiencies in a population of agents arise as a function of the distribution of resources. Although Godin and Keenleyside found that cichlids approximately distribute themselves in accord with the food resources, they also found small but systematic deviations. For the 1:1 and 2:1 ratios of resources, the fish distributed themselves with approximately 1:1 and 2:1 ratios respectively. However, for the 5:1 ratio, the fish distributed with a ratio of approximately 3:1. We will call this pattern of agents distributing themselves to resources in a less extreme (compared to a baseline of uniform distribution) manner than the resources themselves are distributed "undermatching." If undermatching occurs, then an adviser (perhaps a cichlid efficiency consultant) would recommend that a fish partaking of the resource with the lower output rate could increase its resource intake by moving to the resource with the higher output rate.

The second experimental variable that we manipulate is agents' knowledge of their environment and other agents. In Godin and Keenleyside's experiment with cichlids, every cichlid could see the other cichlids as well as the larvae resources at both ends of the tank. Gallistel argues that this information is important for the cichlids to distribute themselves rapidly in accord with the resource distribution. They are learning about the resource distributions by observing events that do not directly involve themselves. However, in standard reinforcement learning situations, an agent only has access to the outcomes of their own actions. They do not have access to the values of options not selected. Both situations occur naturally, and it is possible that the ability of a group to efficiently distribute itself to resources depends on the knowledge at each agent's disposal. It is plausible to hypothesize that as the information available to agents increases, the efficiency with which they can allocate their energy to resources increases, although there are paradoxical cases where more information seems to lead to worse decisions [Gigerenzer 1999].

## 2. Experiment in Group Foraging

We have developed a software system that records the instant-by-instant actions of individuals foraging for resources in a shared environment [Ashpole 2002]. Two resource pools were created with different rates of replenishment. The participants' task was to obtain as many resource tokens as possible during an experiment. An agent obtained a token by being the first to move on top of it. In addition to varying the relative replenishment rate for the two resources (50-50, 65-35, or 80-20), we manipulated whether agents could see each other and the entire food distribution, or had their vision restricted to food at their own location. We were interested in analyzing the resulting data for dynamics and sub-optimalities in the allocation of individuals to resources.

### 2.1 Experimental Methods

One-hundred and sixty-six undergraduate students from Indiana University served as participants in order to fulfill a course requirement. The students were run in 8 groups with 21, 20, 23, 19, 28, 12, 25, and 18 participants. Each student within a group was assigned to a PC computer in a large computer-based classroom with 40 computers. The experimenter controlled the experiment from another computer in the same room. The participants' computers were registered with the experimenter's computer using our developed software for sending messages over the internet to a set of networked computers.

Participants were told that they were being asked to participate in an experiment on group behavior. They were instructed to try to pick up as many "food" resources as possible by moving their icons' position on top of food locations. Participants were told that the food would occur in clumps, and if they learned where the productive clumps were that they could harvest more food. Participants were also told that there would be a lottery at the end of the experiment to win $10, and every piece of food that they collected during the experiment would be worth one lottery ticket. In this manner, participants were motivated to collect as many pieces of food as possible, even if at some point in the experiment they felt that it was improbable that they could collect more food than any other participant. Participants were told to only look at their own computer screen and not to talk with each other at any time.

Participants within a group co-existed in a virtual environment made up of replenishing resource pools and other human-controlled agents. The environment consisted of an 80 X 80 grid of squares. Participants controlled their position within this world by moving up, down, left, and right using the four arrow keys on their computers' keyboard. A participant could *not* walk off one side of the grid and reappear on the other. Each participant was represented by a yellow dot. In the "Visible" condition, all of the other participants' locations were represented by blue dots, and available food resources were represented by green dots. In the "Invisible" condition, each participant only saw their own position on the screen and any food gathered by that participant in the last two seconds. After this time interval, these consumed food pieces disappeared. In both conditions, food was gathered when a participant's position coincided with a piece of food.

Every experiment was divided into six 5-minute sessions. These six games consisted of all combinations of the two levels of knowledge (Visible versus Invisible) and the three levels of resource distribution (50/50, 65/35, 80/20). For each of the three distribution conditions, two resource pools were constructed, with center locations at reflections and rotations of the set of coordinates {40,15} and {15, 65}. A different reflection and rotation was used for each of the six conditions, with the result that the resource centers were approximately equally likely to be in each of eight possible locations, and the two centers within one session always had the same distance from one another. Two opposite orders of the six games were randomly assigned to the eight separate groups of participants. Each order began with a visible condition and then alternated between visible and invisible conditions.

The rate of distribution of food was based on the number of participants, with one piece of food delivered every 4/N seconds, where N is the number of participants. This yields an average of one food piece per participant per four seconds. When a piece of food was delivered, it was assigned to a pool probabilistically based upon the distribution rate. For example, for the 80/20 condition, the food would occur in the more plentiful pool 80% of the time, and in the less plentiful pool 20% of the time. The location of the food within the pool followed a Gaussian distribution with a mean at the center of the pool and a variance of 5 horizontal and vertical positions. Thus, the probability of food occurring in a given location was inversely proportional to the distance between the location and pool's center. Since multiple agents could occupy the same location without colliding, any food placed on such a location would be randomly assigned to one of the agents at that location.

Each of the six sessions lasted 5 minutes. Data were recorded every two seconds that included the positions of all agents, the number of food pieces collected by each agent, and the locations of uncollected food pieces. After all six sessions were completed, a winning participant was selected to receive $10 by an automatic lottery. The probability of a participant winning the lottery was equal to the number of food pieces they collected divided by the total number of food pieces collected by all participants.

## 2.2 Results

As a preliminary analysis of the distribution of agents across resource pools, Figure 1 shows the frequency with which each of the 80 X 80 grid cells was visited by participants broken down by the six experimental conditions. The brightness of a cell increases proportionally with the number of times the cell was visited. The few isolated white specks can be attributed to participants who decided not to move for extended periods of time. In Figure 1, the thick and thin circles show one standard deviation of the food distribution for the more and less plentiful resources, respectively. An inspection of this figure indicates that agents spend the majority of their time within relatively small regions centered on the two resource pools. The concentration of agents in pools' centers is greater for visible than invisible conditions, and is greater for the more plentiful pool. For the invisible conditions, there is substantial diffusion of travel outside of one standard deviation of the pools' centers. A Cochran's test for homogeneity of variances revealed significantly greater variability for the invisible than visible condition ($p<.01$), indicating greater scatter of agents' locations in the invisible condition. The agents approximately distributed themselves in a Gaussian form, with the exception of a small second hump in the frequency distribution in the invisible condition. The cause of this hump is that cells near the edges of the 80 X 80 grid that were close to pools were frequented somewhat more often than cells closer to the pool's center.

The dynamics of the distribution of agents to resources is shown in Figure 2, broken down by the six conditions. In this figure, the proportion of agents in the two pools is plotted over time within a session. Horizontal lines indicate the proportions that would match the distribution of food. An agent was counted as residing in a pool if he/she was within 5 food distribution standard deviations of the pool's center. This created circular pools that were as large as possible without overlapping. Agents who were not in either pool were excluded from Figure 2, and the total number of agents was normalized to exclude these agents. Figure 2 shows that agents roughly match the food distribution probabilities and that asymptotic levels of matching are found within 40 seconds even for the invisible condition. Although fast adaptation takes place, the asymptotic distribution of agents systematically undermatches the optimal probabilities. For the 65/35 distribution the 65% pool attracts an average of 60.6% of the agents in the 50-300 second interval, a value that is significantly different from 65%, one-sample T-test, $t(7)=3.9$, $\underline{p}<.01$. Similarly, for the 80/20 distribution, the 80% pool attracts only 73.5% of the agents, $t(7)=4.3$, $\underline{p}<.01$. For the 65/35 distribution, the asymptotic percentage of agents in the 65% pool in the visible condition (61.3%) was greater than in the invisible condition (60.0%), paired T-test $t(7)=2.4$, $\underline{p} < .05$. Likewise, for the 80/20 distribution, the asymptotic percentage of agents in the 80% pool in the visible condition (74.8%) was greater than in the invisible condition (72.2%), $t(7)=2.9$, $\underline{p} < .05$. Another trend, apparent in Figure 2, is that the proportions of agents in a given pool vary more sporadically with time for the invisible than visible conditions. This is because agents more often move themselves outside of a designated pool in the invisible condition. The percentage of agents falling outside of either pools during the interval 50-300 seconds were 1.2% and 13.4% for visible and invisible conditions respectively, paired $t(7)=8.4$, $\underline{p}<.01$.

A final analysis of interest explores the possibility of periodic fluctuations in resource use. Informal experimental observations suggested the occurrence of waves of overuse and underuse of pools. Participants seemed to heavily congregate at a pool for a period of time, and then became frustrated with the difficulty of collecting food in the pool (due to the large population in the pool), precipitating an emigration from this pool to the other pool. If a relatively large subpopulation within a pool decides at roughly the same time to migrate from one pool to another, then cyclic waves of population change may emerge. This was tested by applying a Fourier transformation of the data shown in Figure 2. Fourier transformations translate a time-varying signal into a set of sinusoidal components. Each sinusoidal component is characterized by a phase (where it crosses the Y-intercept) and a frequency. For our purposes, the desired output is a frequency plot of the amount of power at different frequencies. Large power at a particular frequency indicates a strong periodic response.

Any periodic waves of population change that occur in the experiment would be masked in Figure 2 because the graphs average over 8 different groups of participants. If each group showed periodic changes that occurred at different phases, then the averaged data would likely show no periodic activity. Accordingly, to produce the frequency plot showed in Figure 3, we conducted four steps of analysis. First, we derived a data vector of the proportion of agents in the more plentiful pool across a five-minute session for each of the 8 groups within each of the 6 conditions. Second, we detrended each vector by removing the best straight-line fit from it. If we had not done this, the resulting frequency plot would exhibit inappropriately high power in low frequencies, reflecting slow trends in population growth or decline. Third, we applied a digital Fourier transformation (Fast-Fourier Transform) on each detrended vector. Fourth, we created the frequency plots in Figure 3 by averaging the frequency plots for the 8 groups within a single condition.

The resulting frequency plots show significantly greater power in the low frequency spectra for invisible than visible conditions. In particular, the average power for frequencies up to .05 cycles is 3.4 and 1.1 for invisible and visible conditions, paired T-test $t(7)=4.1$, $\underline{p}<.01$. The power in lower frequencies is particularly high for the invisible condition with an 80/20 distribution. For all three invisible conditions, the peak power is at approximately .02 cycles/second. This means that the agents tend to have waves of relative dense crowding at one pool that repeat about once every 50 seconds. This 50 second period includes both the time to migrate from the first pool to the second pool and to return to the first pool. A pronounced power peak at lower frequencies is absent for the visible condition. One account for the difference in the two visibility conditions is that in the visible condition, each agent can see the whether other agents are closer than themselves to underexploited resource pools. The temptation to leave a dissatisfying pool for a potentially more lucrative pool would be tempered by the awareness that other agents are already heading away from the dissatisfying pool and toward the lucrative pool. However, in the invisible condition, agents may become dissatisfied with a pool populated with many other agents, but as they leave the pool they would not be aware that other agents are also leaving. It is less clear why periodic population waves should be greatest for the most lop-sided, 80/20 distribution, but one speculation is that the slowly replenishing 20% pool has the least power to attract a stable population of agents. A few agents can come into the 20% pool and quickly exhaust all of the fallen food resources. Then, the slow replenishment rate gives all of the agents in the pool little incentive to stay, and they consequently move to the 80% pool, until eventually the 20% pool becomes attractive again because of its low population density and cache of accumulated food.

## 3. Conclusions

The results of the present experiment indicate three systematic inefficiencies in the distribution of human participants to resources over time. First, participants exhibited undermatching in the sense that there were too many participants in the less plentiful resource and too few participants in the more plentiful resource. This undermatching is implied by comparing the distribution of agents to the distribution of resources, and is also directly indicated by the lower rate of food intake for agents in less plentiful compared to more plentiful resource. If this result proves to be general, then advice could be given to participants in similar situations to increase their use of relatively plentiful resources despite the possibly greater population density at those resources. Second, systematic cycles of population change are apparent whereby the migration of people from one pool to another is roughly synchronized. The problem with these synchronized population waves is that competitive crowding results in decreased food intake for those participants moving in the wave. Third, participants were more scattered than were the food resources. Both participants and food were distributed in a roughly Gaussian form, but the positional variance associated with participants was higher.

All three of these systematic inefficiencies were more pronounced for invisible than visible conditions. In fact, of the three inefficiencies described above, the only one that was appreciably large in the visible condition was the systematic undermatching. The influence of visibility suggests that an individual's knowledge of the moment-by-moment state of the environment and other agents can allow the group as a whole to avoid inefficient waves of resource under- and over-exploitation.

A fruitful extension to the current work would be the development of an agent-base computer simulation that models the large quantity of spatial-temporal population information amassed in the experiment. Such a model would need to incorporate a distinction between agents with and without vision. Blind agents might resemble standard reinforcement learning devices if supplemented by biases that undermatch distributions. Incorporating agents with vision is more challenging, raising important issues in generating expectations and planning. However, the current empirical work suggests that developing these more sophisticated agents with knowledge is worth the trouble. Knowledge of food distributions allows an agent to more effectively match those distributions, whereas knowledge of other agents allows an agent to more effectively decouple their responses from others.

## Acknowledgements

The authors wish to express thanks to Jerry Busemeyer, Jason Gold, Nathan Steele, and William Timberlake for helpful suggestions on this work. This research was funded by NIH grant MH56871 and NSF grant 0125287.

## References

Ashpole, B.C., & Goldsone, R.L., 2002, *Interactive Group Participation Experiments Using Java.* under review.

Ballard, D.H. 1997, *An introduction to natural computation*, MIT Press (Cambridge).

Berry, D.A., & Fristedt, B., 1985, *Bandit problems: Sequential allocation of experiments*, Chapman and Hall (London).

Estes W.K. and Straughan J.H., 1954, Analysis of a verbal conditioning situation in terms of statistical learning theory, *Journal of Experimental Psychology*, **47**, 225.

Gallistel, C.R., 1990, *The organization of learning*, MIT Press (Cambridge).

Gigerenzer, G., Todd, P.M., & ABC Research Group, 1999, *Simple heuristics that make us smart*, Oxford University Press (Oxford).

Godin, M.J., & Keenleyside, M.H.A., 1984, Foraging on patchily distributed prey by a cichlid fish (Teleosti, Chichlideae): A test of the ideal free distribution theory, *Animal Behavior*, **32**, 120.

Grant D.A., Hake H.W., and Hornseth J.P., 1951, Acquisition and extinction of verbal expectations in

situation analogous to conditioning, *Journal of Experimental Psychology*, **42**, 1-5.
Harper, D.G.C., 1982), Competitive foraging in mallards: Ideal free ducks, *Animal Behavior*, **30**, 575.
Holland, J.H., 1975, *Adaptation in natural and artificial systems*, MIT Press (Cambridge).
Krebs, J.R., & Davies, N.B., 1978, *Behavioural Ecology: An Evolutionary Approach.* Blackwell (Oxford).
Pleasants, J.M. 1989. Optimal foraging by nectarivores: a test of the marginal-value theorem. *American Naturalist*, **134**, 51.
Pyke, G.H. 1978. Optimal foraging in hummingbirds: testing the marginal value theorem. *American Zoologist*, **18**, 739.

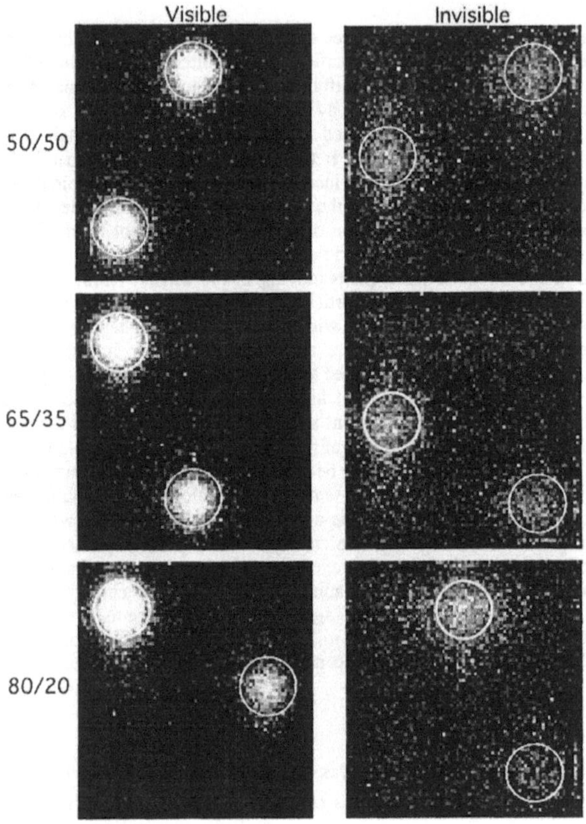


**Figure. 1.** A frequency plot of participants' visits to each grid square.

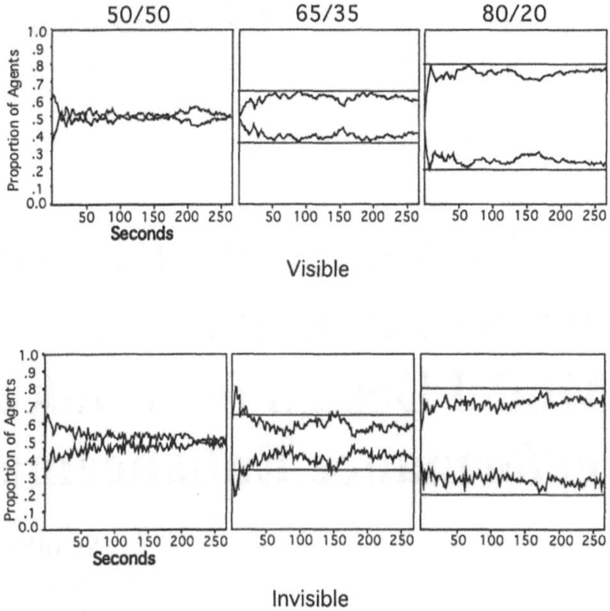


**Figure. 2.** Changes in populations over the course of an experimental session.

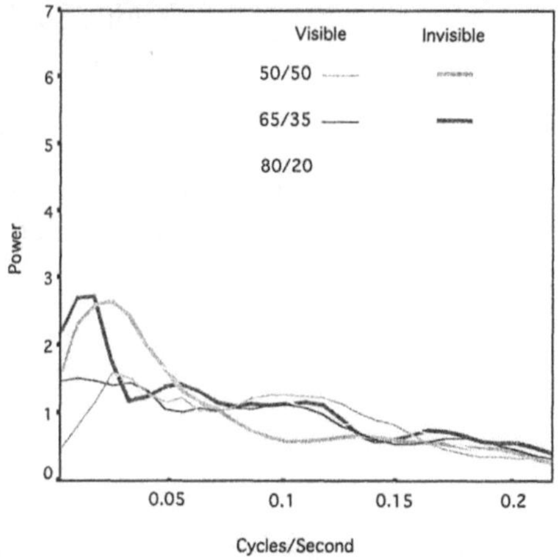


**Figure. 3.** A Fourier analysis of resource populations over time.

# Chapter 11

# Customer Relationship Management in Banking Sector and A Model Design for Banking Performance Enhancement

**Semih Onut**
**Ibrahim Erdem**
Yildiz Technical University
Dept. Of Industrial Engineering
Yildiz, 80750, Istanbul,Turkey
onut@yildiz.edu.tr, erdem@yildiz.edu.tr
**Bora Hosver**
Garanti Bank
Customer Relationship& Marketing Dept.
No :63, Maslak, 80670,Istanbul,Turkey
borah@garanti.com.tr

## 1. Introduction

Today, many businesses such as banks, insurance companies, and other service providers realize the importance of Customer Relationship Management (CRM) and its potential to help them acquire new customers, retain existing ones and maximize their lifetime value. At this point, close relationship with customers will require a strong coordination between IT and marketing departments to provide a long-term retention of selected customers. This paper deals with the role of Customer Relationship Management in banking sector and the need for Customer Relationship Management to increase customer value by using some analitycal methods in CRM applications.

CRM is a sound business strategy to identify the bank's most profitable customers and prospects, and devotes time and attention to expanding account relationships with those customers through individualized marketing, repricing, discretionary decision making, and customized service-all delivered through the various sales channels that the bank uses. Under this case study, a campaign management in a bank is conducted using data mining tasks such as dependency analysis, cluster profile analysis, concept description, deviation detection, and data visualization. Crucial business decisions with this campaign are made by extracting valid, previously unknown and ultimately comprehensible and actionable knowledge from large databases. The model developed here answers what the different customer segments are, who more likely to respond to a given offer is, which customers are the bank likely to lose, who most likely to default on credit cards is, what the risk associated with this loan applicant is. Finally, a cluster profile analysis is used for revealing the distinct characteristics of each cluster, and for modeling product propensity, which should be implemented in order to increase the sales.

## 2. Customer Relationship Management

In literature, many definitions were given to describe CRM. The main difference among these definitions is technological and relationship aspects of CRM. Some authors from marketing background emphasize technological side of CRM while the others considers IT perspective of CRM. From marketing aspect, CRM is defined by [Couldwell 1998] as ".. a combination of business process and technology that seeks to understand a company's customers from the perspective of who they are, what they do, and what they are like". Technological definition of CRM was given as ".. the market place of the future is undergoing a technology-driven metamorphosis" [Peppers and Rogers 1995]. Consequently, IT and marketing departments must work closely to implement CRM efficiently. Meanwhile, implementation of CRM in banking sector was considered by [Mihelis et al. 2001]. They focused on the evaluation of the critical satisfaction dimensions and the determination of customer groups with distinctive preferences and expectations in the private bank sector. The methodological approach is based on the principles of multi-criteria modeling and preference disaggregation modeling used for data analysis and interpretation. [Yli-Renko et al. 2001] have focused on the management of the exchange relationships and the implications of such management for the performance and development of technology-based firms and their customers. Spesifically the customer relationships of new technology-based firms has been studied. [Cook and Hababou, 2001] was interested in total sales activities, both volume-related and non-volume related. They also developed a modification of the standard data envelope analysis (DEA) structure using goal programming concepts that yields both a sales and service measures. [Beckett-Camarata et al. 1998] have noted that managing relationships with their customers (especially with employees, channel partners and strategic alliance partners) was critical to the firm's long-term success. It was also emphasized that customer relationship management based on social exchange and equity significantly assists the firm in developing collaborative, cooperative and profitable long-term relationships. [Yuan and Chang 2001] have presented a mixed-initiative synthesized learning approach for better understanding of customers and the provision of clues for improving customer relationships based on different sources of web customer data.

They have also hierarchically segmented data sources into clusters, automatically labeled the features of the clusters, discovered the characteristics of normal, defected and possibly defected clusters of customers, and provided clues for gaining customer retention. [Peppers 2000] has also presented a framework, which is based on incorporating e-business activities, channel management, relationship management and back-office/front-office integration within a customer centric strategy. He has developed four concepts, namely Enterprise, Channel management, Relationships and Management of the total enterprise, in the context of a CRM initiative. [Ryals and Knox 2001] have identified the three main issues that can enable the development of Customer Relationship Management in the service sector; the organizational issues of culture and communication, management metrics and cross-functional integration-especially between marketing and information technology.

## 3. CRM Objectives in Banking Sector

The idea of CRM is that it helps businesses use technology and human resources gain insight into the behavior of customers and the value of those customers. If it works as hoped, a business can: provide better customer service, make call centers more efficient, cross sell products more effectively, help sales staff close deals faster, simplify marketing and sales processes, discover new customers, and increase customer revenues.It doesn't happen by simply buying software and installing it. For CRM to be truly effective, an organization must first decide what kind of customer information it is looking for and it must decide what it intends to do with that information. For example, many financial institutions keep track of customers' life stages in order to market appropriate banking products like mortgages or IRAs to them at the right time to fit their needs. Next, the organization must look into all of the different ways information about customers comes into a business, where and how this data is stored and how it is currently used. One company, for instance, may interact with customers in a myriad of different ways including mail campaigns, Web sites, brick-and-mortar stores, call centers, mobile sales force staff and marketing and advertising efforts. Solid CRM systems link up each of these points. This collected data flows between operational systems (like sales and inventory systems) and analytical systems that can help sort through these records for patterns. Company analysts can then comb through the data to obtain a holistic view of each customer and pinpoint areas where better services are needed. In CRM projects, following data should be collected to run process engine: 1) Responses to campaigns, 2) Shipping and fulfillment dates, 3)Sales and purchase data, 4) Account information, 5) Web registration data, 6) Service and support records, 7) Demographic data, 8) Web sales data.

## 4. A Model Design for CRM At Garanti Bank

Garanti Bank, one of the leading banks in Turkey were looking at new ways to enhance its customer potential and service quality. Electronic means of banking have proved a success in acquiring new customer groups until the end of 2001. After then, a strategic decision was made to re-engineer their core business process in order to enhance the bank's performance by developing strategic lines. Strategic lines were given in order to meet the needs of large Turkish and multinational corporate customers, to expand commercial banking business, to focus expansion in retail banking and small business banking, to use different delivery channels while growing, and to enhance operating efficiency though investments in technology and human resources

To support this strategy Garanti Bank has implemented a number of projects since 1992 regarding branch organization, processes and information systems. The administration burden in the branches has been greatly reduced and centralized as much as possible in order to leave a larger room to marketing and sales. The BPR projects have been followed by rationalizing and modernizing the operational systems and subsequently by the introduction of innovative channels: internet banking, call center and self-servicing. In parallel, usage of technology for internal communication: intranet, e-mail, workflow and management reporting have become widespread.

**4.1.CRM Development**

To be prepared to the changing economic conditions and, in particular, to a rapidly decreasing inflation rate scenario Garanti Bank has started timely to focus on developing a customer relationship management (CRM) system. The total number of customers is presently around two millllions, but an increase to roughly three millions is foreseen as mergings with Osmanli Bank and Koferzbank are achieved and the present growth targets are reached.

The importance for the bank of managing the relationhips with their customers has been the drive of the joint projects that have been developed with IBM in the last three years. During the projects a number of crucial technological and architecture choices have been made to implement the entire process. Realizing the importance of customer information availability the first of these projects has focussed on the problem of routinely collecting and cleansing data. The project has been undertaken by the bank with the spirit that has characterized the whole CRM development. The project has promoted a massive involvment of the branches, namely of the portfolio managers and campaigns have been launched for popularizing among branch staff the importance of gathering and maintaining reliable customer data. Another set of methods have been tested for customer not included in portfolios (pool customers), such as mailing or distributing questionnaires in the branches or using automatic teller machines (ATM) and the call center. Methods for data checking and testing have been developed to be routinely employed by the bank's staff. Results obtained are very good: for portfolio customers data available are respectively 98% for the commercial ones and 85% for the retail ones. For pool customers availability goes down to 65%: this is a well-known phenomenon due to the loose relationship with the latter customers.

**4.2. Data Warehouse and Data Mining**

The Data warehouse is the core of any decision suppport system and hence of the CRM. In implementing its Data Warehouse Garanti Bank has selected an incremental approach, where the development of information systems is integrated with the business strategy. Instead of developing a complete design of a corporate Data Warehouse before implementing it, the bank has decided to develop a portion of the Data Warehouse to be used for customer relationship management and for the production of accurate and consistent management reports. Here we are not concerned with the latter goal, but are concentrating on the former.
The Data Warehouse has been designed according to the IBM BDW (Banking Data Warehouse) model, that has been developed as a consequence of the collaboration between IBM and many banking customers. The model is currently being used by 400 banks worldwide. The Garanti Bank Data Warehouse is regularly populated both from operational systems and from intermediate sources obtained by partial preprocessing of the same raw data.

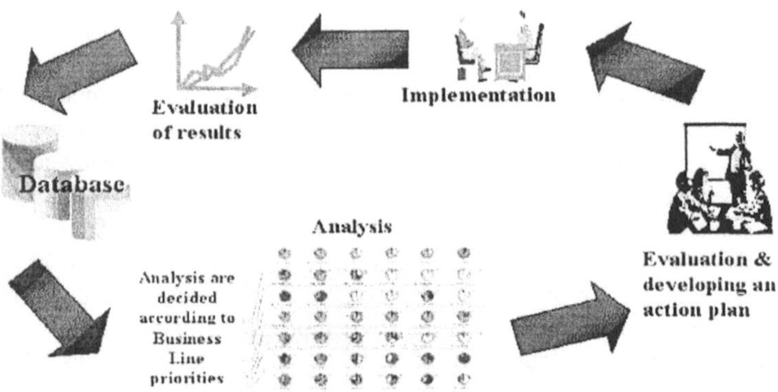


Figure 1. The process of Relational Marketing

It includes customers' demographic data, product ownership data and transaction data or, more generally product usage data as well as risk and profitability data. Most data are monthly averages and today's historical depth is 36 months starting from 1/1/1999 to 12/31/2001. As new data are produced they are placed temporarily in an intermediate, from which they are preprocessed and transferred to the warehouse. The importance of the Data Warehouse stems from the analysis of Figure 1. As a result of strategic decisions customer analysis is carried out by using data continuously updated as well the analytical methods and tools to be described later on. The CRM group analyzes results obtained and designs action plans, such as campaigns, promotions, special marketing initiatives, etc. Plans developed are then implemented by means of the several channels used by the bank to reach cusomers. Evaluation or results completes the cycle. The results become an integral part of the description of the bank-customer relationship in the warehouse. The learning cycle is thus complete and results obtained can be reused in future analyses and in future marketing plans. It is easy to understand that the Data Warehouse cannot actually be built 'once for all' but is a kind of living structure continuously enriched and updated as the Relational Marketing activity developes. OLAP (On Line Application Programming) analyses are developed by means of Business Object in its web version. CRM analysts use this tool to issue complex SQL queries on the Data Warehouse or on the Analytical Datamart and carry out mono and bivariate statistics on the whole customers' population or on selected groups. Figure 2 shows general structure of Relational Marketing Activity.

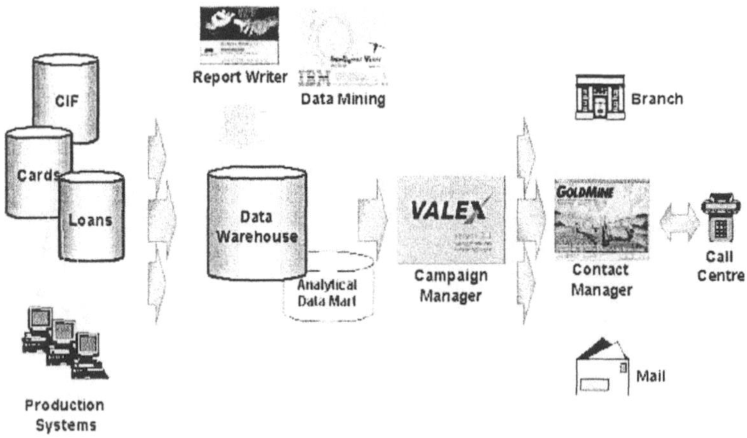


Figure 2.The Relational Marketing process is supported by a computing infrastructure where many software packages are integrated with the bank's information system.

Data Mining analyses are not carried out directly on the Data Warehouse, but on the Analytical Datamart by means of the software package IBM Intelligent Miner [Cabena et.al. 1999], using as a computing and data server the same mainframe where the Data Warehouse resides. Garanti Bank believes these tools and methodologies are a powerful competitive weapon and is investing heavily in the human resources needed to develop these analyses.

The Analytical Datamart is derived from the Data Warehouse through the following steps: 1)*Raw data processing:* data selection, data extraction, and data verification and rectification *2) Data modelling and variable preprocessing:* variable selection, new variable creation, variable statistics, variable discretization. The above processing, based on traditional data analysis, is strictly dependent on the investigated process; new variable creation, for instance, is intended to aggregate information contained in the raw data into more expressive variables. A simple example is the number of credit transaction on current account, that contains much of the information contained in the individual transactions, but is easier to analyze and represent. Variable discretization, based on the distribution of the original variables, is intended to generate categorical variables that better express the *physical reality* of the problem under investigation. The Analytical Datamart is customer centric and contains the following data:

1. demographic (age, sex, cultural level, marital status, etc.)
2. ownership of bank's product/services
3. product/services usage (balance, transactions, etc.)
4. global variables : profit, cost, risk, assets, liabilities
5. relationship with the bank: segment, portfolio, etc.

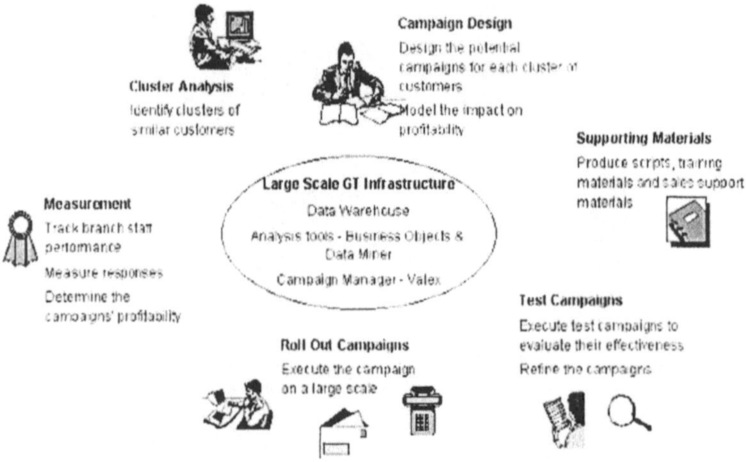


Figure 3. The marketing campaign process and the software supporting it.

## 4.3.Marketing Campaigns

After analyzing strategic and analytical CRM we concentrate here on the equally important operational aspects. Marketing Campaigns is the first method that Garanti Bank has used to test the above described analyses and techniques. The overall campaign process is reported in Figure 3, that shows that propensity determination and targeting are the first step of the whole activity.
A number of experimental campaigns have been designed and carried out to test the soundness of the approach before attempting a large scale roll-out. Experimental campaigns have addressed about 900 customers selected within six branch offices in Instanbul. An education process has been started by meeting sales forces in the branch offices, by distribution of an explanation booklet and by publishing on the Intranet a note explaining the whole process.
System interfaces have been modified in order to track the customers under promotion, as well as to enable salespeople in the branch office to complete the sales on promoted customers as well as to record the fact that the sale was a consequence of the promotion. The bank has so far used for promotion two channels: the salespeople in the branches and the call center. Each channel was used in four different campaigns.

| RETAIL | CHANNEL | SALES (%) |
|---|---|---|
| 1st Mutual Fund Campaign with 4 branches | BRANCH | 26 |
| 2nd Mutual Fund Campaign with 6 branches | BRANCH | 44 |
| Credit Card Campaign with 4 branches | BRANCH | 61 |
| E.L.M.A Campaign with Gayrettepe branch | BRANCH | 47 |
| Mutual Fund Campaign with 6 branches | TELE-MARKETING | 6 |
| Home Insurance Campaign | TELE-MARKETING | 21 |
| **COMMERCIAL** | | |
| Alternative Delivery Channels | TELE-MARKETING | 22 |
| Internet Branch Activation Campaign | TELE-MARKETING | 67 |

Figure 4. Summary table of experimental promotions deployed by Garanti Bank.

The activity of the call center was supported by the Goldmine software package, while the overall campaign management was achieved through Valex. This product used customer data stored in the Data Warehouse and at the same time manages itself a smaller local database, where campaign data are temporarily stored (list of customers, date of promotion, responses, etc.). These data must be *copied* manually into the Data Warehouse when the campaign has been completed.
Table in Figure 4 reports a summary of the campaigns deployed and the results thereof. By considering that the maximum response rate obtained by campaigns run by using traditional methods is normally around 1-2%, we see that the benefit of using targeted campaigns is very remarkable. As precise measurements on results obtained by the bank by using traditional methods are not available, we cannot accurately quantify the improvement. An indirect measurement of this improvement comes from the observation that in the timeframe where investment funds campaign were run, the branch offices included in the promotion record an increase in product purchase of about 214%, against an average 6% throughout the Bank.
Table in Figure 4 shows also that the results obtained by using the call center as a promotion channel are equally satisfactory, with the single exception of a low result obtained on the first campaign on investment funds.

## 5.Conclusions

Results obtained by extensive usage of customer data to develop and apply Relational Marketing have convinced the Garanti Bank to proceed along the line undertaken. As lists of customers eligible for four very important banking product/services are available, as above described, the following actions are now being deployed:

1. extension of promotions to a larger customer population by having sales people in the branches contacting progressively 15,000 customers
2. targeted campaigns through Internet and the call center for customers actively using one or both of these innovative channels for their banking operations.

The same approach is now being extended to small and medium businesses and to commercial customers. Moreover the analytical and strategic CRM cycle is being completed by developing an application analyzing customers' attrition and deploying strategies to reduce it.

**References**

Alis, O.F., Karakurt, E., & Melli, P., 2000, *Data Mining for Database Marketing at Garanti Bank, Proceeding of the International Seminar "Data Mining 2000"*, WIT Publications.

Beckett-Camarata, E.J., Camarata, M.R., Barker, R.T., 1998, *Integrating Internal and External Customer Relationships Through Relationship Management: A Strategic Response to a Changing Global Environment, Journal of Business Research*, 41, 71-81.

Berry, M.J.A. & Linoff G.S., 2000, Mastering Data Mining: The Art and Science of Customer Relationship Management, John Wiley & Sons, Inc.

Cabena, P., Choi H.H., Kim I.S., Otsuka S., Reinschmidt J. & Saarenvirta G., 1999, Intelligent Miner for Data Applications Guide, IBM Redbooks, SG24-5252-00.

Cook, W.D., & Hababou, M., 2001, *Sales Performance Measurement in Bank Branches, Omega, 29*, 299 –307.

Couldwell, C., 1998, *A Data Day Battle, Computing*, 21 May, 64–66.

Hosking, J.R.M., Pednault, E. P. D. & Sudan, M., 1997, *A statistical perspective on data minin, Future Generation Computer Systems*, 13, 17-134.

Mihelis, G., Grigoroudis, E., Siskos, Y., Politis, Y., & Malandrakis, Y., 2001, *Customer Satisfaction Measurement in the Private Bank Sector, European Journal of Operational Research*, 347-360.

Peppard, J., 2000, *Customer Relationship Management (CRM) in Financial Services, European Management Journal*, Vol. 18, No. 3, pp. 312–327,

Peppers, D., & Rogers, M., 1995, *A New Marketing Paradigm, Planning Review*, 23(2), 14–18.

Ryals, L., & Knox, S., 2001, *Cross-Functional Issues in the Implementation of Relationship Marketing Through Customer Relationship Management, European Management Journal*, Vol. 19, No. 5, pp. 534–542.

Yli-Renko, H., Sapienza, H.J., Hay, M., 2001, *The Role of Contractual Governance Flexibility in Realizing the Outcomes of Key Customer Relationships, Journal of Business Venturing*, 16, 529–555.

Yuan, S.T., & Chang, W.L., 2001, *Mixed-Initiative Synthesized Learning Approach For Web-Based CRM, Expert Systems with Applications*, 20, 187-200.

# Chapter 12

# How Complex Systems Studies Could Help in Identification of Threats of Terrorism?

**Czeslaw Mesjasz**
Cracow University of Economics
Cracow, Poland
mesjaszc@ae.krakow.pl

## 1. Introduction

The terrorist attacks of 11 September 2001 have once again reminded that limitations of prediction constitute the key factor in security theory and policy. In addition to fundamental epistemological limitations of prediction, one specific reason must be taken into account. Threats are often unpredictable, not because of objective barriers of their predictability, but due to the impact of social context and subsequent mental constraints, which make perception biased and eventually limit validity of prediction.

The main aim of the paper is to assess what could be the use of complex systems studies in improving predictive instruments of security theory.[1] The concepts of securitization of threats and vulnerability of social systems are used as a point of departure of analysis. Special stress is put upon analogies and metaphors drawn from complexity studies which are used in security theory and policy. Attention is focused on the threats of terrorism yet conclusions can be extended to other domains of security studies.

## 2. Concepts of Security

In the realist, and later, neorealist approach, military security is an attribute of relations of a state, a region or a grouping of states (alliance) with other state(s), regions, groupings of states. Security is viewed as an absence of threat or a situation in which occurrence of consequences of that threat could be either prevented or state (region, alliance) could be made isolated from that.

---

[1] According to the author's views, the term "complex systems science" seems premature.

Broadening the neorealist concept of security means inclusion of a wider range of potential threats, beginning from economic and environmental issues, and ending with human rights and migrations. Deepening the agenda of security studies means moving either down to the level of individual or human security or up to the level of international or global security, with regional and societal security as possible intermediate points. Parallel broadening and deepening of the concept of security has been proposed by the constructivist approach associated with the works of the Copenhagen School [Buzan *et al.* 1998].

Prediction, or identification of threats is undoubtedly the core issue in analytical approaches to security. It should make possible subsequent future actions ("emergency measures").

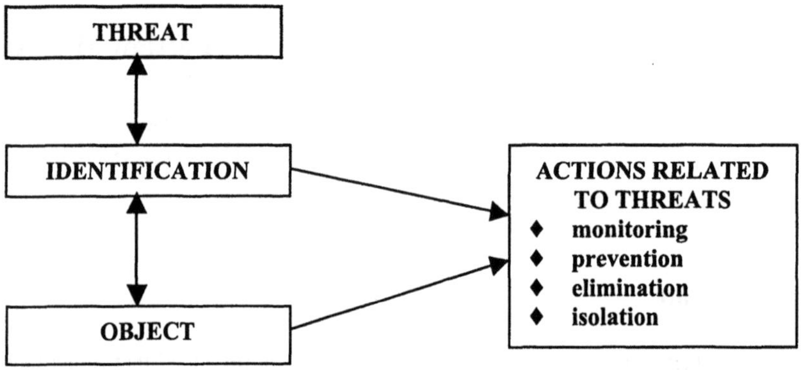


Fig. 1. The core of the concept of security

In order to preserve and develop analytical properties of the concept of security, a specific "middle-of-the-road", eclectic approach is proposed. It combines at least declarative objective value of widened neorealist broadened security concept with the constructivist, and at the same time "deepened" idea of security viewed as an "act of speech" [Buzan *et al.* 1998].

In the eclectic approach security is referred to the following sectors: military, economic, political, environmental and societal. Following Buzan *et al.* (1998] the concepts of existential threat and securitization are used. Any public issue can be securitized (meaning the issue is presented as an existential threat, requiring emergency measures and justifying actions outside the normal limits of political procedure). Security is thus a self-referential practice, because it is in this practice that the issue becomes a security issue - not necessarily because a real existential threat exists, but due to the fact that the issue is depicted as a threat [Buzan *et al* 1998].

A mirror concept of desecuritization can be defined as a process in which a factor (threat) which under one "speech act" compels extraordinary measures in another "speech act" is presented as not requiring such measures [Wæver 1995].[2]

The proposed approach to security allows to find a compromise between a neorealist assumption of predictability of objective threats, and constructivism's denial of any possibilities of prediction. Inspiration for solution of this dilemma can be found in other normative social sciences, especially in economics and management. Possibility of reconceptualisation of prediction in those disciplines was mainly resulting from abandonment of mechanistic views of social processes. Instead of refining extrapolations, computer models, scenarios and forecasts, stress is being put on the mechanisms of learning which result in making predictions, like in management [van der Heijden 1996], or in refining methods applied in forecasting like in the future studies [Glenn & Gordon 2006].

## 2. Applications of Complex Systems Studies in Security Theory and Policy

Some of the founders of the concept of "systems thinking" or "systems approach", etc. were also engaged in various domains of security-related studies - peace research - Anatol Rapoport, Kenneth Boulding, international relations - Karl W. Deutsch [1964], Morton Kaplan [1957]. Similarly many modern works on security exposes the links with systems thinking and complexity studies, for example:

1. Direct references - [Rosenau 1990, 1997], [Snyder & Jervis 1993], and indirect, introductory references [Kauffman 1993, 1995].
2. The links between broadly defined security and "complexity paradigm" presented frequently in a different manner, for example in a book edited by Alberts & Czerwinski [2002] reflecting the interest of the US military in universal issues of security and in specific, combat and command solutions; similarly the interest of the US policy making centers can be quoted, e.g. the RAND Corporation [RAND Workshop 2000].
3. Studies of specific security oriented issues with non-linear mathematical models [Saperstein 1984, 1991, 1999], [Alberts & Czerwinski 2002], [Center for Naval Analyses 2002].
4. Studies of specific security oriented issues with analogies and metaphors deriving from various non-linear mathematical models (complexity models) - see, for example, an interesting discussion on Clausewitz and complexity initiated by Beyerchen [1992], chapters in the book of Alberts & Czerwinski [2002], [Center for Naval Analyses 2002].
5. Applications of systems thinking and complexity studies in research on terrorism which have been given special attention after 11 of September 2001 [Ilachinski 2002].

Ideas originated in systems thinking and complexity studies are used in social sciences as models, analogies and metaphors - the term "models" is used herein only for mathematical structures. The main attention in theory, and particularly in practice,

[2] These concepts are supplemented with complacency - nonsecuritization of apparent threats (Buzan *et al.* 1998, p. 57).

is paid to analogies and metaphors deriving from systems thinking and complexity studies. They are treated as "scientific" and obtain supplementary political influence resulting from "sound" normative (precisely prescriptive) approach. In applications of models, analogies and metaphors the following approaches can be identified: descriptive, explanatory, predictive, normative, prescriptive, retrospective, retrodictive, control and regulatory.

A question is thus arising: How systems thinking and complexity studies can be applied in theory and policy of security? Focusing attention upon prevention of terrorism, the following problem has to be taken into account. Since security is an outcome of social discourse (securitization), it is necessary to ask how the ideas drawn from systems thinking and complexity studies can be used in all aspects of securitization - identification of threats and in their prevention.

The uses of ideas taken from systems thinking and complexity studies include mathematical models and analogies and metaphors. Bearing in mind the state-of-the-art of systems thinking and complexity studies their mathematical apparatus can be only partly helpful in anti-terrorist research. The recent review of writings on that topic only reaffirms this assertion [Ilachinsky 2002]. Some methods are useful - models taken from operations research and agent based modeling, specific computer based information systems for data storage and processing, e.g. fuzzy reasoning applied in cross-referencing, more advanced methods of "data mining", etc. could be helpful in enriching descriptions and analyzes. However, one can hardly expect that they will substantially help in predictions.

The second area of applications of systems thinking and complexity studies in security theory and policy is associated with the analogies and metaphors. Similarly as in management, in addition to description and analysis, they can be used even for prediction and identification of threats as well as in actions allowing to eliminate those threats.

Since anti-terrorist actions require separate research, the following problems of prediction must be studied:

1. Identification of threats of terrorism.
2. Securitization and desecuritization of threats of terrorism.
3. Methods of prediction of potential terrorist attacks - strategic and operational, day-to-day basis.

## 3. Terrorism and Vulnerability of Social Systems

### 3.1. How to Define Terrorism?

Definitions of terrorism vary widely and are contested as inadequate. Frequently the basic U.S. Department of State definition of terrorism is quoted "Premeditated, politically motivated violence perpetrated against noncombatant targets by subnational groups or clandestine agents, usually intended to influence an audience."

In another interpretations a terrorist action is the calculated use of unexpected, shocking, and unlawful violence against noncombatants (including, in addition to civilians, off-duty military and security personnel in peaceful situations) and other symbolic targets perpetrated by a clandestine member(s) of a subnational group or a clandestine agent(s) for the psychological purpose of publicizing a political or religious cause and/or intimidating or coercing a government(s) or civilian population into accepting demands on behalf of the cause [The sociology and psychology of terrorism.....1999].

Three categories of terrorism may be discerned:

- unexpected attacks against military forces in the combat situations under the conditions of an open military conflict - for the supporters, the "terrorists" are "guerillas", "patriots", "freedom fighters", etc.,
- the surprising terrorist attacks against unprepared civilians or members of armed forces out of combat ,
- a new category - mega-terrorism and genocide - the attacks on New York and Washington.

Terrorism understood herein solely as an attack on unprepared civilians or members of any kind of armed forces in out of combat situations, usually exploits vulnerabilities existing in social systems. The more open and complex are the social systems, the more vulnerable they are.

### 3.2. Vulnerability of Social Systems

The vulnerabilities the terrorist use result from inadequate securitization and/or from implementation of irrelevant preventive and/or protective measures. The attacks of 11 September 2001 could be accomplished not because of the lack of potential adequate measures of prevention but because of the insufficient securitization, and especially because of inadequate identification of vulnerabilities and impossibility of subsequent prediction.

The eclectic approach to security can be made more specific by narrowing the sense of securitization to identification of vulnerabilities. Securitization relates to a system and its environment while in a long-term prevention of terrorism it is of a special importance to identify weak points (loopholes) of the system which could be prone to any threats from inside and from outside.

According to the New Webster's Dictionary and Thesaurus of the English Language [1992] vulnerable means "...open to attack, hurt or injury; a vulnerable position - capable of being hurt or wounded (either because insufficiently protected or because sensitive and tender)".

Vulnerability of social systems, similarly as security, can be described in three ways: objective vulnerability, vulnerability as an effect of social discourse, and as a result of the eclectic interpretation

### 3.3. Identification and Possible Prevention of Threats of Terrorism

As an introduction to the discussion on the links between unpredictability and security, the case of the lack of prediction of the end of the Cold War can be quoted. Scholarly disputes after the collapse of the USSR, and especially the works by Gaddis [1992], reflected methodological weaknesses of social sciences. In the writings by Hopf [1993] and by Singer [1999], an opinion was expressed that unpredictability of the end of the Cold War was resulting from a social context of research and policy making.

As to show the limits of securitization treated as a social discourse, it is worthwhile to recall a quotation from Hopf [1993, p. 207]: "Can anyone imagine a senior international relations scholar applying to the Carnegie Endowment in 1972 for a research grant to investigate the conditions under which Moscow would most likely voluntarily relinquish control over Eastern Europe? Predicting the end of the Cold War was an unimaginable research question during the Cold War. But, had it been asked, no methodological or theoretical barriers would have stood in the way of formulating an answer".

By the same token, the terrorist attacks of 11 September 2001 can be reminded. Any security analyst trying to make a study of feasibility of a scenario of a mega-terrorist genocide attack before that date would likely have had problems with preserving his/her professional reputation. Attempts to sell a similar screenplay to the Hollywood filmmakers would have likely been vain as going beyond any acceptable limits of imagination.

Mechanisms of counterproductive self-imposed limits on prediction and/or rationality are not unknown. Tuchmann [1992] in her "March of Folly" showed several historical examples of that kind. Maybe the time is ripe to conduct study what social phenomena lead to the situations when societies and individuals blind themselves against major security threats.

It can be argued that systems thinking and complexity studies may provide some insights how to avoid that kind of inertia in perception and in prediction. They allow to enrich a broadly defined language of research with mathematical models as well as with analogies and metaphors.

In the attacks of 11 September the terrorists have exploited the vulnerabilities which generally resulted from inadequate securitization. It was a mixture of systemic properties of the US society - openness, purposive decision of relevant US institutions responsible for flight security, and likely, everyday negligence. Mistakes made by the US intelligence services and law enforcement agencies also were of a great importance.

The situation is easy to grasp in terms of securitization, desecuritization and vulnerabilities. An implicit decision was made by all parts concerned, who consciously left the room for a threat of a single hijacking and subsequent suicidal aggressive use by a terrorist or by a mentally impaired person. It was thus implicitly accepted that there could have been an incident involving a single passenger plane. But nobody expected four simultaneous suicidal attacks against the places of vital importance - it was beyond imagination in any potential securitization discourse.

## 4. Vulnerability of Social Systems, Terrorism and Complex Systems Studies

The most important problem with terrorism is that some vulnerabilities of social systems cannot be identified due to social and mental constraints of creative thinking (thinking about the yet non-existing, or even unthinkable), which are relatively well-known in social sciences.

The question is thus arising - how systems thinking and complexity studies could be helpful in predicting the unpredictable in the process of securitization of threats of terrorism?

Since applications of mathematical models in security studies have still limited usefulness attention must be paid to the use of concepts deriving from systems thinking and complexity studies as heuristic instruments - metaphors and analogies.

Securitization and identification of vulnerabilities are learning processes in which ideas taken from systems thinking and complexity studies have been already used in the micro-scale in management theory and practice, e.g. [Senge 1990], [van der Heijden 1996] and other following writings. Perhaps experiences stemming from preparation of objectivized scenarios will be equally useful [Glenn & Gordon 2006]. Using this inspiration it seems reasonable to ask what are the barriers hampering the social learning processes, or in other words, what are the limitations of securitization of vulnerabilities?

## 5. Barriers of Social Learning

Securitization of vulnerabilities of social systems should be helpful in better prediction and prevention of threats of terrorism. For the use of analogies and metaphors taken from systems thinking and complexity studies in the securitization, a specific condition must be fulfilled. Source field and target field have to be properly defined. It means that in addition to a trivial demand that ideas of complexity must be well understood by the users, there should not be any limitations in access to information concerning social systems (target field of metaphor). For mechanistic metaphor of social systems, the information about those systems could be limited to those features which could be used in building analogy/metaphor possessing the facets of a machine.

For the metaphor of learning system referring to the concepts of complex adaptive systems any limitations of access to information about the target field of metaphor make the reasoning worthless. Although information is always incomplete yet securitization of vulnerabilities (social discourse) is impeded by the specific barriers: political correctness and tabooization, and secrecy.

### 5.1. Political Correctness and Tabooization

In a simplified approach political correctness will be treated as attempts to limit normative description of a person, a group, an institution or a social phenomenon. The key issue with political correctness is that the source of the norms is frequently not well-known and properly identified. The rules of political correctness may frequently create their eigendynamik, independent from the initial intentions of the authors. Tabooization as understood herein goes even further, by making evaluation of some aspects of social life not only forbidden but even unthinkable.

It must be stressed that political correctness and tabooization are not necessarily connected with ethnicity or nationality. The group or the person must always have an equal chance for self-defense. The point is that it should not be accomplished by imposing new rules limiting language and eventually, thinking and behavior of the others, but rather through topical argument. Experience from complex systems studies tell us that systems evolve in a spontaneous manner. Why then some issues cannot be discussed openly, especially in the democratic countries which cherish the freedom of speech?

Tabooization and political correctness are methods of control of the language and subsequently of the actions. It is well-known that control of spontaneity of the language and of the actions is counterproductive in learning social systems. Complex adaptive systems provide additional arguments for this assertion even with the simplest metaphors of bifurcations and "organization at the edge of chaos".

Political correctness and tabooization are resulting from a static (mechanistic) view of society since some views are to be rejected without reflection upon their consequences. It is rooted in simplifying applications of mechanistic metaphors in which equilibrium and stability are viewed as dominant characteristics of social systems.

As to illustrate this conclusion it is easy to guess that anybody who would have earlier produced and publicized a scenario of 11 September would have been criticized both because of offending an ethnic group, and because of undermining reputation of the airlines and of the US government agencies.

Of course one may raise an argument that terrorist propaganda can also exploit the freedom of speech under the claims for adequate securitization discourse. The argument is however not symmetric. Terrorism by definition acts in a hidden form. Arguments supporting terrorism as a method of struggle for no matter what cause are never treated as legitimate. Therefore abandonment of limitations of open discussions caused by political correctness will not lead to open promotion of terrorism as a method of gaining political goals.

The lessons of 11 September 2001 can be reduced to the following conclusion. As to facilitate learning of social systems, the process of securitization/desecuritization of any threats must be free of any artificial semantic barriers. The discourse on vulnerabilities among the public, in the media and among the specialists must be moved to a higher level of thoroughness, specificity and objectivity.

**5.2. Secrecy**

Unrestricted access to information is a precondition of effective social learning. There exists a deficiency of discussion on terrorism conducted by any outsider who uses publicly available sources. Truly significant information, which could be important in anti-terrorist policy is obviously unavailable to any outsider who is but a careful "media watcher". Therefore any studies of terrorism with applications of complex systems could become a GI-GO (garbage in - garbage out) enterprise. It is obvious that this remark concerns all security studies, both technical and more general, accomplished by the less informed "analysts(?)".

Assuming hypothetically that, for instance, it is known to the government agencies in various countries that terrorist groups have successfully attempted to obtain anthrax or weapons of mass destruction from different sources, e.g. from the inventory of post-Soviet arsenals.

Those who have such knowledge can produce more relevant predictions yet frequently not for the public use. For the analysts who do not have access to this kind of information, the research on the links between complexity studies and security, including terrorism, provides but an inspiration for better informed analysts and policy makers.

The following example mixes up political correctness and secrecy. After the 11 September 2001 attacks the officials in NATO countries have warned against new terrorist attacks, especially chemical, biological and even nuclear ones.

So what happened to the voices about incomplete inventories of Russian nuclear arsenals, biological and chemical weapons after the collapse of the Soviet Union? Maybe in the early 1990s the terrorist groups exploited havoc in Russia and obtained what they needed? Maybe they keep mass destruction weapons as a means of the ultimate resort?

The fate of the post-Soviet mass destruction weapons after 1991 has become a textbook example of desecuritization of an objective security threat. Some writings on that topic were neglected for the sake of maintaining good relations with Russia and for the sake of not spreading panic among the Western societies, e.g. [Allison *et al.* 1996].

## 6. Vulnerability of Social Systems and Implicit Contracting

Since ordinary negligence, institutional problems and typical barriers to creative thinking are well-known in political science, security theory and policy, attention is paid to a specific phenomenon leading to negligence of some terrorist threats.

Humans and social systems are limitlessly vulnerable to various unpredictable and unthinkable threats. However, in some cases vulnerabilities to terrorist attacks are well-known but due to physical limitations and/or conscious decisions a kind of "implicit contracts" are made with potential terrorists.

As an example the protection of airports can be quoted. Before the 11 September 2001 it could not be ruled out that some day a suicide terrorist attack with the use of an airplane (maybe even spectacular) could take place. It was implicitly assumed, however, that such an accident could be isolated. The terrorists of 11 September have broken the implicit contract firstly by a group suicide (irrational behavior), and secondly, by exploiting well-known vulnerabilities of the US aviation security system, which was more liberal in comparison with that one in Europe.

It may be easily guessed that similar situational implicit contracting is still quite common in most of the airports although at present more security protection measures have been introduced, and direct actions against the known terrorist organizations and individual terrorists have been taken. Many other situations of implicit contracts still exist in social systems and we have to search for them ("securitize" them) as to prevent new sophisticated terrorist attacks.

Bearing in mind the above it can be proposed that incomplete contracting should be regarded as an element of vulnerabilities of social systems which should be implemented in mathematical models of collective behavior used in security studies, e.g. agent based modeling. The models should be supplemented with game theory models of implicit contracting. Similarly, awareness of the role of implicit contracting in prediction of threats of terrorism should be supported with relevant analogies and metaphors, such as for example, the core metaphor of implicit contracting, the common knowledge - "I know that you know, you know that I know that you know......, etc., *ad infinitum*".

## 7. Conclusions

The ideas presented in the paper are but preliminary answers to the questions about security theory and policy after 11 September 2001. The main conclusion is that systems thinking and complexity studies can and should become instruments of security theory and policy in the 21st Century. It concerns both prediction associated with the social mechanisms of securitization/desecuritization, as well as actions aiming at preventing and eliminating threats, not only the already known ones. This assertion concerns also new threats such as mega-terrorist genocide, and "cyber-wars" and emergence of "intelligent machines" [Joy 2000], [Kurzweil 2000], casting shadow on the future of the world social and political system. The ideas presented in the paper can be treated solely as an inspiration for further discussion on the links between complex systems studies, and security theory and policy.

Perhaps they could be helpful in the anti-terrorist policy making and in building future anti-terrorist strategy, as well as in elaborating a new security theory and policy recommendations meeting the challenges of yet unthinkable threats of the 21st Century.

## References


Alberts, David. S. & Thomas J. Czerwinski, eds., 2002. *Complexity, Global Politics and National Security,* University Press of the Pacific, Honolulu.

Allison, G. T. *et al.*, 1996, *Avoiding Nuclear Anarchy. Containing the Threat of Loose Russian Nuclear Weapons and Fissile Material*, The MIT Press (Cambridge, Mass.)

Bell, D. E. Raiffa, H., Tversky, A., 1988, *Decision Making: Descriptive, Normative, and Prescriptive Interactions*, Cambridge University Press (Cambridge).

Beyerchen, A., 1992, Clausewitz, Nonlinearity and the Unpredictability of War, *International Security*, **17**, 3.

Center for Naval Analyses, 2002, http://www.cna.org/isaac/on-line-papers.html, retrieved 6 September 2006.

Buzan, B., , 1991, *People, States and Fear,* 2nd edition, Harvester Wheatsheaf (New York).

Buzan, B., Wæver, O., de Wilde, J., 1998, *Security. A New Framework for Analysis,* Lynne Rienner Publishers (Boulder-London).

Deutsch, K., W., 1964, *The Nerves of Government. Models of Political Communication and Control*, Free Press, New York.

Gaddis, J. L., 1992, International Relations Theory and the End of the Cold War, *International Security*, **17**, 3.

Glenn, J. C., Gordon, T. J., 2006, *2002 State of the Future*, The Millennium Project, American Council for the United Nations University (Washington, D.C.).

Heijden, K., van der, 1996, *Scenarios. The Art of Strategic Conversation,* John Wiley & Sons (New York).

Hopf, T., 1993, Getting the Cold War Wrong (correspondence), *International Security*, **18**, 2.

Ilachinski, A., 2002, *Terrorism, Non-Linearity and Complex Adaptive Systems*, Compiled by Center for Naval Analyses, http://www.cna.org/isaac/terrorism_and_cas.htm#C, retrieved 6 September 2006.

Joy, B., 2000, Why the Why the Future Doesn't Need Us, *Wired*, 8.04 - April 2000, http://www.wired.com/wired/wrchive/8.04/joy.html, retrieved 6 September 2006.

Kaplan, M., A., 1957, *System and Process in International Relations,* Wiley (New York).

Kauffman, S. A., 1993, *The Origins of Order: Self-Organization and Selection in Evolution*, Oxford University Press (New York./Oxford).

Kauffman, S. A., 1995, *At Home in the Universe. The Search for Laws of Self-Organization and Complexity*, Oxford University Press (New York/Oxford).

Kurzweil, R., 2000, *The Age of Spiritual Machines: When Computers Exceed Human Intelligence*, Penguin Press (New York).

Mesjasz, C., 1988, Applications of Systems Modelling in Peace Research, *Journal of Peace Research,* **25**, 3.

Mesjasz, C., 1993, International Stability: What Can We Learn from Systems Metaphors and Analogies, *Working Paper,* **3**, Centre for Peace & Conflict Research (Copenhagen).

Mesjasz, C., 1994, Systems Metaphors, Systems Analogies, and Present Changes in International Relations, *Cybernetics and Systems*, **25**, 6.

Morgan, G. 1997, *Images of Organization,* 2nd edition, Sage, London.

Morgan, G., 1997a., *Imaginization. New Mindsets for Seeing, Organizing and Managing*, Berrett-Koehler Publishers/Sage Publications (San Francisco/Thousand Oaks).

*New Webster's Dictionary and Thesaurus of the English Language,* 1992, Lexicon Publications, Inc. (Danbury, CT).

*RAND Workshop on Complexity and Public Policy, Complex Systems and Policy Analysis: New Tools for a New Millennium,* 27 and 28 September 2000, RAND, Arlington, Virginia, http://www.rand.org/scitech/stpi/Complexity/index.html, retrieved 6 September 2006

Rosenau, J. N., 1990, *Turbulence in World Politics. A Theory of Change and Continuity*, Princeton University Press, (Princeton).

Rosenau, J.N., 1997, *Along the Domestic-Foreign Frontier: Exploring Governance in a Turbulent World*, Cambridge University Press (Cambridge).

Saperstein, A. M., 1984, Chaos - A Model for the Outbreak of War, *Nature*, **309**.

Saperstein, A. M., 1991, The "Long Peace" - Result of A Bipolar Competitive World, *Journal of Conflict Resolution*, **35**, 1.

Senge, P. M., 1990, *The Fifth Discipline. The Art and Practice of the Learning Organization,* Doubleday (New York).

Singer, J. D., 1999, Prediction, Explanation, and the Soviet Exit from the Cold War, *International Journal of Peace Studies*, **4**, 2.

Snyder, J. & Jervis, R., (eds.), 1993, *Coping with Complexity in the International System*, Westview Press (Boulder, CO).

The sociology and psychology of terrorism: Who becomes a terrorist and why? 1999, *A Report prepared under an Interagency Agreement by the Federal Research Division*, September, Library of Congress (Washington D.C.).

Tuchmann, B. W., 1992, *March of Folly: From Troy to Vietnam*, Reissue edition, Ballantine Books (New York).

Wæver, O., 1995, Securitization and Desecuritization, in Lipschutz, R. D., (ed.), *On Security*, Columbia University Press, (New York).

## Index of authors

MIX
Papier aus verantwortungsvollen Quellen
Paper from responsible sources
FSC® C105338

If you have any concerns about our products,
you can contact us on
**ProductSafety@springernature.com**

In case Publisher is established outside the EU,
the EU authorized representative is:
**Springer Nature Customer Service Center GmbH**
**Europaplatz 3, 69115 Heidelberg, Germany**

Printed by Libri Plureos GmbH
in Hamburg, Germany